Fired Heater
Theory and Practice

Fired Heater
Theory and Practice

발행일	2018년 2월 28일			
지은이	장 구 수			
펴낸이	손 형 국			
펴낸곳	(주)북랩			
편집인	선일영	편집	권혁신, 오경진, 최예은, 최승헌	
디자인	이현수, 김민하, 한수희, 김윤주	제작	박기성, 황동현, 구성우, 정성배	
마케팅	김회란, 박진관, 유한호			
출판등록	2004. 12. 1(제2012-000051호)			
주소	서울시 금천구 가산디지털 1로 168, 우림라이온스밸리 B동 B113, 114호			
홈페이지	www.book.co.kr			
전화번호	(02)2026-5777	팩스	(02)2026-5747	
ISBN	979-11-5987-681-3 93550(종이책)	979-11-5987-682-0 95550(전자책)		

이 도서의 국립중앙도서관 출판예정도서목록(CIP)은 서지정보유통지원시스템 홈페이지(http://seoji.nl.go.kr)와 국가자료공동목록시스템(http://www.nl.go.kr/kolisnet)에서 이용하실 수 있습니다.

(주)북랩 성공출판의 파트너
북랩 홈페이지와 패밀리 사이트에서 다양한 출판 솔루션을 만나 보세요!
홈페이지 book.co.kr • **블로그** blog.naver.com/essaybook • **원고모집** book@book.co.kr

Fired Heater

Theory
and
Practice

Gu Su Jang

북랩 **book** Lab

Preface

A fired heater is a combustion reactor and large heat exchanger where fuel and air react by combustion and that energy is transferred to process energy. Most fired heaters are comprised of two sections - a radiant section and a convection section. The radiant section receives heat from the burner primarily by radiation, while the convection section recovers additional heat by convection. Heat recovered from both sections is then transferred to pipes that carry process fluids, which ultimately generates process energy. Finally, hot flue gases are collected and discharged into the atmosphere through a vertical conduit, called stacks.

Fired heaters are vital in the refining and petrochemical industries, as they are crucial components to key refinery process units. Generally a significant amount of energy is provided to the unit through the fired heater by the combustion of fuel, usually in the form of gas or oil. Frequently the successful operation of the entire unit is directly influenced by how the fired heater works. Therefore, learning about the design, operation, maintenance and revamp of the fired heater can maximize the process unit profitability in many ways. Proper application of knowledge in the fired heater can increase the heater's capacity, decrease the operational costs, achieve more stringent emissions limits, increase reliability and flexibility of the process unit operation and provide tools and devices to optimize the unit control.

Despite the importance of the fired heater, I was not able to find a book that compiled the theory and practice of a fired heater systematically into one book. Therefore, I decided to prepare a book about the fired heater with my 30 years of experience in energy saving and environmental emission reduction as a chemical engineer. For over 10 years, I have collected a vast amount of data from my own work and other journals. This book was published by summarizing and editing all those data and photos.

This published book contains 4 chapters: chapter 1. Basic theory of design, chapter 2. Operation practice, chapter 3. General methods of maintenance, chapter 4. Exemplary cases of heater revamp. The outlined contents of each chapter are as follows.

Chapter 1 contains a basic concept necessary to design an optimized fired heater: terms & definition, classification of fired heater, configuration and the method of air supply & flue gas removal, introduction of processes with a fired heater, construction materials, combustion conditions, film temperature control, heat recovery, and economics, and etc. Chapter 2 includes a proper guide in operation and combustion: types of fuel, various burners, higher efficiency, operation & emission control, operating variables, and best practice, and etc. The chapter 3 demonstrates the problems and counter measures in operating a fired heater: coke accumulation, plugging & leak, flame impingement, corrosion, accidents, T/A inspection, safety practice, and accident troubleshooting, and etc. Lastly chapter 4 presents various methods and case studies of revamped fired heater in refinery and petrochemical processes.

I hope this book can help many people who work with the fired heater. I also hope that the book contributes to the energy saving and reduction of environmental emission. Lastly I would like to thank Book Lab Co. Ltd for the support in publishing this book.

Gu-Su Jang

Seoul, Korea
February 2018

Grand Contents

PART 1

Design

Sub Contents
for Design

1. Definition of Fired Heater

Fired heaters are large heat exchangers. The source of heat is primarily the heat of reaction provided by the combustion of fuel, usually gas or oil. Heat is transferred indirectly from the combustion gases to fluid contained within tubes. The rating involves a heat balance between heat releasing and absorbing streams within constraints of furnace boundaries. Much of the heat is transferred by radiation instead of convection in contrast to the usual heat exchanger.

The radiant section is commonly known as the firebox. It is within this area that the burners are situated and combustion is completed. The primary mode of heat transfer within the radiant section is radiation. Some convective heat transfer occurs as recirculating gases contact the tubes. Perhaps 60-90% of the total heat absorbed within in a radiant-convection furnace is absorbed within the radiant section. Flue gas temperatures exiting the radiant section may range from 704 to 816°C in typical refinery furnaces. Hydrogen and ethylene furnaces, among others, may have considerably higher flue gas temperatures exiting the radiant section.

The convection section recovers additional heat from the flue gas exiting the radiant section. The primary mode of heat transfer in the convection section is convection. Some radiative heat transfer occurs as the polar molecules (CO_2, H_2O, SO_2, etc.) in the flue gas radiate heat to the tubes.

A stack is provided to provide the differential necessary to establish the flow of air and flue gases through the furnace.

In most cases, a fired heater comprises of a radiant section receiving heat which is transferred to process fluid by combustion of fuel via the burners and a convection section recovering additional heat from hot flue gas leaving the radiant section before discharging to atmosphere.

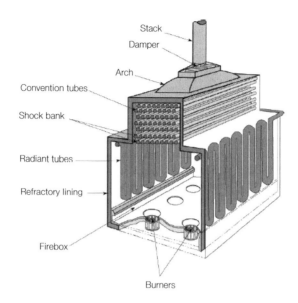

Stack
Damper
Arch
Convention tubes
Shock bank
Radiant tubes
Refractory lining
Firebox
Burners

Fig. 1 Typical heater

2. Terms Definition

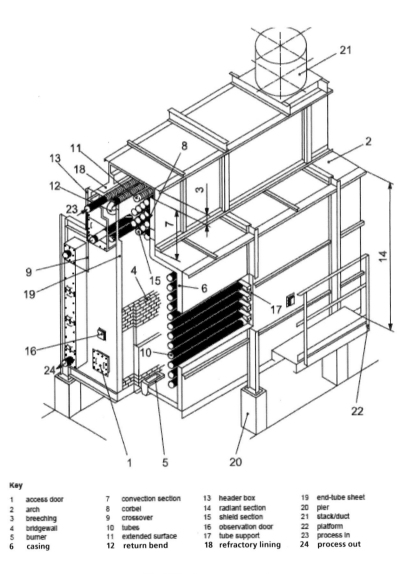

Key

1	access door	7	convection section	13	header box	19	end-tube sheet
2	arch	8	corbel	14	radiant section	20	pier
3	breeching	9	crossover	15	shield section	21	stack/duct
4	bridgewall	10	tubes	16	observation door	22	platform
5	burner	11	extended surface	17	tube support	23	process in
6	casing	12	return bend	18	refractory lining	24	process out

Fig.2 Heater components

Air preheat system

A collection of ductwork, fans and heat exchange equipment used to heat the combustion air prior to its entry into the burners. The use of an APH improves to overall efficiency of the heater.

Air heater or air preheater is a heat transfer apparatus through which combustion air is passed and heated by a medium of higher temperature such as the products of combustion, steam, or other fluid. It has three basic types: regenerative, recuperative, heat pipe.

Direct air preheater is an exchanger which transfers heat directly between the flue gas and the combustion air. A regenerative air preheater uses heated rotating elements and a recuperative design uses stationary tubes, plates, or cast iron elements to separate the two heating media.

Indirect-type air preheater is a fluid-to-air heat transfer device. The heat transfer can be accomplished by using a heat transfer fluid, process stream or utility stream which has been heated by the flue gas or other means. A heat pipe air preheater uses a vaporizing/condensing fluid to transfer heat between the flue gas and air.

Air registers and dampers
Air registers and dampers are used to vary and control the amount of air flowing through the burner. If the registers or dampers are not adjustable, the targeted level of excess oxygen at the burner cannot be maintained over the desired range of operation.

Arch
Arch is a flat or sloped portion of the heater radiant section opposite the floor.

Atomizer
Atomizer is device used to reduce a liquid fuel oil to fine mist. Atomization media are either steam, air or mechanical.

Anchor or tieback
Anchor or tieback is a metallic or refractory device that retains the refractory or insulation in place.

Backup layer
Backup layer is any refractory layer behind the hot face layer.

Balanced draft
Balanced draft heater is a unit in which the combustion air is supplied by a fan and the flue gases are removed by a fan.

Box

The burners and tubes are enclosed in the furnace box, which consists of a structure, refractory lining and tube supports. The "Box" may be rectangular or cylindrical in design

Buoyancy

The tendency for warmer fluids to rise due to density differences compared to surrounding fluids

Breeching (Breech)

The enclosure which collects the flue gas at the convection section exit, for transmission to the stack or connecting ductwork.

Bridge wall, division or gravity wall

Bridge wall, division or gravity wall is a wall separating two adjacent heater zones.

Bridge wall temperature

Bridge wall temperature is the flue gas temperature leaving the radiant section. The term comes from the old horizontal box furnace, where a bridge wall physically is separated the radiant and convection sections.

Bulk temperature

The average temperature of the process fluid at any tube cross section.

Burner

Burner is a device for the introduction of fuel and air into a heater at the desired velocities, turbulence, and concentration to establish and maintain proper ignition and combustion. Burners are classified by the types of fuel fired, such as oil, gas, or combination of gas and oil. A secondary consideration in classifying burners is the means by which combustion air is mixed with the fuel.

Casing

A steel sheathing which encloses the furnace box and makes it essentially air- tight.

Castable

Castable is an insulating concrete poured or gunned in place to form a rigid refractory shape or structure.

Ceramic fiber

Ceramic fiber is a fibrous refractory insulation composed primarily of silica and alumina. Applicable forms include blanket, board, module, rigidized blanket, and vacuum-formed shapes.

Coil

A series of tube lengths forming a continuous path through which a fluid passes and is heated. The coil may be straight lengths connected by 180° return bends, helical, wickets, or of arbor design.

Coil pressure drop

Coil pressure drop is the difference between the coil inlet pressure and the coil outlet pressure between terminals, excluding the effect of static head.

Combination burner

Combination burner can burn liquid fuel, gaseous fuel, or both simultaneously.

Convection section

The portion of a furnace, consisting of a bank of tubes or coils, which receives heat from the hot flue gases, mainly by convection. The convection section is typically located above the radiant section.

Corbel

Narrow ledges extending from the convection section side walls to prevent flue gas from flowing preferentially up the side of the convection section, between the wall and the nearest tubes, thereby by-passing the tube bank.

Corrosion allowance

Corrosion allowance is the additional material thickness added to allow for material loss during the design life of the component. It is the corrosion rate time's tube design life, expressed in inches (millimeters).

Corrosion rate

Corrosion rate is the reduction in the material thickness due to the chemical attack from the process fluid and/or flue gas expressed in millimeters per year.

Crossover

Piping which transfers the process fluid between any two heater sections. Typically, it is from the convection section outlet to the radiant section inlet.

Damper

A device to regulate flow of gas through a stack or duct and to control draft in a furnace. A typical damper consists of a flat plate connected to a shaft which can be rotated, similar to a butterfly valve.

Butterfly-type is a single blade damper pivoted about its center.

Louver-type is a damper consisting of several blades each pivoted about its center and linked together for simultaneous operation.

Down-fired burner

Burner located in the roof of the furnace firing down; often used in reforming furnaces

Draft

The negative pressure (vacuum) of the flue gas at a given point inside the furnace, usually expressed as inches of water column.

Draft loss

Draft loss is the pressure drop through duct conduits or across tubes and equipment in air and flue gas systems.

Duct

Duct is a conduit for air or flue gas flow.

Excess air

Excess air is the amount of air above stoichiometric requirement for complete combustion in furnace, expressed as a percentage.

Efficiency

The ratio of heat absorbed to the heat fired in furnace.

Fuel efficiency

Fuel efficiency refers to the heat absorbed divided by the net heat of combustion of the fuel as heat input, expressed as a percentage.

Thermal efficiency

Thermal efficiency refers to the total heat absorbed divided by the total heat input, expressed as a percentage.

Erosion

Erosion is the reduction in the material thickness due to mechanical attack from a fluid, expressed in inches (millimeters).

Extended surface

Surface added to the outside of bare tubes in the convection section to provide more heat transfer area. This usually consists of fins continuously wound around and welded to the tube.

Extension ratio

Extension ratio is the ratio of total outside exposed surface to the outside surface of the bare tube.

Film

A thin fluid layer adjacent to a pipe wall which remains in laminar flow, even when the bulk flow is turbulent. The velocity profile in the film is approximately linear, with zero velocity existing at the wall.

Film coefficient

The convective heat transfer coefficient of the film.

Film temperature

The maximum temperature in the film, at the tube wall.

Fire box (Radiant section)

A term used to describe the structure which surrounds the radiant tubes and into which the burners protrude.

Flashback

Velocity of a flammable mixture is slower than the flame speed, causing the mixture to burn back (flash back) toward the source.

Flat flame
Narrow, rectangular flame shape

Flue gas
A mixture of gaseous products resulting from combustion of the fuel, including the excess air.

Forced draft heater
Forced draft heater is a unit in which the combustion air is supplied by a fan or other mechanical means.

Fouling allowance
Fouling allowance is for a layer of residue that increases pressure drop, usually a buildup of coke or scale on the inner surface of a coil, expressed as inches (millimeters). This value shall be used in calculating the fouled pressure drop.

Fouling resistance
Fouling resistance is a factor used to calculate the overall heat transfer coefficient. The inside fouling resistance shall be used to calculate the maximum metal temperature for design. The external fouling resistance is used to compensate the loss of performance due to deposits on the external surface of the tubes or extended surface.

Fuel staging
Some of the fuel is injected downstream from the primary flame zone

Gross fuel liberation
The total fuel fired in the furnace, including all losses (usually expressed in lb/hr).

Guillotine blind
Guillotine blind is a device consisting of a single blade that is used to isolate equipment or heaters.

Gross heating value
See higher heating value.

Hanger-tube support
Device that holds a tube hanging from the wall or roof of the furnace

Header (Return bends)

The fitting which connects two tubes in a coil. Strictly speaking, the removable plug type fitting ("plug headers" or "cleanout plugs") into which the tube is either rolled or welded. In common usage, "header" refers to cast or forged 180° "U- bends" (return "bends").

Header box

The compartment at the end of the convection section where the headers are located. There is no flue gas flow in the header box, since it is separated from the inside of the furnace by an insulated tube sheet. It is used to enclose a number of headers or manifolds.

Hearth burner

Burner mounted in the floor (hearth) of a furnace

Liftoff

Phenomenon that can occur when the velocity of a flammable fuel-air mixture is much higher than the flame speed.

Heat flux

Energy transfer per unit time per unit area

Heat available

The heat absorbed from the products of combustion (flue gas) as they are cooled from the flame temperature to a given flue gas temperature.

Heat absorption

Heat absorption is the total heat absorbed by the coils excluding any combustion air preheat.

Heat density (Heat flux)

The rate of heat transfer per unit area of a tube, usually based on total outside surface area. Typical units are Btu/hr ft^2 also called "heat flux". Average heat flux for an extended surface tube shall be indicated on a bare surface basis with extension ratio noted. Maximum heat flux is the maximum local heat transfer rate in the coil section.

Heat release

Heat release is the total heat liberated from the specified fuel, using the lower heating value (LHV) of the fuel, usually expressed in Btu/hr.

Heat duty

The total heat absorbed by the process fluid, usually expressed in Million Btu/hr (MMBTU/hr). Total furnace heat duty is the sum of heat transferred to all process streams, including auxiliary services, such as steam super heaters and drier coils.

Higher Heating Value (HHV)

The theoretical heat of combustion of a fuel, when the water formed is considered as a liquid (credit taken for its heat of condensation). Also called gross heating value.

Hot face layer

Hot face layer is the refractory layer exposed to the highest temperatures in a multi-layer or multi-component lining.

Hot face temperature

Hot face temperature is the temperature of the refractory surface in contact with the flue gas or combustion air. The hot face temperature is used to determine refractory or insulation thickness and heat transmitted. The design temperature is used to specify the service temperature limit of the refractory materials.

Induced draft heater

Induced draft heater is a unit in which a fan is used to remove flue gases and maintain a negative pressure in the heater to induce combustion air without a forced draft fan.

IPS

Abbreviation for "Iron Pipe Size", used to refer to standard nominal pipe sizes.

Lower Heating Value (LHV)

The theoretical heat of combustion of a fuel, when no credit is taken for the heat of condensation of water in the flue gas. That is the higher heating value minus the latent heat of vaporization of the water formed by combustion of hydrogen in the fuel. Also called net heating value. Usually expressed in Btu/lb.

Manifold

A pipe connected to several parallel passes and used to distribute or collect fluid into or out of these passes.

Metal fiber reinforcement

Metal fiber reinforcement is stainless steel needles added to castable for improved toughness and durability.

Monolithic lining

Monolithic lining is a single component lining system without joints or interfaces.

Mortar

Mortar is a refractory material preparation used for laying and bonding refractory bricks.

Multi-component lining

Multi-component lining is a refractory system consisting of two or more layers of different refractory types; for example, castable and ceramic fiber.

Multi-layer lining

Multi-layer lining is a refractory system consisting of two or more layers of the same refractory type.

Mass velocity

The mass flow rate per unit of flow area through the tubes. Typical units are lb/ sec/ ft^2.

Natural draft

System in which the draft required to move combustion air into the furnace and flue gas through the furnace and out the stack is provided by stack effect alone.

Normal heat release

Normal heat release is the design heat absorption of the heater divided by the calculated fuel efficiency expressed in MM Btu/hr.

Net fuel

The fuel which would be required in the furnace if there were no radiation losses. Usually expressed in lb/hr.

Observation doors
Viewing ports placed at selected points along the furnace walls, to permit viewing of furnace tubes, supports and burners. Also call "peep holes".

Pass or stream
A coil which transports the process fluid from furnace inlet to outlet. The total process fluid can be transported through the furnace by one or more parallel passes.

Pilot burner
A smaller burner that provides ignition energy to light the main burner.

Plenum or wind box
Plenum or wind box is a chamber surrounding the burners that is used to distribute air to the burners or reduce combustion noise.

Plug-type header
Plug-type header is a cast return bend which is provided with one or more openings for the purpose of inspection, mechanical tube cleaning, or draining.

Primary air
Primary air is that portion of the total combustion air that first mixes with the fuel.

Premix burner
At least some, if not all, of the fuel and combustion air are mixed inside the burner.

Protective coating
Protective coating is a corrosion resistant material applied to a metal surface; such as on casing plates behind porous refractory materials to protect against sulfur in the flue gases.

Radiant section
The portion of the furnace in which heat is transferred to the furnace tubes primarily by radiation from the flame and high temperature flue gas.

Radiation or setting loss
Radiation or setting loss is the heat lost to the surroundings from the casing of the heater and the ducts and auxiliary equipment when heat recovery systems are used, expressed as percent of heat release.

Refractory

Material that lines the roof, walls, and floor of the firebox. Refractory re-radiates energy, thus heating the tubes more evenly than flames alone could. With other insulating layers behind it, the refractory liner limits heat loss and protects the heater structure and operators from high temperature. It is used in the convection section, ducts, and stack. Bridge walls are also refractory.

Regen tile

Inner refractory tile used to add heat to the oil stream inside the burner to improve oil vaporization

Secondary air

Secondary air is the air supplied to the fuel to supplement primary air.

Service factor

A measure of the continuity of operation, generally expressed as the ratio of total running days for a given time period to the total calendar days in the period.

Setting

The refractory insulation on the inside of the furnace box and any ducting. That is the heater casing, brickwork, refractory and insulation including the tiebacks or anchors.

Shield section (Hip section)

The first two tube rows of the convection section. These tubes are exposed to direct radiation from the radiant section and usually receive about half of their heat in this manner. They are usually made of more resistant material than the rest of the tubes in the convection section.

Shock or Shield tubes

First two rows of convection tubes (in the direction of flue gas flow) that may be exposed to radiation from the fire box as well as to convection heating

Soot blower

Soot blower is a mechanical device for discharging steam or air to clean heat absorbing surfaces.

Stack

A cylindrical steel, concrete or brick shell (a vertical conduit) which carries flue gas to the atmosphere and provides necessary draft.

Stack effect

The difference (buoyancy) between the weight of a column of high-temperature gases inside the furnace and/ or stack and the weight of an equivalent column or external air, usually expressed in inches of water per foot of height.

Stack temperature

The temperature of the flue gas as it leaves the convection section. If an APH system is used, it is the flue gas temperature after it leaves the air preheater.

Staged air

Some of the combustion air is added downstream of the primary flame zone.

Staged fuel

Some of the fuel is added downstream of the primary flame zone.

Strakes or spoilers

Strakes or spoilers are metal stack attachments that reduce wind-induced vibration.

Target wall

Target wall is a vertical refractory firebrick wall which is exposed to direct flame impingement on one or both sides.

Temperature allowance

Temperature allowance is the number of degrees Celsius (°C) to be added to the process or flue gas temperature for flow maldistribution and operating unknowns. The temperature allowance is added to the calculated maximum tube metal temperature or the equivalent tube metal temperature to obtain the design metal temperature.

Terminal

Terminal is a flanged or welded projection from the coil providing for inlet and outlet of fluids.

Tube guide

Tube guide is a device used with vertical tubes to restrict horizontal movement while allowing the tube to expand axially.

Tube retainer

Tube retainer is a device used to restrain horizontal radiant tubes from lifting off the intermediate tube supports during operation.

Tube sheet/Supports

A large tube support plate located at the end of a horizontal tube and supports a number of tubes. The end supports are usually carbon steel or low-alloy steel and make up one side of the header box. The inside of the end supports, which is exposed to the flue gas, is insulated. Intermediate supports sometimes required and are exposed to flue gas on both sides and are fabricated of a suitable alloy material.

Turndown

Ratio of the highest design firing rate to the lowest design firing rate

Vapor barrier

Vapor barrier is a metallic foil placed between layers of refractory as a barrier to flue gas flow.

Volumetric heat release

Volumetric heat release is the heat released divided by the net volume of the radiant section, excluding the coils and refractory dividing walls, expressed in MM Btu/ ft^3 hr.

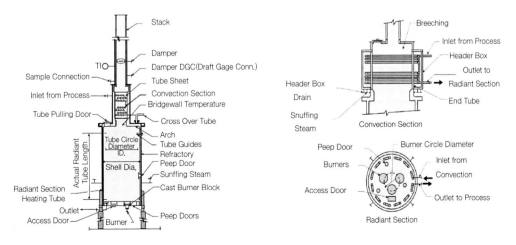

Fig. 3 Vertical cylindrical heater

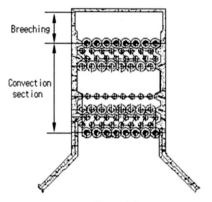

a. Breeching

b. Ceramic fiber(Castable)

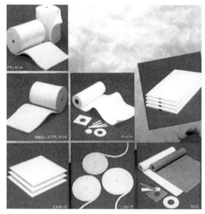

c. Ceramic fiber(Blanket)

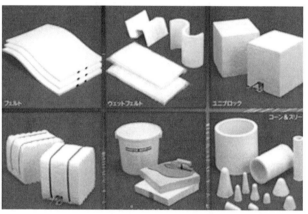

d. Ceramic fiber(folded/monolithic module)

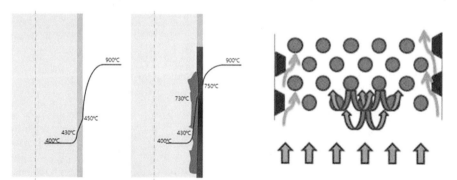

e. Fouling resistance

f. Corbel

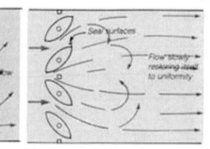

g. Louver damper(parallel & combination blade)

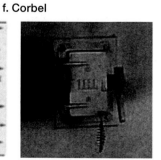

h. Inspection door

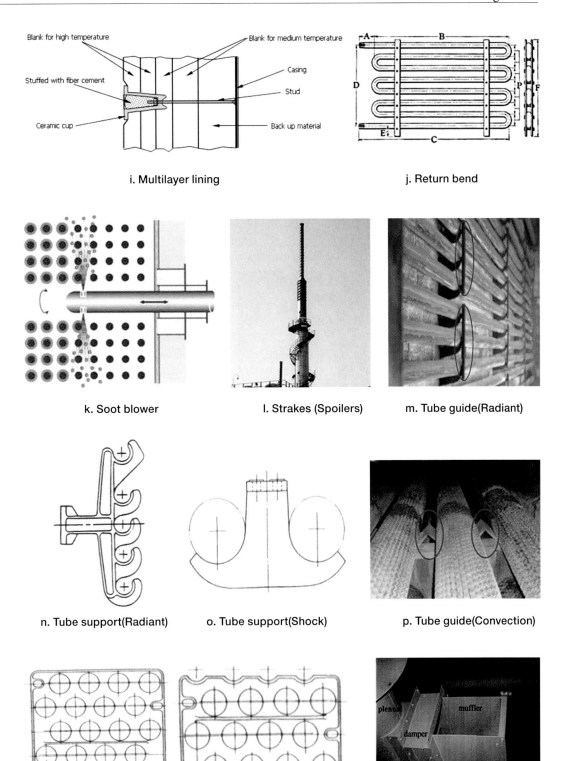

i. Multilayer lining

j. Return bend

k. Soot blower

l. Strakes (Spoilers)

m. Tube guide(Radiant)

n. Tube support(Radiant)

o. Tube support(Shock)

p. Tube guide(Convection)

q. Tube sheet(Lower, Upper)

r. Plenum damper & Muffler assembly

s. Damper detail

t. Burner air register

u. Stack damper

v. Tube skin thermocouple

w. Draft gauge

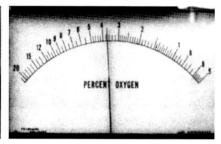

x. Excess air indicator

y. Explosion door

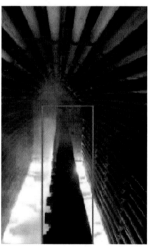

z. Target wall

aa. Air preheater

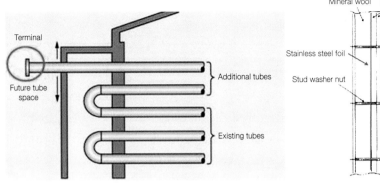

bb. Terminal & future tube space

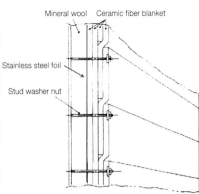

cc. Vapor barrier

dd. Inspection door

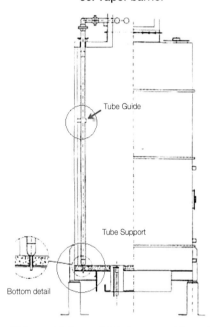

ee. Tube guide(upper) & tube upport(lower)

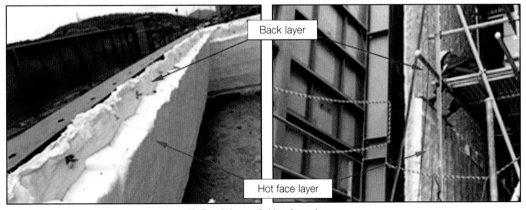

ff. Back layer & Hot face layer

Fig. 4 Other parts photos

3. Process Utilization

Heaters can be classified according to method of utilization in the process, described as follows. The process function of heater is as follows: used for fractionation, upstream of reactors, black oil cracking and miscellaneous specialty.

3.1 Process heaters

These generally are used for heating only in which there is little or no decomposition of the feed stream. The principal uses are for feed preheating, reboiling, reheating or for raising the temperature of a hot oil which exchanges heat with other process streams.

3.1.1 Fractionating column feed preheaters

Fired heaters in this service tend to be the workhorses of many process operations. The charge stock (usually all liquid although some feeds may contain a nominal amount of vapor at the inlet) is sent to the fired heater following upstream preheating in unfired equipment. In the fired heater, the fluid temperature is usually raised high enough to achieve partial vaporization of the charge stock. A typical example of this service is the feed heater for an atmospheric distillation column in the crude oil unit of a petroleum refinery. Here, crude oil entering the fired heater as a 232℃ liquid might exit near 371℃ with about 60% of the charge stock vaporized. When the crude oil enters the refinery, it must be heated to 360℃ to separate the products.

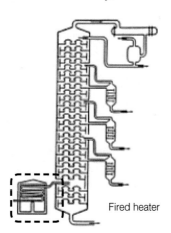

Fired heater

Fig. 5 Fractionating feed heater

3.1.2. Reactor feed preheaters

Many catalytic processes require fired heaters to obtain the proper reactor temperature. That is to raise the charge stock temperature to a level necessary for controlling a chemical reaction taking place in an adjoining reactor vessel. The nature of the charge stock and the heater operating temperatures and pressures can vary considerably, depending on the process. The following examples illustrate the diversity of the applications performed by reactor feed preheaters. Single phase/ single component heating such as steam superheating in the reaction of styrene manufacturing processes. In this service, the fluid temperature across the fired heater increases from an inlet temperature of about 371℃ to an exit temperature of approximately 866℃.

Single phase/multicomponent heating such as the heating of mixtures of vaporized hydrocarbons and recycle hydrogen gas prior to catalytic reforming in a refinery. In this service, the charge stock enters the fired heater at about 427℃ and exits at approximately 538℃. In reformers, the fluid pressure may range from about 250 to 600 psig. Severe restrictions on fluid pressure drop are normally associated with this service. Regeneration heaters are used to regenerate catalyst beds and molecular sieve driers.

Mixed phase/multicomponent heating such as the heating of mixtures of liquid hydrocarbons and recycle hydrogen gas for reaction in a refinery hydrocracker. Fluid temperatures typically run from 371℃ at the inlet to 454℃ at the outlet. Operating pressures may reach 3,000 psig, depending on the process.

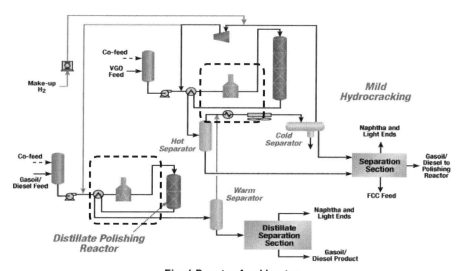

Fig. 6 Reactor feed heater

3.1.3 Column reboilers

This is normally considered one of the mildest and least critical of fired heater applications. The charge stock taken from a distillation column is a recirculating liquid that is partially vaporized in the fired heater. The mixed vapor liquid stream reenters the column, where the vapor condenses and releases the heat of vaporization. Reboiler applications are characterized by relatively small differentials between the inlet and outlet fluid temperatures across the fired heater, and by substantial vaporization (typically, 50% or more of the charge stock is vaporized). Depending on the particular application, reboiler heater outlet temperatures generally fall in the range of 204℃ to 288℃.

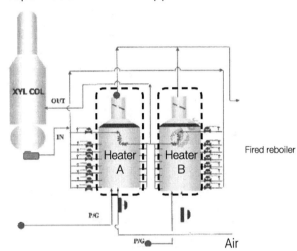

Fig. 7 Column reboiler

3.2 Steam reforming heaters

Heaters in which a catalytic chemical reaction takes place in the furnace tubes. Heaters of this type are steam methane reformers used for production of hydrogen, ammonia, methanol and ketene reactors. In case of steam hydrocarbon reformer heaters, in which the tubes of the combustion chamber function individually as vertical reaction vessels filled with nickel-bearing catalyst. In reformers that yield hydrogen, fluid outlet temperatures range from 788℃ to 899℃.

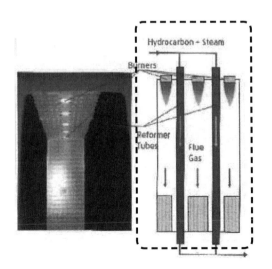

Fig. 8 Steam reforming heater

3.3 Reaction heaters

High temperature conversion furnaces in which the required decomposition of the process fluid occurs in the furnace coil, such as ethylene cracking heaters, or ethylene dichloride cracking heaters. Heaters used in refinery processing for Visbreakers, Delayed Coker and Thermal Cracking (pyrolysis) service are also classified as mild cracking heaters(low severity) as chemical reaction and thermal cracking of heavy molecules take place.

Pyrolysis heaters are used to produce olefins from gaseous feedstocks such as ethane and propane and from liquid feedstocks such as naphtha and gasoil. In cracking heaters, where chemical reactions occur in the coil, the tubes and burners are arranged so as to assure pinpoint firing control. Fluid outlet temperatures in heaters designed for liquid feedstocks are in the 816°C to 899°C range.

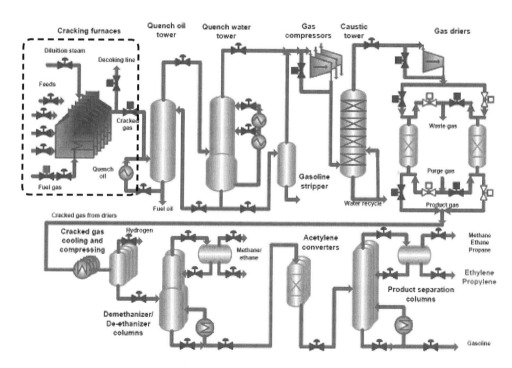

Fig. 9 Pyrolysis heater

The Delayed Coker process uses a fired heater to heat vacuum bottoms to the 491°C range. Most of the thermal reaction occurs in the coke drum.

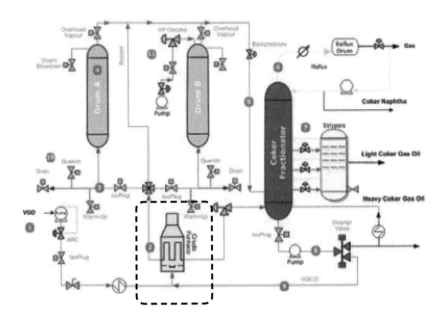

Fig. 10 Delayed Coker heater

3.4 Other heaters

Heat supplied to viscous fluids. Often heavy oil must be pumped from one location to another for processing. At low temperatures, where the oil may have so high a viscosity as to render pumping infeasible, a fired heater is employed to warm the oil to a temperature that will facilitate pumping.

Heat supplied to heat transfer media. Many plants furnish heat to individual users via an intermediate heat transfer medium. A fired heater is generally employed to elevate the temperature of the recirculating medium, which as typically a heating oil, Dowtherm, Therminol, molten salt, etc. Fluids flowing through the fired heater in these systems almost always remain in the liquid phase from inlet to outlet.

4. The Type of Heaters

There are many variations in the layout, design, and detailed construction of the fired heaters. A consequence of this flexibility is that virtually every fired heater is custom-engineered for its particular application. In addition to the radiant section, most modern fired heaters include a separate convection section. The residual heat of the flue gases leaving the radiant section is recovered in this section, primarily by convection. Using this heat for preheating the charge stock, or for other supplementary heating services, increases the thermal efficiency of the fired heater. The first few rows of tubes in the convection section are subject to radiant heat transfer, in addition to convective transfer from the hot flue gases as they flow across the tubes. Because these tube rows are usually being subjected to the highest heat transfer rates in the fired heater, they are aptly termed "shield" or "shock" tubes. The principal classification of fired heaters, however, relates to the orientation of the heating coil in the radiant section: i.e., where the tubes are vertical or horizontal. Vertical arrangements are shown in Fig. 11 and horizontal arrangements in Fig. 12.

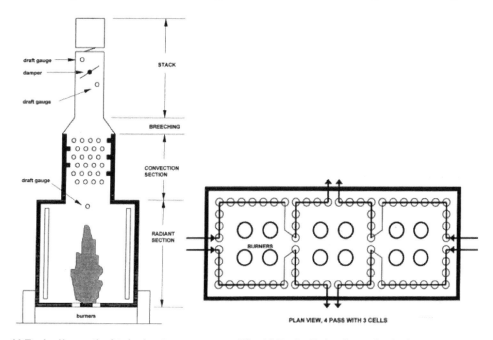

Fig. 11 Typically vertical tube heater   **Fig. 12 Typically horizontal tube heater**

4.1 Vertical-cylindrical

Vertical-cylindrical is with crossflow convection. These heaters, also fired vertically from the floor, feature both radiant and convection sections. The radiant section tube coil is disposed in a vertical arrangement along the walls of the combustion chamber. The convection section tube coil is arranged as a horizontal bank of tubes positioned above the combustion chamber. This configuration provides an economical, high efficiency design that requires a minimum of plot area. The majority of new, vertical tube fired heater installations fall into this category. Typical duty range is 2.52 to 50.4MM Kcal/hr.

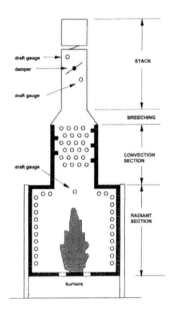

Fig. 12 Vertical cylindrical heater

4.2 Horizontal tube cabin

The radiant section tube coils of these heaters are arranged horizontally so as to line the sidewalls of the combustion chamber and the sloping roof or "hip". The convection section tube coil is positioned as a horizontal bank of tubes above the combustion chamber. Normally the tubes are fired vertically from the floor, but they can also be horizontally fired by sidewall-mounted burners located below the tube coil. This economical, high efficiency design currently represents the majority of new, horizontal tube fired heater installations. Duties run 2.52 to 25.2MM Kcal/hr.

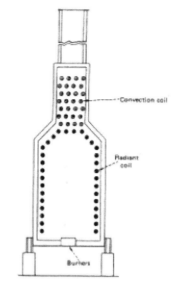

Fig. 13 Horizontal tube cabin heater

4.3 Arbor (wicket)

This is a specialty design in which the radiant heating surface is provided by U tubes connecting the inlet and outlet terminal manifolds. This type is especially suited for heating large flows of gas under conditions of low pressure drop. Typical applications are found in petroleum refining, where this design is often employed in the catalytic reformer charge heater, and in various reheat services. Firing modes are usually vertical from the floor, or horizontal between the riser portions of the U tubes. This design type can be expanded to accommodate several arbor coils within one structure. Each coil can be separated by dividing walls so that individual firing control can be attained. In addition, a crossflow convection section is normally installed to provide supplementary heating capacity for chores such as steam generation. Typical duties for each arbor coil of this design are about 12.6 to 25.2MM Kcal/hr.

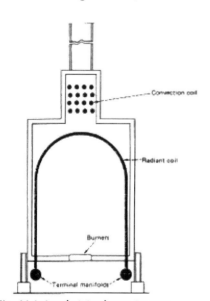

Fig. 14 Arbor (wicket) type heater

4.4 Vertical tube, double fired

In these units, vertical radiant tubes are arranged in a single row in each combustion cell (there are often two cells) and are fired from both sides of the row. Such an arrangement yields a highly uniform distribution of heat transfer rates (heat flux) about the tube circumference. Another variation of these heaters uses multilevel side wall firing, which gives maximum control of the heat flux profile along the length of the tubes. Multilevel sidewall firing units are often employed in fired reactor services and in critical reactor feed heating services. In addition to the twin cell furnaces to be mentioned, single cell models are available for smaller duties. As a group, these represent the most expensive fired heater configuration. The typical duty range for each cell runs from about 5.04 to 31.5MM Kcal/hr.

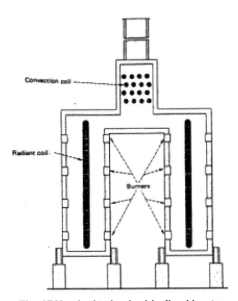

Fig. 15 Vertical tube double fired heater

4.5 Horizontal tube, double fired

Horizontal radiant tubes are arranged in a single row and are fired from both sides to achieve a uniform distribution of heat transfer rates around the rube circumference. Such heaters are normally fired vertically from the floor. They are often selected for critical reactor feed heating services. For increased capacity, the concept can be expanded to provide for a dual combustion chamber. A typical duty range for each cell of this design is about 5.04 to 12.6MM Kcal/hr.

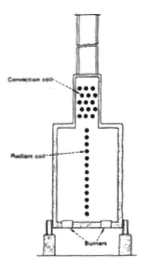

Fig. 16 Horizontal tube double fired heater

4.6 Two cell horizontal tube box

Here the radiant section tube coil is deployed in a horizontal arrangement along the sidewalls and roof of the two combustion chambers. The convection section tube coil is arranged as a horizontal bank of tubes positioned between the combustion chambers. Vertically fired from the floor, this is again an economical high efficiency design. Typical duties range from 25.2 to 63MM Kcal/hr. For increased capacity, the basic concept can be expanded to include three or four radiant chambers.

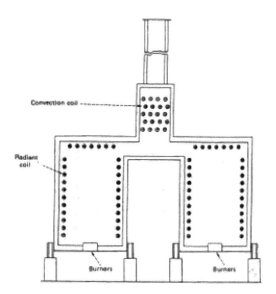

Fig. 17 Two cell Horizontal tube box heater

4.7 Horizontal tube cabin with dividing bridge wall

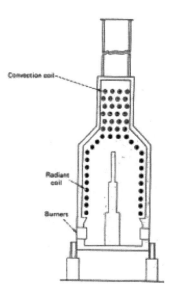

Again the radiant section tube coil is arranged horizontally along the sidewalls of the combustion chamber, and along the hip. The convection section tube coil takes the form of a horizontal bank of tubes positioned above the combustion chamber. A dividing bridge wall between the cells allows for individual firing control over each cell in the combustion chamber. Available options permit horizontal firing with sidewall-mounted burners, or vertical firing from the floor along both sides of the bridge wall. A typical duty range for this design is 5.04 to 25.2MM Kcal/hr.

Fig. 18 Horizontal tube cabin with dividing bridge wall heater

4.8 End fired horizontal tube box

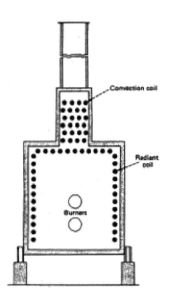

The radiant section tube coil is disposed in a horizontal arrangement along the sidewalls and roof of the combustion chamber. The convection section tube coil is arranged as a horizontal bank of tubes positioned above the combustion chamber. These furnaces are horizontally fired by burners mounted in the end walls. Typical duty range for this design is 1.26 to 12.6MM Kcal/hr.

Fig. 19 End fired horizontal tube box heater

4.9 Horizontal tube and vertical tube heater

Most old heater designs are horizontal tube cabin heaters. Horizontal tube heaters take up more plot area and are 15-20% more expensive than a vertical tube heater. Most new heaters today are vertical cylindrical to save cost and plot space. The vertical often has a taller radiant section and is much better suited for the long flames of low NO_x burners. A common problem with vertical cylindrical heaters is that the burner to tube clearance is not adequate to keep the flames off the tubes.

4.10 Selection type of heater to be used

The type of heater to be used for a given application is frequently specified by the client. The first consideration must be whether to use a vertical cylindrical or cabin type firebox. Cylindrical heaters are used when available plot space is limited. It requires much less plot space than the equivalent box type furnace. For a vertical cylindrical heater, the possibilities are a vertical type of all radiant heater and one with a horizontal convection section mounted on top of the radiant section. All radiant heaters are used when the furnace duty is small, and no great economies could be made by increasing the efficiency by heat recovery in a convection section. They should be considered for duties up to about 1.26MM Kcal/hr for continuously operating heaters and for higher duties where the furnace is to be operated only intermittently and for short periods. Vertical tube, all radiant, cylindrical heaters are used when decoking facilities are required. When a cylindrical heater with an overhead convection section is required, the radiant section tubes are usually vertically disposed. This type of heater usually proves to be a good choice when the furnace duty is greater than 1.512MM Kcal/hr.

Cabin and box type heaters are usually recommended for larger furnace duties. This type of heater invariably has a convection section. Horizontal tube cabin heater requires free area alongside the heater for tube removal. The tube removal area should be at least equal to the overall length of any tubes there may be in the convection section.

The radiant tubes can either be horizontally or vertically disposed, the latter being adopted when a large number of process fluid passes are required and when even heat distribution is essential. Horizontally arranged radiant tubes allow the use of very long tubes (up to about 24.384m) in both the radiant and convection section, thereby reducing the number of return bends or fittings and their associated welds.

Cabin and box type heaters are also recommended for processes requiring uniform heat distribution to the tubes. The radiant tubes are arranged in the center of heater with burners on both sides of the tubes. This finds application in high temperature services (about 760°C fluid outlet temperature), such as steam super heaters, reforming and cracking furnaces. When the radiant tubes are arranged along the furnace wall and are horizontally disposed, then the radiant section may be arranged so that there is a gradual convergence to the convection section, with tubes along the sloping walls.

It should be emphasized that the layout of a furnace radiant section should be made in conjunction with the layout of the convection section tubes. The convection section layout should be compatible with a good flue gas mass velocity for the type of draft available in the heater.

4.11 Flow of fired side and process side

The process fluid enters the top of the convection section and flows down countercurrent to the flue gas flow. Most of the heat transfer occurs in the radiant section. Approximately 70% of the process duty is absorbed in the radiant section and approximately 30% in the convection section.

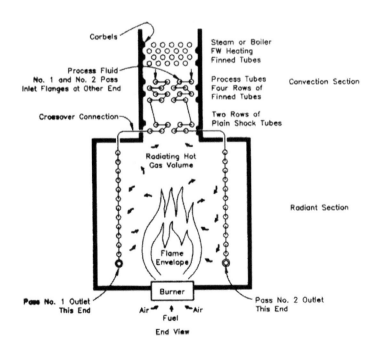

4.12 Typical heat transfer of fired heater

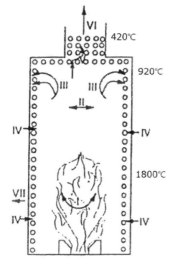

I	Direct heat transfer through flame radiation	15%
II	Direct heat transfer through gas radiation	30%
III	Direct heat transfer through gas convection	5%
IV	Indirect heat transfer through refractory radiation	10%
	Total heat transfer in radiation section	60%
V	Direct heat transfer through convection in convection bank	25%
	Total useful heat	85%
VI	Flue gas loss	13%
VII	Radiation loss	2%
	Total	100%

4.13 Typical specification of refinery heater

Process		Radiant flux	Mass velocity	Process △P	Coil metallurgy	Heater type
		Btu/hr ft^2	lb/sec ft^2	psi		
Crude		10,000	200 min	150	5 Cr or 9 Cr	Vertical Tube Box, Cabin, or V.C.
Vacuum		8,000	250 min	75	5 Cr or 9 Cr	Vertical Tube Box, Cabin, or V.C.
Visbreaking		7,000-8,000	350 min (5-6 fps cold oil)	250-300 (clean)	9 Cr	Cabin with brick Center wall
Delayed Coking		9,000	350 min (6 fps cold oil)	300 (clean)	9 Cr	Cabin with Brick Center Wall or Double Fired Horizontal Tube
Semi Regen Platformer		10,000	40-50	7	2-1/4 Cr	V.C., Cabin, Wicket
CCR Platformer		10,000 SFU 15,000 DFU	23-35	3 per cell	9 Cr	Single Fired U-tube, Double Fired U-tube
Reboiler		10,000	150-300	50	C. S.	V.C.
Hot Oil		10,000	400-600	65	C. S.	V.C.
Naphtha Hydrotreater	Max 15 mils/yr	10,000	70-180	50	9 Cr	V.C.
	With cracked stock				347	
Kerosene Hydrotreater		10,000	150-250	50	347H SS	V.C.
Diesel Hydrotreater		10,000	150-250	50	347H SS	V.C., or Double Fired Vertical Tube
Hydrocracker	Gas only	10,000-13,000	50-80	30-50	347H SS	Double Fired Vertical Tube
	Gas & liquid	15,000	180-300	50-80	347H SS	Double Fired Vertical Tube
Catalytic Cracker Feed Heater		10,000-11,000	300-400	-	-	-
Steam Superheater		9,000-13,000	30-75	-	-	-

Note) V.C. (Vertical Cylindrical)

4.14 Typical heater and surrounding parts

(1) Vertical cylindrical heater

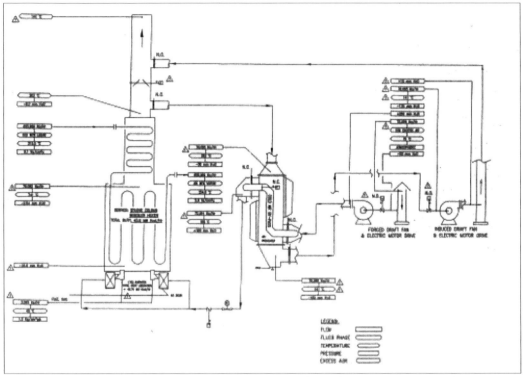

Note: From left ① Fired heater ② Air Pre-Heater ③ Forced Draft Fan (FD Fan) & Electric Motor Drive ④ Induced Draft Fan (ID Fan) & Electric Motor Drive

(2) Two cell horizontal cabin heater

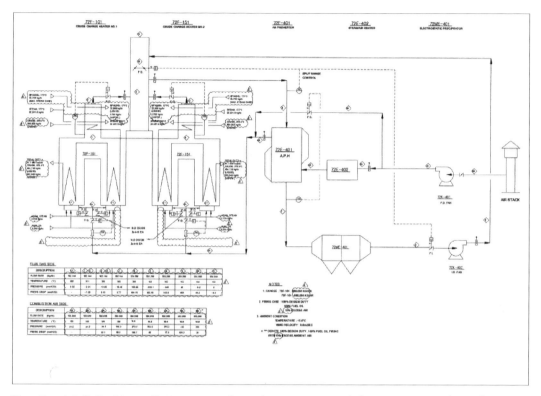

Note: From left ① Fired heater ② Air PreHeater ③ E.P. (Electric Precipitator) ④ Forced Draft Fan (FD Fan) & Electric Motor Drive ⑤ Induced Draft Fan (ID Fan) & Electric Motor Drive

5. Air Supply and Flue Gas Removal

Aside from the major classifications according to service and configuration, fired heaters can also be grouped according to their methods of combustion air supply and flue gas removal.

The capability for inducing the flow of combustion air into a fired heater exists when hot flue gas of relatively low density is confined in a structure and isolated from higher density air at ambient temperature. The buoyancy of the hot flue gas contained in the fired heater creates "draft" (less than atmospheric pressure) which induces flow or air into the combustion chamber.

5.1 Natural draft heater

The majority of fired heaters use natural draft. A stack of appropriate height provides the draft needed to draw in air at the burners; move flue gas through the fire box, convection section and stack; and keep all parts of the heater under negative pressure. Natural draft uses the natural driving force of the hot low density gases in the stack to pull the combustion air into the heater through the burners. Natural draft units have no fans.

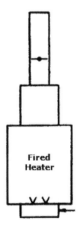

5.2 Induced draft heater

Heaters which have a fan pulling hot flue gases out of the top. An induced draft fired heater incorporates an induced draft fan, instead of a stack, to maintain a negative pressure and to induce the flow of combustion air and the removal of flue gas.

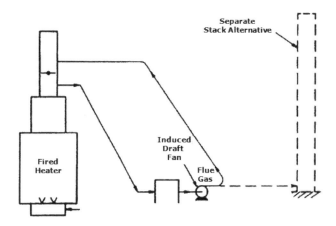

5.3 Forced draft heater

Heaters with a fan to supply combustion air are called forced draft. A forced draft fired heater is one wherein the combustion air is supplied under positive pressure by means of a forced draft fan. It is to be noted that even with air supplied under positive pressure, the combustion chamber and all other parts of the fired heater are maintained under negative pressure, and the flue gas is removed by stack effect.

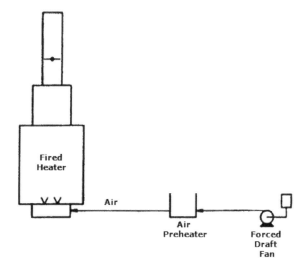

5.4 Balanced draft heater

Heaters which have both fans. A forced draft/induced draft heater uses a forced draft fan to supply combustion air under positive pressure: an induced draft fan maintains the combustion chamber and all other parts of the fired heater under negative pressure and removes the flue gas. Most fired heaters equipped with air preheaters are of the forced draft/induced draft type.

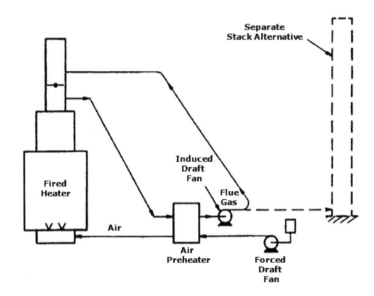

6. Standard Petroleum Refining Units

The below is a schematic flow diagram of a typical petroleum refinery that describes the various refining processes and the flow of intermediate product streams that occurs between the inlet crude oil feedstock and the final end-products.

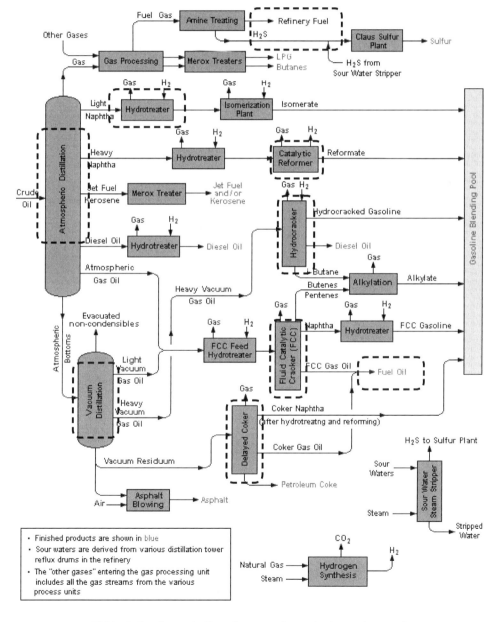

Table 2. A schematic flow diagram of a typical petroleum refinery

6.1 Crude oil distillation unit

The crude oil distillation unit is the first processing unit in virtually all petroleum refineries. The CDU distills the incoming crude oil into various fractions of different boiling ranges, each of which are then processed further in the other refinery processing units. The CDU is often referred to as the atmospheric distillation unit (CDU) because it operates at slightly above atmospheric pressure.

Below is a schematic flow diagram of a typical crude oil distillation unit. The incoming crude oil is preheated by exchanging heat with some of the hot, distilled fractions and other streams. It is then desalted to remove inorganic salts (primarily sodium chloride). Following the desalter, the crude oil is further heated by exchanging heat with some of the hot, distilled fractions and other streams. It is then heated in a fuel-fired furnace (Fired Heater) to a temperature of about 398°C and routed into the bottom of the distillation unit. The cooling and condensing of the distillation tower overhead is provided partially by exchanging heat with the incoming crude oil and partially by either an air-cooled or water-cooled condenser. Additional heat is removed from the distillation column by a pump around system.

The overhead distillate fraction from the distillation column is naphtha. The fractions removed from the side of the distillation column at various points between the column top and bottom are called side cuts. Each of the side cuts (i.e., the kerosene, light gas oil and heavy gas oil) is cooled by exchanging heat with the incoming crude oil. All of the fractions (i.e., the overhead naphtha, the side cuts and the bottom residue) are sent to intermediate storage tanks before being processed further.

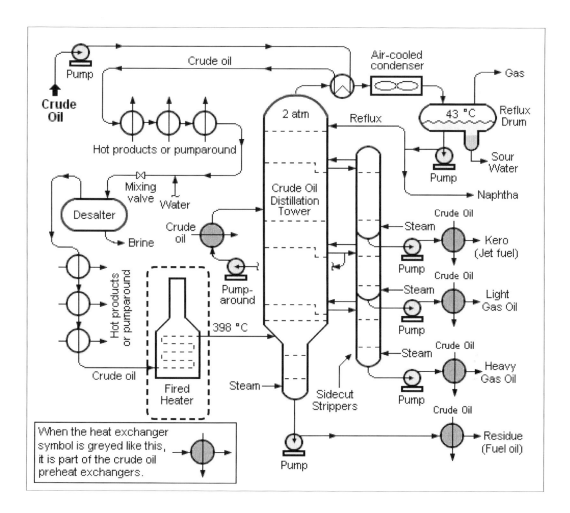

6.2 Vacuum distillation unit

Vacuum distillation has several advantages. Close boiling mixtures may require many equilibrium stages to separate the key components. One tool to reduce the number of stages needed is to utilize vacuum distillation. Vacuum distillation increases the relative volatility of the key components in many applications. The higher the relative volatility, the more separable are the two components; this denotes fewer stages in a distillation column in order to effect the same separation between the overhead and bottoms products. Lower pressures increase relative volatilities in most systems.

A second advantage of vacuum distillation is the reduced temperature requirement at lower pressures. For many systems, the products degrade or polymerize at elevated temperatures. Vacuum distillation can improve a separation by: 1) Prevention of product degradation or polymer formation because of reduced pressure leading to lower tower bottoms temperatures 2) Reduction of product degradation or polymer formation because of reduced mean residence time especially in columns using packing rather than trays 3) Increasing capacity, yield, and purity.

In distilling the crude oil, it is important not to subject the crude oil to temperatures above 370 to 380°C because the high molecular weight components in the crude oil will undergo thermal cracking and form petroleum coke at temperatures above that. Formation of coke would result in plugging the tubes in the furnace that heats the feed stream to the crude oil distillation column. Plugging would also occur in the piping from the furnace to the distillation column as well as in the column itself.

The constraint imposed by limiting the column inlet crude oil to a temperature of less than 370 to 380°C yields a residual oil from the bottom of the atmospheric distillation column consisting entirely of hydrocarbons that boil above 370 to 380°C. To further distill the residual oil from the atmospheric distillation column, the distillation must be performed at absolute pressures as low as 10 to 40mmHg (also referred to as Torr) so as to limit the operating temperature to less than 370 to 380°C.

The 10 to 40mmHg absolute pressure in a vacuum distillation column increases the volume of vapor formed per volume of liquid distilled. The result is that such columns have very large diameters. Distillation columns may have diameters of 5m or more, heights ranging up to about 50m, and feed rates ranging up to about 25,400m³/day (160,000 barrels/day).

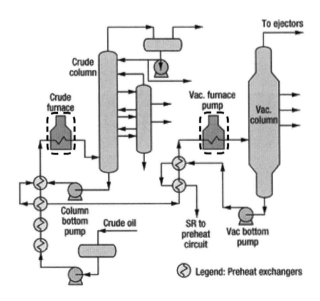

6.3 Hydrotreater process unit

Hydrodesulfurization (HDS) is a catalytic chemical process widely used to remove sulfur (S) from natural gas and from refined petroleum products such as gasoline or petrol, jet fuel, kerosene, diesel fuel, and fuel oils.

The industrial hydrodesulfurization processes include facilities for the capture and removal of the resulting hydrogen sulfide (H_2S) gas. In petroleum refineries, the hydrogen sulfide gas is then subsequently converted into byproduct elemental sulfur or sulfuric acid (H_2SO_4). An HDS unit in the petroleum refining industry is also often referred to as a hydrotreater.

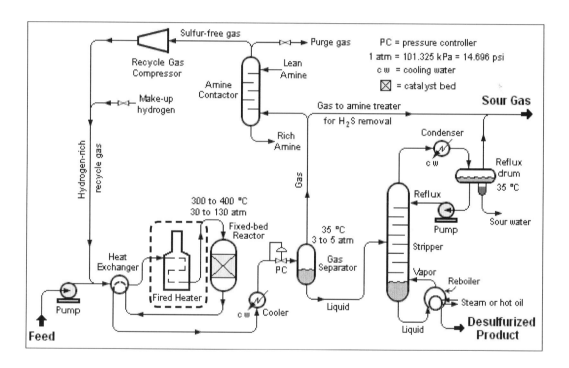

The liquid feed (at the bottom left in the diagram) is pumped up to the required elevated pressure and is joined by a stream of hydrogen rich recycle gas. The resulting liquid gas mixture is preheated by flowing through a heat exchanger. The preheated feed then flows through a fired heater where the feed mixture is totally vaporized and heated to the required elevated temperature before entering the reactor and flowing through a fixed bed of catalyst where the hydrodesulfurization reaction takes place.

The hot reaction products are partially cooled by flowing through the heat exchanger where the reactor feed was preheated and then flows through a water cooled heat exchanger before it flows through the pressure controller (PC) and undergoes a pressure reduction down to about 3 to $5kg_f/m^2$. The resulting mixture of liquid and gas enters the gas separator vessel at about 35°C and 3 to $5kg_f/m^2$.

Most of the hydrogen rich gas from the gas separator vessel is recycle gas, which is routed through an amine contactor for removal of the reaction product H_2S that it contains. The H_2S free hydrogen rich gas is then recycled back for reuse in the reactor section. Any excess gas from the gas separator vessel joins the sour gas from the stripping of the reaction product liquid. The liquid from the gas separator vessel is routed through a reboiled stripper distillation tower. The bottoms product from the stripper is the final desulfurized liquid product from hydrodesulfurization unit.

The overhead sour gas from the stripper contains hydrogen, methane, ethane, hydrogen sulfide, propane, and, perhaps, some butane and heavier components. That sour gas is sent to the refinery's central gas processing plant for removal of the hydrogen sulfide in the refinery's main amine gas treating unit and through a series of distillation towers for recovery of propane, butane and pentane or heavier components. The residual hydrogen, methane, ethane, and some propane is used as refinery fuel gas.

The refinery HDS feedstocks (naphtha, kerosene, diesel oil, and heavier oils) contain a wide range of organic sulfur compounds, including thiols, thiophenes, organic sulfides and disulfides, and many others. When the HDS process is used to desulfurize a refinery naphtha, it is necessary to remove the total sulfur down to ppm or lower in order to prevent poisoning the noble metal catalysts in the subsequent catalytic reforming of the naphthas.

When the process is used for desulfurizing diesel oils, the latest environmental regulations in the United States and Europe, requiring what is referred to as ultralow sulfur diesel (ULSD), in turn requires that very deep hydrodesulfurization is needed.

6.4 Catalytic reforming unit

Catalytic reforming is a chemical process used to convert petroleum refinery naphthas distilled from crude oil (typically having low octane ratings) into high octane liquid products called reformates, which are premium blending stocks for high octane gasoline. The process converts low octane linear hydrocarbons (paraffins) into branched alkanes (isoparaffins) and cyclic naphthenes, which are then partially dehydrogenated to produce high octane aromatic hydrocarbons. The dehydrogenation also produces significant amounts of byproduct hydrogen gas, which is fed into other refinery processes such as hydrocracking. A side reaction is hydrogenolysis, which produces light hydrocarbons of lower value, such as methane, ethane, propane and butanes.

In addition to a gasoline blending stock, reformate is the main source of aromatic bulk chemicals such as benzene, toluene, xylene and ethylbenzene which have diverse uses, most importantly as raw materials for conversion into plastics. There are many chemical reactions that occur in the catalytic reforming process, all of which occur in the presence of a catalyst and a high partial pressure of hydrogen. Depending upon the type or version of catalytic reforming used as well as the desired reaction severity, the reaction conditions range from temperatures of about 495 to 525°C and from pressures of about 5 to 45kg$_f$/cm^2.

The commonly used catalytic reforming catalysts contain noble metals such as platinum and/or rhenium, which are very susceptible to poisoning by sulfur and nitrogen compounds. Therefore, the naphtha feedstock to a catalytic reformer is always pre-processed in a hydrodesulfurization unit which removes both the sulfur and the nitrogen compounds. Most catalysts require both sulfur and nitrogen content to be lower than 1ppm.

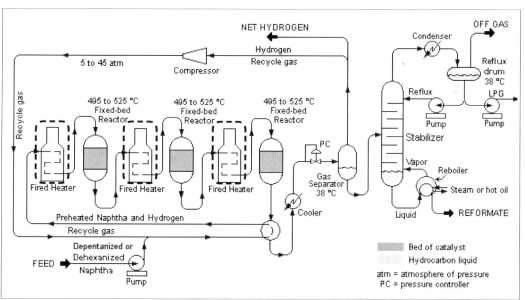

The most commonly used type of catalytic reforming unit has three reactors, each with a fixed bed of catalyst, and all of the catalyst is regenerated in-situ during routine catalyst regeneration shutdowns which occur approximately once each 6

to 24 months. Such a unit is referred to as a semi-regenerative catalytic reformer (SRR). The latest and most modern type of catalytic reformers are called continuous catalyst regeneration (CCR) reformers. Such units are characterized by continuous in-situ regeneration of part of the catalyst in a special regenerator, and by continuous addition of the regenerated catalyst to the operating reactors. The process flow diagram above describes a typical semi-regenerative catalytic reforming unit.

6.5 Fluid catalytic cracking unit

Fluid catalytic cracking (FCC) is one of the most important conversion processes used in petroleum refineries. It is widely used to convert the high boiling, high molecular weight hydrocarbon fractions of petroleum crude oils to more valuable gasoline, olefinic gases, and other products. Cracking of petroleum hydrocarbons was originally done by thermal cracking, which has been almost completely replaced by catalytic cracking because it produces more gasoline with a higher octane rating. It also produces byproduct gases that are more olefinic, and hence more valuable, than those produced by thermal cracking.

The feedstock to an FCC is usually that portion of the crude oil that has an initial boiling point of 340°C or higher at atmospheric pressure and an average molecular weight ranging from about 200 to 600°C or higher. This portion of crude oil is often referred to as heavy gas oil or vacuum gas oil (HVGO). The FCC process vaporizes and breaks the long chain molecules of the high boiling hydrocarbon liquids into much shorter molecules by contacting the feedstock, at high temperature and moderate pressure, with a fluidized powdered catalyst.

The reactor and regenerator are considered to be the heart of the fluid catalytic cracking unit. The schematic flow diagram of a typical modern FCC unit in Fig. below is based upon the "side-by-side" configuration. The preheated high boiling petroleum feedstock (at about 315 to 430°C) consisting of long chain hydrocarbon molecules is combined with recycle slurry oil from the bottom of the distillation column and injected into the catalyst riser where it is vaporized and cracked into smaller molecules of vapor by contact and mixing with the very hot powdered catalyst from the regenerator. All of the cracking reactions take place in the catalyst riser within a period of 2-4 seconds. The hydrocarbon vapors "fluidize" the powdered catalyst and the mixture of hydrocarbon vapors and catalyst flows upward to enter the reactor at a temperature of about 535 °C and a pressure of about 1.72 barg.

The reactor is a vessel in which the cracked product vapors are (a) separated from the so called spent catalyst by flowing through a set of two stage cyclones within the reactor and (b) the spent catalyst flows downward through a steam stripping section to remove any hydrocarbon vapors before the spent catalyst returns to the catalyst regenerator. The flow of spent catalyst to the regenerator is regulated by a slide valve in the spent catalyst line.

Since the cracking reactions produce some carbonaceous material (referred to as catalyst coke) that deposits on the catalyst and very quickly reduces the catalyst reactivity, the catalyst is regenerated by burning off the deposited coke with air blown into the regenerator. The combustion of the coke is exothermic and it produces a large amount of heat that is partially absorbed by the regenerated catalyst and provides the heat required for the vaporization of the feedstock and the endothermic cracking reactions that take place in the catalyst riser. For that reason, FCC units are often referred to as being 'heat balanced'.

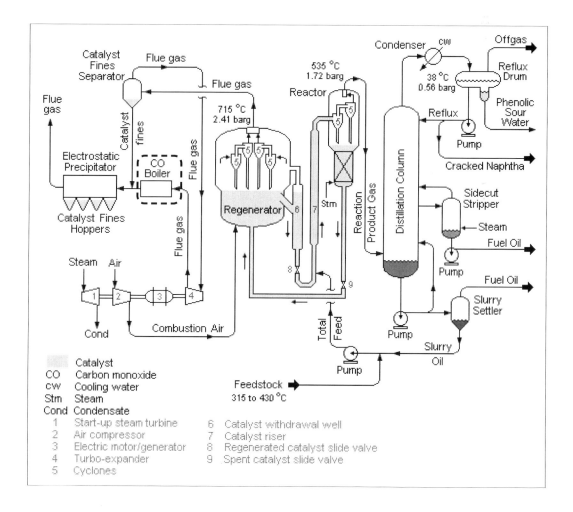

6.6 Hydrocracking process unit

Hydrocracking is a catalytic cracking process assisted by the presence of added hydrogen gas. Unlike a hydrotreater, where hydrogen is used to cleave C-S and C-N bonds, hydrocracking uses hydrogen to break C-C bonds (hydrotreatment is conducted prior to hydrocracking to protect the catalysts in a hydrocracking).

The products of this process are saturated hydrocarbons; depending on the reaction conditions (temperature, pressure, catalyst activity) these products range from ethane, LPG to heavier hydrocarbons consisting mostly of isoparaffins. Hydrocracking is normally facilitated by a bifunctional catalyst that is capable of rearranging and breaking hydrocarbon chains as well as adding hydrogen to aromatics and olefins to produce naphthenes and alkanes. The major products from hydrocracking are jet fuel and diesel, but low sulfur naphtha fractions and LPG are also produced. All these products have a very low content of sulfur and other contaminants.

The hydrocracking process depends on the nature of the feedstock and the relative rates of the two competing reactions, hydrogenation and cracking. Heavy aromatic feedstock is converted into lighter products under a wide range of very high pressures (70-140kg$_f$/cm^2) and fairly high temperatures (399-816°C), in the presence of hydrogen and special catalysts.

The primary function of hydrogen is, thus: a) If feedstock has a high paraffinic content, the primary function of hydrogen is to prevent the formation of polycyclic aromatic compounds. b) Reduced tar formation c) Reduced Impurities d) Prevent buildup of coke on the catalyst. e) High cetane fuel is achieved.

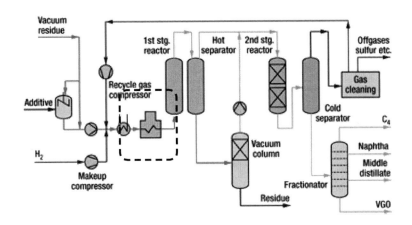

6.7 Visbreaker process unit

A visbreaker is a processing unit in an oil refinery whose purpose is to reduce the quantity of residual oil produced in the distillation of crude oil and to increase the yield of more valuable middle distillates (heating oil and diesel) by the refinery. A visbreaker thermally cracks large hydrocarbon molecules in the oil by heating in a furnace to reduce its viscosity and to produce small quantities of light hydrocarbons (LPG and gasoline). The process name of "visbreaker" refers to the fact that the process reduces (i.e., breaks) the viscosity of the residual oil. The process is non-catalytic.

The term coil (or furnace) visbreaking is applied to units where the cracking process occurs in the furnace tubes (or "coils"). Material exiting the furnace is quenched to halt the cracking reactions: frequently this is achieved by heat exchange with the virgin material being fed to the furnace, which in turn is a good energy efficiency step, but sometimes a stream of cold oil (usually gas oil) is used to the same effect. The gas oil is recovered and re-used. The extent of the cracking reaction is controlled by regulation of the speed of flow of the oil through the furnace tubes. The quenched oil then passes to a fractionator where the products of the cracking (gas, LPG, gasoline, gas oil and tar) are separated and recovered.

In Soaker Visbreaking, the bulk of the cracking reaction occurs not in the furnace but in a drum located after the furnace called the soaker. Here the oil is held at an elevated temperature for a pre-determined period of time to allow cracking to occur before being quenched. The oil then passes to a fractionator. In soaker visbreaking, lower temperatures are used than in coil visbreaking. The comparatively long duration of the cracking reaction is used instead.

Visbreaker tar can be further refined by feeding it to a vacuum fractionator. Here additional heavy gas oil may be recovered and routed either to catalytic cracking, hydrocracking or thermal cracking units on the refinery. The vacuum-flashed tar (sometimes referred to as pitch) is then routed to fuel oil blending. In a few refinery locations, visbreaker tar is routed to a delayed coker for the production of certain specialist cokes such as anode coke or needle coke.

The yields of the various hydrocarbon products will depend on the "severity" of the cracking operation as determined by the temperature the oil is heated to in the visbreaker furnace. At the low end of the scale, a furnace heating to 425°C would crack only mildly, while operations at 500°C would be considered as very severe.

Arabian light crude residue when visbroken at 450°C would yield around 76% (by weight) of tar, 15% middle distillates, 6% gasolines and 3% gas and LPG.

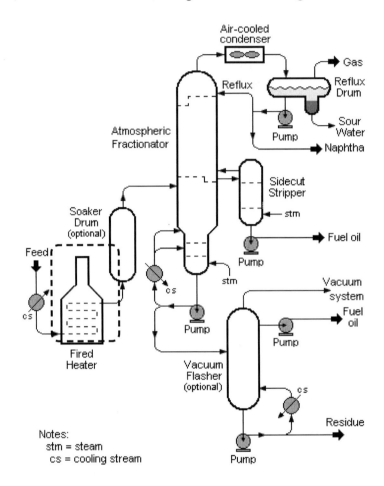

6.8 Delayed coker process unit

A delayed coker is a type of coker whose process consists of heating a residual oil feed to its thermal cracking temperature in a furnace with multiple parallel passes. This cracks the heavy, long chain hydrocarbon molecules of the residual oil into coker gas oil and petroleum coke. Delayed coking is one of the unit processes used in many oil refineries. The yield of coke from the delayed coking process ranges from about 18 to 30 wt% of the feedstock residual oil, depending on the composition of the feedstock and the operating variables.

The flow diagram and description are based on a delayed coking unit with a single pair of coke drums and one feedstock furnace.

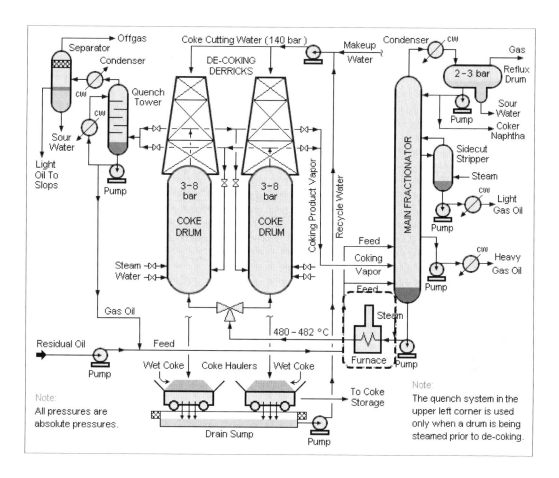

Residual oil from the vacuum distillation unit (sometimes including high boiling oils from other sources within the refinery) is pumped into the bottom of the distillation column called the main fractionator. From there it is pumped, along with some injected steam, into the fuel-fired furnace and heated to its thermal cracking temperature of about 480°C. Thermal cracking begins in the pipe between the furnace and the coke drums, and finishes in the coke drum that is onstream. The injected steam helps to minimize the deposition of coke within the furnace tubes.

Pumping the incoming residual oil into the bottom of the main fractionator, rather than directly into the furnace, preheats the residual oil by having it contact the hot vapors in the bottom of the fractionator. At the same time, some of the hot vapors condense into a high boiling liquid which recycles back into the furnace along with the hot residual oil.

As cracking takes place in the drum, gas oil and lighter components are generated in vapor phase and separate from the liquid and solids. The drum effluent is vapor except for any liquid or solids entrainment, and is directed to main fractionator where it is separated into the desired boiling point fractions. The solid coke is deposited and remains in the coke drum in a porous structure that allows flow through the pores. Depending upon the overall coke drum cycle being used, a coke drum may fill in 16 to 24 hrs.

After the drum is full of the solidified coke, the hot mixture from the furnace is switched to the second drum. While the second drum is filling, the full drum is steamed out to reduce the hydrocarbon content of the petroleum coke, and then quenched with water to cool it. The top and bottom heads of the full coke drum are removed, and the solid petroleum coke is then cut from the coke drum with a high pressure water nozzle, where it falls into a pit, pad, or sluiceway for reclamation to storage.

6.9 Steam cracker (petrochemical)

Steam cracking is a petrochemical process in which saturated hydrocarbons are broken down into smaller, often unsaturated, hydrocarbons. It is the principal industrial method for producing the lighter alkenes (or commonly olefins), including ethene (or ethylene) and propene (or propylene). Steam cracker units are facilities in which a feedstock such as naphtha, liquefied petroleum gas (LPG), ethane, propane or butane is thermally cracked through the use of steam in a bank of pyrolysis furnaces to produce lighter hydrocarbons. The products obtained depend on the composition of the feed, the hydrocarbon-to-steam ratio, and on the cracking temperature and furnace residence time.

In steam cracking, a gaseous or liquid hydrocarbon feed like naphtha, LPG or ethane is diluted with steam and briefly heated in a furnace without the presence of oxygen. Typically, the reaction temperature is very high, at around 850°C, but the reaction is only allowed to take place very briefly. In modern cracking furnaces, the residence time is reduced to milliseconds to improve yield, resulting in gas velocities faster than the speed of sound. After the cracking temperature has been reached, the gas is quickly quenched to stop the reaction in a transfer line heat exchanger or inside a quenching header using quench oil.

Light hydrocarbon feeds such as ethane, LPGs or light naphtha give product streams rich in the lighter alkenes, including ethylene, propylene, and butadiene. Heavier hydrocarbon (full range and heavy naphthas as well as other refinery products) feeds give some of these, but also give products rich in aromatic hydrocarbons and hydrocarbons suitable for inclusion in gasoline or fuel oil.

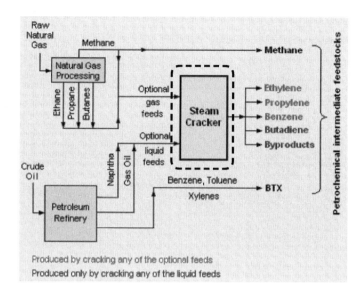

A higher cracking temperature (also referred to as severity) favors the production of ethene and benzene, whereas lower severity produces higher amounts of propene, C_4 hydrocarbons and liquid products. The process also results in the slow deposition of coke, a form of carbon, on the reactor walls. This degrades the efficiency of the reactor, so reaction conditions are designed to minimize this. Nonetheless, a steam cracking furnace can usually only run for a few months at a time between decokings. Decokes require the furnace to be isolated from the process and then a flow of steam or a steam/air mixture is passed through the furnace coils (steam/air decoking). This converts the hard solid carbon layer to carbon monoxide and carbon dioxide. Once this reaction is complete, the furnace can be returned to service.

6.10 Steam reforming unit

Steam reforming is a method for producing hydrogen, carbon monoxide or other useful products from hydrocarbon fuels such as natural gas. This is achieved in a processing device called a reformer which reacts steam at high temperature with the fossil fuel. The steam methane reformer is widely used in industry to make hydrogen. There is also interest in the development of much smaller units based on similar technology to produce hydrogen as a feedstock for fuel cells. Small scale steam reforming units to supply fuel cells are currently the subject of research and development, typically involving the reforming of methanol, but other fuels are also being considered such as propane, gasoline, autogas, diesel fuel, and ethanol.

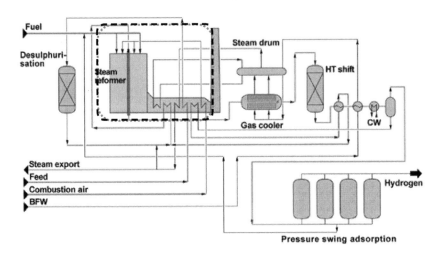

Steam reforming of natural gas- sometimes referred to as steam methane reforming (SMR)- is the most common method of producing commercial bulk hydrogen. Hydrogen is used in the industrial synthesis of ammonia and other chemicals. At high temperatures (700-1,100°C) and in the presence of a metal based catalyst (nickel), steam reacts with methane to yield carbon monoxide and hydrogen.

$$CH_4 + H_2O \rightleftharpoons CO + 3H_2$$

Additional hydrogen can be recovered by a lower temperature gas shift reaction with the carbon monoxide produced. The reaction is summarized by

$$CO + H_2O \rightleftharpoons CO_2 + H_2$$

The first reaction is strongly endothermic (consumes heat, ΔH_r= 49.3kcal/mol), the second reaction is mildly exothermic (produces heat, ΔH_r= -9.8 kcal/mol).

This SMR process is quite different from and not to be confused with catalytic reforming of naphtha, an oil refinery process that also produces significant amounts of hydrogen along with high octane gasoline. SMR is approximately 65-75% efficient.

In contrast to conventional steam reforming, the process is operated at lower temperatures and with lower steam supply, allowing a high content of methane (CH_4)

in the produced fuel gas. The main reactions are

Steam reforming: $C_nH_m + nH_2O \leftrightarrow (n + m/2)H_2 + nCO$

Methanation: $CO + 3H_2 \leftrightarrow CH_4 + H_2O$

Water gas shift: $CO + H_2O \leftrightarrow H_2 + CO_2$

6.11 Alkylation unit

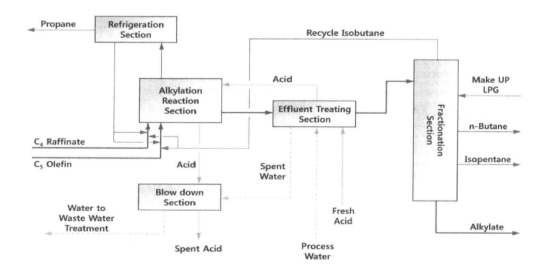

Alkylation of alkenes (shown here is propene) by isobutane is a major process in refineries. It is catalyzed by strong acids such as HF and sulfuric acid.

The product is called alkylate and is composed of a mixture of high octane, branched chain paraffinic hydrocarbons (mostly isoheptane and isooctane). Alkylate is a premium gasoline blending stock because it has exceptional antiknock properties and is clean burning. Alkylate is also a key component of jet fuels. The octane number of the alkylate depends mainly upon the kind of alkenes used and upon operating conditions. For example, isooctane results from combining butylene with isobutane and has an octane rating of 100 by definition. There are other products in the alkylate, so the octane rating will vary accordingly.

6.12 Isomerization unit

Isomerization in hydrocarbon cracking is usually employed in organic chemistry, where fuels, such as pentane, a straight chain isomer, are heated in the presence of a platinum catalyst. The resulting mixture of straight and branched chain isomers then have to be separated. An industrial process is also the isomerization of n-pentane into isobutane.

pentane   2-methylbutane 2,2-dimethylpropane

6.13 Merox unit

Merox is an acronym for mercaptan oxidation. It is a proprietary catalytic chemical process developed by UOP used in oil refineries and natural gas processing plants to remove mercaptans from LPG, propane, butanes, light naphthas, kerosene and jet fuel by converting them to liquid hydrocarbon disulfides.

The Merox process requires an alkaline environment which, in some process versions, is provided by an aqueous solution of sodium hydroxide (NaOH), a strong base, commonly referred to as caustic. In other versions of the process, the alkalinity is provided by ammonia, which is a weak base. The catalyst in some versions of the process is a water-soluble liquid. In other versions, the catalyst is impregnated onto charcoal granules.

UOP has developed many versions of the Merox process for various applications. In all of the Merox versions, the overall oxidation reaction that takes place in converting mercaptans to disulfide is

$$4RSH + O_2 \rightarrow 2RSSR + 2H_2O$$

Process flow diagrams and descriptions of the two conventional versions of the Merox process are presented as follows.

The conventional Merox process for extraction and removal of mercaptans from liquefied petroleum gases (LPG), such as propane, butanes and mixtures of propane and butanes, can also be used to extract and remove mercaptans from light naphthas. It is a two step process. In the first step, the feedstock LPG or light naphtha is contacted in the trayed extractor vessel with an aqueous caustic solution containing UOP's proprietary liquid catalyst. The caustic solution reacts with mercaptans and extracts them. The reaction that takes place in the extractor is

$$2RSH + 2NaOH \rightarrow 2NaSR + 2H_2O$$

In the above reaction, RSH is a mercaptan and R signifies an organic group such as a methyl, ethyl, propyl or other group. For example, the ethyl mercaptan (ethane thiol) has the formula C_2H_5SH.

The second step is referred to as regeneration and it involves heating and oxidizing of the caustic solution leaving the extractor. The oxidations results in converting the extracted mercaptans to organic disulfides (RSSR) which are liquids that are water-insoluble and are then separated and decanted from the aqueous caustic solution. The reaction that takes place in the regeneration step is

$$4NaSR + O_2 + 2H_2O \rightarrow 2RSSR + 4NaOH$$

After decantation of the disulfides, the regenerated "lean" caustic solution is recirculated back to the top of the extractor to continue extracting mercaptans.

The net overall Merox reaction covering the extraction and the regeneration step may be expressed as

$$4RSH + O_2 \rightarrow 2RSSR + 2H_2O$$

The feedstock entering the extractor must be free of any H_2S. Otherwise, any H_2S entering the extractor would react with the circulating caustic solution and interfere with the Merox reactions. Therefore, the feedstock is first "prewashed" by flowing through a batch of aqueous caustic to remove any H_2S. The reaction that takes place in the prewash vessel is

$$H_2S + NaOH \rightarrow NaSH + H_2O$$

The batch of caustic solution in the prewash vessel is periodically discarded as "spent caustic" and replaced by fresh caustic as needed.

The flow diagram below describes the equipment and the flow paths involved in the process. The LPG (or light naphtha) feedstock enters the prewash vessel and flows upward through a batch of caustic which removes any H_2S that may be present in the feedstock. The coalescer at the top of the prewash vessel prevents caustic from being entrained and carried out of the vessel.

The feedstock then enters the mercaptan extractor and flows upward through the contact trays where the LPG intimately contacts the down flowing Merox caustic that extracts the mercaptans from the LPG. The sweetened LPG exits the tower and flows through: a caustic settler vessel to remove any entrained caustic, a water wash vessel to further remove any residual entrained caustic and a vessel containing a bed of rock salt to remove any entrained water. The dry sweetened LPG exits the Merox unit.

The caustic solution leaving the bottom of the mercaptan extractor ("rich" Merox caustic) flows through a control valve which maintains the extractor pressure needed to keep the LPG liquified. It is then injected with UOP's proprietary liquid catalyst, flows through a steam-heated heat exchanger and is injected with compressed air before entering the oxidizer vessel where the extracted mercaptans are converted to disulfides. The oxidizer vessel has a packed bed to keep the aqueous caustic and the water-insoluble disulfide well contacted and well mixed.

The caustic disulfide mixture then flows into the separator vessel where it is allowed to form a lower layer of "lean" Merox caustic and an upper layer of disulfides. The vertical section of the separator is for the disengagement and venting of excess air and includes a Raschig ring section to prevent entrainment of any disulfides in the vented air. The disulfides are withdrawn from the separator and routed to fuel storage or to a hydrotreater unit. The regenerated lean Merox caustic is then pumped back to the top of the extractor for reuse.

The conventional Merox process for the removal of mercaptans (i.e., sweetening) of jet fuel or kerosene is a one step process. The mercaptan oxidation reaction takes place in an alkaline environment as the feedstock jet fuel or kerosene, mixed with compressed air, flows through a fixed bed of catalyst in a reactor vessel. The catalyst consists of charcoal granules that have been impregnated with UOP's proprietary catalyst. The oxidation reaction that takes place is

$$4RSH + O_2 \rightarrow 2RSSR + 2H_2O$$

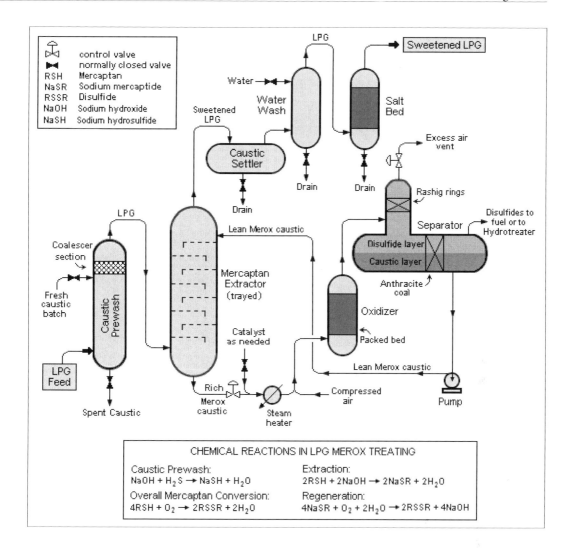

control valve
normally closed valve
RSH   Mercaptan
NaSR   Sodium mercaptide
RSSR   Disulfide
NaOH   Sodium hydroxide
NaSH   Sodium hydrosulfide

LPG

Sweetened LPG

Water

Water
Wash

Salt
Bed

Sweetened
LPG

Excess air
vent

Caustic
Settler

Rashig rings

Drain

Drain

Disulfides to
fuel or to
Hydrotreater

Separator

LPG

Drain

Lean Merox caustic

Coalescer
section

Disulfide layer

Mercaptan
Extractor
(trayed)

Caustic layer

Anthracite
coal

Fresh
caustic
batch

Oxidizer

Packed bed

Catalyst
as needed

LPG
Feed

Lean Merox caustic

Rich
Merox
caustic

Compressed
air

Spent Caustic

Steam
heater

Pump

CHEMICAL REACTIONS IN LPG MEROX TREATING

Caustic Prewash:
$NaOH + H_2S \rightarrow NaSH + H_2O$

Extraction:
$2RSH + 2NaOH \rightarrow 2NaSR + 2H_2O$

Overall Mercaptan Conversion:
$4RSH + O_2 \rightarrow 2RSSR + 2H_2O$

Regeneration:
$4NaSR + O_2 + 2H_2O \rightarrow 2RSSR + 4NaOH$

As is the case with the conventional Merox process for treating LPG, the jet fuel or kerosene sweetening process also requires that the feedstock be prewashed to remove any H_2S that would interfere with the sweetening. The reaction that takes place in the batch caustic prewash vessel is

$$H_2S + NaOH \rightarrow NaSH + H_2O$$

The Merox reactor is a vertical vessel containing a bed of charcoal granules that have been impregnated with the UOP catalyst. The charcoal granules may be impregnated with the catalyst in-situ or they may be purchased from UOP as pre-impregnated with the catalyst. An alkaline environment is provided by caustic being pumped into reactor on an intermittent, as needed basis.

The jet fuel or kerosene feedstock from the top of the caustic prewash vessel is injected with compressed air and enters the top of the Merox reactor vessel along with any injected caustic. The mercaptan oxidation reaction takes place as the feedstock permeates downward over the catalyst. The reactor effluent flows through a caustic settler vessel where it forms a bottom layer of aqueous caustic solution and an upper layer of water-insoluble sweetened product.

The caustic solution remains in the caustic settler so that the vessel contains a reservoir for the supply of caustic that is intermittently pumped into the reactor to maintain the alkaline environment.

The sweetened product from the caustic settler vessel flows through a water wash vessel to remove any entrained caustic as well as any other unwanted water-soluble substances, followed by flowing through a salt bed vessel to remove any entrained water and finally through a clay filter vessel. The clay filter removes any oil-soluble substances, organometallic compounds (especially copper) and particulate matter, which might prevent meeting jet fuel product specifications.

The pressure maintained in the reactor is chosen so that the injected air will completely dissolve in the feedstock at the operating temperature.

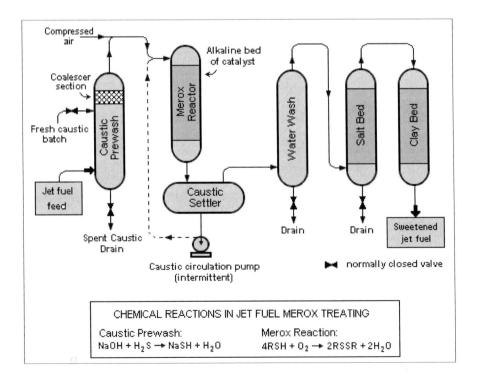

6.14 Amine gas treating unit

Amine gas treating, also known as amine scrubbing, gas sweetening and acid gas removal, refers to a group of processes that use aqueous solutions of various alkylamines (commonly referred to simply as amines) to remove hydrogen sulfide (H_2S) and carbon dioxide (CO_2) from gases. It is a common unit process used in refineries, and is also used in petrochemical plants, natural gas processing plants and other industries.

Processes within oil refineries or chemical processing plants that remove hydrogen sulfide are referred to as "sweetening" processes because the odor of the processed products is improved by the absence of hydrogen sulfide. An alternative to the use of amines involves membrane technology. However, membranes are less attractive since the relatively high capital and operation costs as well as other technical factors.

Many different amines are used in gas treating: Diethanolamine (DEA), Monoethanolamine (MEA), Methyldiethanolamine (MDEA), Diisopropanolamine (DIPA), Aminoethoxyethanol (Diglycolamine) (DGA),

The most commonly used amines in industrial plants are the alkanolamines DEA, MEA, and MDEA. These amines are also used in many oil refineries to remove sour gases from liquid hydrocarbons such as liquified petroleum gas (LPG).

Gases containing H_2S or both H_2S and CO_2 are commonly referred to as sour gases or acid gases in the hydrocarbon processing industries. The chemistry involved in the amine treating of such gases varies somewhat with the particular amine being used. For one of the more common amines, monoethanolamine (MEA) denoted as RNH_2, the chemistry may be expressed as

$$RNH_2 + H_2S \Longleftrightarrow RNH_3^+ + SH^-$$

A typical amine gas treating process (as shown in the flow diagram below) includes an absorber unit and a regenerator unit as well as accessory equipment. In the absorber, the down flowing amine solution absorbs H_2S and CO_2 from the up flowing sour gas to produce a sweetened gas stream (i.e., a gas free of hydrogen sulfide and carbon dioxide) as a product and an amine solution rich in the absorbed acid gases. The resultant "rich" amine is then routed into the regenerator (a stripper with a reboiler) to produce regenerated or "lean" amine that is recycled for reuse in the absorber. The stripped overhead gas from the regenerator is concentrated H_2S and CO_2.

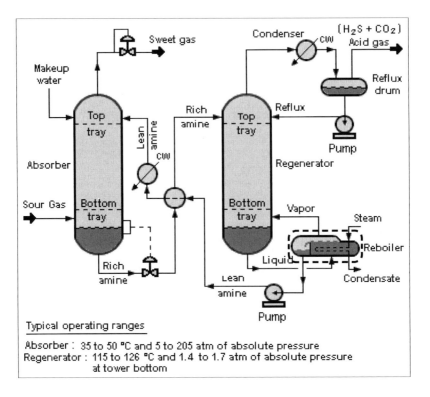

Typical operating ranges

Absorber : 35 to 50 °C and 5 to 205 atm of absolute pressure
Regenerator : 115 to 126 °C and 1.4 to 1.7 atm of absolute pressure
at tower bottom

The choice of amine concentration in the circulating aqueous solution depends upon a number of factors and may be quite arbitrary. It is usually made simply on the basis of experience. The factors involved include whether the amine unit is treating raw natural gas or petroleum refinery by-product gases that contain relatively low concentrations of both H_2S and CO_2 or whether the unit is treating gases with a high percentage of CO_2 such as the off gas from the steam reforming process or the flue gases from power plants.

Both H_2S and CO_2 are acid gases and hence corrosive to carbon steel. However, in an amine treating unit, CO_2 is the stronger acid of the two. H_2S forms a film of iron sulfide on the surface of the steel that acts to protect the steel. When treating gases with a high percentage of CO_2, corrosion inhibitors are often used and that permits the use of higher concentrations of amine in the circulating solution.

In oil refineries, that stripped gas is mostly H_2S, much of which often comes from a sulfur-removing process called hydrodesulfurization. This H_2S-rich stripped gas stream is then usually routed into a Claus process to convert it into elemental sulfur.

6.15 Sulfur recovery unit (Claus process)

The Claus process is the most significant gas desulfurizing process, recovering elemental sulfur from gaseous hydrogen sulfide. First patented in 1883 by the scientist Carl Friedrich Claus, the Claus process has become the industry standard.

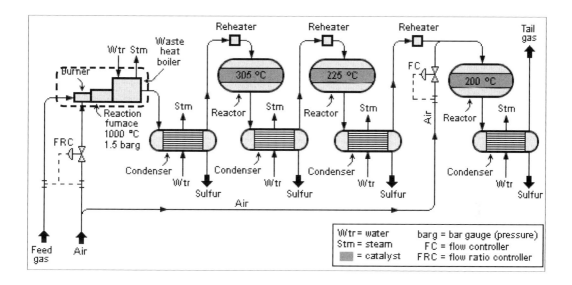

The Claus technology can be divided into two process steps, thermal and catalytic. In the thermal step, hydrogen sulfide-laden gas reacts in a substoichiometric combustion at temperatures above 850°C such that elemental sulfur precipitates in the downstream process gas cooler.

The H_2S content and the concentration of other combustible components (hydrocarbons or ammonia) determine the location where the feed gas is burned. Claus gases (acid gas) with no further combustible contents apart from H_2S are burned in lances surrounding a central muffle by the following chemical reaction.

$$2H_2S + 3O_2 \rightarrow 2SO_2 + 2H_2O \; (\Delta H = -992 \text{ kcal/mol})$$

This is a strongly exothermic free flame total oxidation of hydrogen sulfide generating sulfur dioxide that reacts away in subsequent reactions. The most important one is the Claus reaction.

$$2H_2S + SO_2 \rightarrow 3S + 2 H_2O$$

The overall equation is

$$10H_2S + 5O_2 \rightarrow 2H_2S + SO_2 + 7/2S_2 + 8H_2O$$

This equation shows that in the thermal step alone two thirds of the hydrogen sulfide can be converted to sulfur.

The Claus reaction continues in the catalytic step with activated aluminum or titanium oxide, and serves to boost the sulfur yield. More hydrogen sulfide (H_2S) reacts with the SO_2 formed during combustion in the reaction furnace in the Claus reaction, and results in gaseous, elemental sulfur.

$$2H_2S + SO_2 \rightarrow 3S + 2H_2O \ (\Delta H = -279 \ kcal/mol)$$

This sulfur can be S_6, S_7, S_8 or S_9.

The catalytic recovery of sulfur consists of three sub-steps: heating, catalytic reaction and cooling plus condensation. These three steps are normally repeated a maximum of three times. Where an incineration or tail gas treatment unit (TGTU) is added downstream of the Claus plant, only two catalytic stages are usually installed.

In the sulfur condenser, the process gas coming from the catalytic reactor is cooled to between 150 and 130°C. The condensation heat is used to generate steam at the shell side of the condenser.

Before storage, liquid sulfur streams from the process gas cooler, the sulfur condensers and from the final sulfur separator are routed to the degassing unit, where the gases (primarily H_2S) dissolved in the sulfur are removed.

The tail gas from the Claus process still containing combustible components and sulfur compounds (H_2S, H_2 and CO) is either burned in an incineration unit or further desulfurized in a downstream tail gas treatment unit.

7. Design Basis for Fired Heater Rating

The main parameters utilized to design a fired heater are basically two: one is the required efficiency, the other is the heat absorbed in one hour per square meter of transfer surface (heat flux).

Obviously higher heat flux means lower installed surface and therefore a more economical design. Normally the design heat flux depends on which process fluid is considered and its maximum acceptable film temperature. The heat flux raises the fluid temperature in contact with the internal tube wall as follows:

$$\Delta T = \Phi / h_{io}$$

Where, ΔT differential temperature bulk/film

Φ heat flux

h_{io} internal heat transfer coefficient of process fluid

In addition to film ΔT, the tube temperature rises for the heat transfer resistance through the metal of the coil as follows:

$$[\Delta T = \Phi \times th]/k$$

Where, th tube thickness

k metal thermal conductivity

ΔT differential metal temperature between the internal and external tube surfaces

Higher heat flux directly results in higher film and metal temperature which might be a limit for the process fluid stability and/or for the tube mechanical properties.

The heat flux used in the fired heater design influences the average flue gas temperature in the radiant section and bridge wall temperature and therefore the radiant efficiency. For example, the higher the radiant flux rate and the more economical the radiant design, the lower the radiant section efficiency, and vice versa.

High overall efficiency allows recovery of more heat in the convection section. High overall efficiency means a lower approach temperature between the inlet process fluid temperature and flue gases leaving the convection section, or pinch temperature, requiring more installed surface. The additional cost for this extra surface in the convection sections is normally well covered by the fuel saving.

8. Construction Materials, Mechanical Features, Performance Monitoring

A variety of process, structural and environmental factors influence the choice of materials and mechanical design features used in fired heaters. For example, high operating temperatures, or poor fuel quality (excessive trace metals and ash) may force the selection of costly, highly alloyed materials. Environmental considerations may necessitate extremely tall stack heights, and available plot plan area may restrict dimensions.

An economic factor that has assumed greater importance in recent years is the shift toward shop assembly and fabrication in order to reduce field construction time and costs. A consequence of this shift is that provision for shipping clearance along the route from fabricator to jobsite has assumed the importance of a primary design criterion.

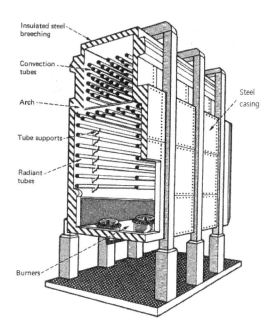

Fig. shows a schematic for a typical fired heater depicting the major structural components and their relationship to each other. These components will be taken up here.

8.1 Casing and structural framework

Typically, the outer wall or casing of the heater is fabricated from 4.76mm steel plate, reinforced against warping. However, for vertical cylindrical heaters, in which the shell itself serves as a load carrying structural member, the normal plate thickness can be as low as 6.35mm. Floor plates too, are normally designed for a thickness of 6.35mm. Prevailing design guide lines call for sealing the heater casing plate by welding, in order to prevent air and water infiltration.

The heater's structural steel framework provides load carrying members, which permit lateral and vertical expansion of all parts of the fired heater. The framework also supports the tube coil, independently of the refractory.

Common design practice calls for fireproofing the main structural columns to a specified height above grade, as well as the main floor beams. When header boxes are provided to receive return bend fittings, the minimum plate thickness is normally 4.76mm, reinforced against warping. Where appurtenances such as ladders and platforms are provided or anticipated, the structural design must be adjusted to carry such loads.

8.1.1 Refractories
The casing described above is lined internally with insulating materials. Aside from the basic function of preventing the steel structure from overheating, the insulation also serves to contain the fire box heat at high temperature by reradiating it to the tube coil. In addition, the internal insulation serves to minimize casing heat loss, and also functions as a barrier to prevent flue gas particle migration to the steel casing. Such migration, in the case of sulfur bearing fuels, may lead to acid corrosion of the steel plate. In order to properly select and design a refractory lining for a fired heater, concern must be given to several important factors:

Extreme temperature. Exposure to temperatures beyond the design limitation of refractory material can cause melting or fusion, and failure under load.

Thermal shock. Extreme or frequent temperature fluctuations can cause disintegration and spalling of refractory linings.

Mechanical stress. Abnormal vibration can contribute to the deterioration of some materials. Stresses due to the expansion and contraction of the structure can cause the loss of lining integrity unless proper allowance is made in the mechanical design.

Erosion. Extremely fine particles such as flyash or catalyst being carried at high velocity in a flue gas stream can cause erosion of the refractory material.

Chemical attack. Some fuels contain impurities that can react with various refractory constituents, causing slagging and failure of the refractory lining. Alkalis and acids, depending on the temperature and dew point of the flue gases, can attack the components of a refractory lining, causing corrosion and deterioration

Cost. The economic evaluation of refractory materials and construction types is complicated because materials having the best insulating properties often fall short on mechanical strength. Thus most refractory choices represent a compromise between insulating value and mechanical durability.

8.1.2 Insulation

Insulating systems for modern fired heaters fall into three basic categories:
Insulating firebrick (IFB). This is a porous brick with good insulation characteristics, manufactured by firing mixtures of sawdust, coke and high alumina fireclays. Design temperature ratings of IFB range from 871 to 1,538°C.

Typical IFB walls are suspended from the heater casing, and supported by horizontal steel angles. With this arrangement, heavy loading on the lower bricks of a wall are avoided. Several concepts are employed to retain the wall in place. One method anchors individual bricks, at frequent intervals, to the steel casing with steel hooks or rods. Another method utilizes long steel rods placed through holes in the bricks, with the rods then anchored at intervals to the steel casing.

When IFB walls are designed for vertical cylindrical casing, the "keying" effect of the bricks on the curved wall normally holds the wall in place without the use of tieback hardware such as hooks or rods. Improved insulating effectiveness of IFB walls can be achieved by incorporating a backup layer of mineral wool block insulation. This "brick and block" construction serves in countless operating installations.

With the current emphasis on shop preassembly of heater sections in modular form, the problems associated with shipping brick lined modules have discouraged their use in new installations. By far, the more popular setting employed in conventional heaters today is the castable refractory wall construction. Nevertheless, brick and block settings remain the standard for high temperature specialty units such as steam hydrocarbon reformer heaters and pyrolysis heaters.

Castable refractory. Castable refractory used in heater setting is normally an insulating castable applied by pouring or gunning. For shop assembled modules, gunning under controlled conditions has been shown to be a very economical method of application. However, Pneumatic placement of castable is a skilled craft, and the techniques of the applicator can mean the difference between success and failure of the installation.

Lumnite-Haydite-Vermiculite insulating castable (1:2:4 mix by volume) is an inexpensive material having excellent insulation characteristics. Because of its low cost, LHV is used extensively in heater applications. In addition, it has a very low expansion coefficient and is therefore used on large wall areas, without expansion joints.

Fig. Convection section

Fig. Radiant section

A maximum temperature limit of 982 to 1,038℃ precludes the use of LHV on exposed walls (unprotected by tubes) in close proximity to the flame burst. For such applications, proprietary castable refractory mixes are available. It is to be noted that as the service temperature and density of the castable increase, the insulating effectiveness decreases, resulting in a need for additional thickness to achieve the same cold face temperature. In many cases, dual layer constructions are employed in which the high temperature, high density material is exposed to the flame, and the lower grade, better insulating material is provided as a backup layer. Below describes a duel layer insulating castable at 26.7℃ ambient air temperature and still wind.

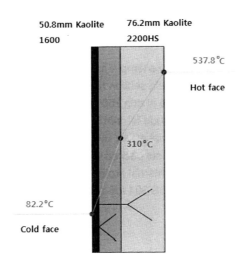

The thicknesses are typically 127mm for convection section walls and radiant section walls protected by tubes, and 152.4 to 203.2mm for exposed radiant walls, arches and floors. It is now fairly common to provide a facing of first quality firebrick over castable areas on the floor, particularly when the heater is designed for liquid fuels.

As important as the choice of insulating material is, the key to a properly applied castable refractory wall is the anchoring system. The most popular system employs "V" clips or modifications thereof, welded to the steel casing. These clips are normally 3.175 to 4.763mm dia. with anchor heights usually not less than 70% of the castable thickness. The preferred anchor material is austenitic stainless steel. Typical anchor spacings are a maximum of twice the lining thickness but not exceeding 304.8mm on a square pattern for walls, and 228.6mm a square pattern for arches.

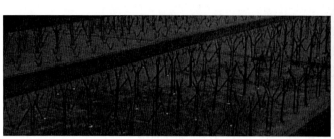

Fig. Anchor for dual layer

Ceramic fiber. Ceramic fiber construction is the most current development in the field of fired heater insulation. These linings normally consist of a hot face layer followed by one or more layers of backup material. For the hot face layer, ceramic fiber or blanket with a minimum density of 128.3kg/m^3 is recommended, with a minimum thickness of 25.4mm. Backup layers should also have a minimum thickness of 25.4mm and a minimum density of 64.1kg/m^3. Mineral wool block insulation can also be employed as backup material, provided that the fuel does not contain more than 1wt% sulfur if it is fuel oil, or 1.5vol% H$_2$S if it is gas.

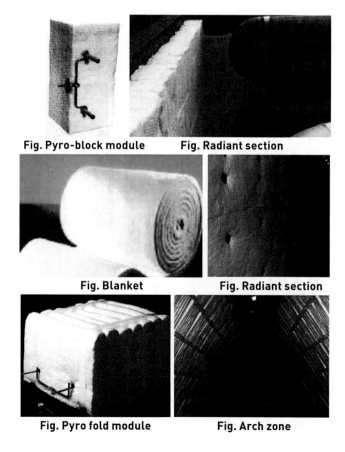

Fig. Pyro-block module   **Fig. Radiant section**

Fig. Blanket   **Fig. Radiant section**

Fig. Pyro fold module   **Fig. Arch zone**

Advantages of ceramic fiber installations derive primarily from their light weight, which permits a reduction of structural steel and from their immediate availability for operation without special startup procedures such as curing, dry out or cold weather precautions. Because ceramic fiber is more porous than insulating castable, it is desirable to provide an internal protective coating on the casing plate to prevent corrosion. Ceramic fiber construction should not be employed in convection sections when soot blowing or steam lancing is considered.

In brief, refractory type and characteristics are summarized as following table.

Refractory types	Characteristics
Firebrick	-. Preformed bricks -. Easier to install at some locations -. Good load bearing strength -. Have much higher erosion resistance
Insulating castable	-. Somewhat like concrete mix, mix with water, apply and let it cure -. Needs dry-out time -. Application typically by casting or gunning -. Various types of anchors used to keep castable in place -. Can be applied to all parts of a heater
Ceramic fiber (Blanket, Module- pyro block & pyro fold)	-. Somewhat like wool blanket -. Much better insulation property than castable, but much lower strength -. Excellent resistance to thermal shock -. Not good for heavy fuel oil firing applications -. Anchors provided to hold blanket or modules -. Use of rigidizer

The basic components of refractory are mostly alumina(Al_2O_3) and silica(SiO_2). Also others contain ferric oxide(Fe_2O_3), titanium oxide(TiO_2), calcium oxide(CaO), magnesium oxide (MgO) etc.

8.2 The tube coil

The tube coil of a fired heater is the most important component of a heater installation. It is also a major cost contributor to the overall heater investment. Normally, the tube coil consists of a number of tubes connected in series by 180° return bends. In the event of internal coke deposition in the tubes, all welded coils can usually be cleaned by a steam/air decoking procedure, or by a relatively new process using an abrasive (i.e., "pig" device) propelled by a high velocity gas stream. In many older installations as well as in current high temperature designs, wherein heavy oils having the potential for substantial coke deposition are heated-plug-type headers which are employed to connect the tubes. In this construction, the tubes are rolled into, or welded to the headers. The tubes can be cleaned by reaming or turbining. Occasionally, all-welded coils are provided with plug type headers at certain key locations, simply to permit internal tube inspection.

8.2.1 Tube design

Principal factors affecting the selection of tubing material for elevated temperatures are service life, environment, and cost. Specified service life varies substantially for individual heating applications and even for the same application within different operating companies. For example, one company might use Type 304 stainless steel as a coil material with an expected life of 8 to 11 years for a particular service. Another company might select a chromium-molybdenum steel, expecting to make tube replacements after about five years and perhaps replace all the tubes within seven years. In the case of the Type 304 steel, the initial investment is substantially greater, as is the cost of the replacement of any tubes lost through faulty operation. More continuous operation, however, with less downtime for tube replacement, may make Type 304 a more economical long- term choice.

The temperature and stresses to which the tubing is subjected are as crucial as the media to which it is being exposed. In fired heaters, the tube metal temperature is always higher than the bulk fluid temperature at a given location. In addition, the temperature differential between tube wall and bulk fluid may increase as coke or scale is deposited on the inner wall. Therefore, consideration must be given not only to the initial tube wall temperature but also to the maximum metal temperature that may result at the end of a run.

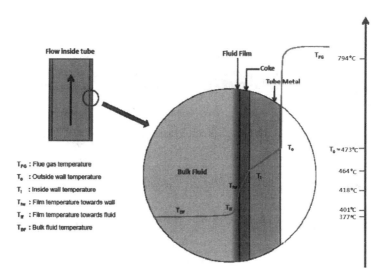

$$\text{Total Metal Temperature} = T_{bulk} + \triangle T_{film} + \triangle T_{coke} + \triangle t_{metal}$$

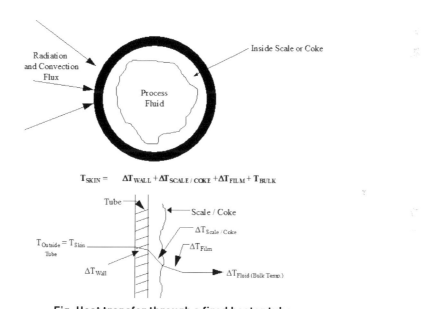

Fig. Heat transfer through a fired heater tube

The estimated tube wall temperature (TWT) in process is the process outlet temperature + X (X: 80℃ for platforming heaters, 110℃ for vacuum heater outlet, 30-55℃ for all other heaters). Also important thing is whether the stress is constant or cyclic. Thermal stresses that are created when a unit is starting up or shutting down can result in failure of steel components. Under certain conditions, such stresses can be of greater magnitude than the steady state operating stresses. The medium to which the tubing is exposed affects the oxidation or corrosion behavior of the steel.

If anticipated oxidation levels are severe, tube material selection must provide a high resistance to scaling. As for corrosion activity, the selection should reflect, where possible, data obtained from actual operating conditions at commercial units or pilot plants.

Probably the most important factor in the selection of tubing material is cost. A steel with excellent elevated temperature properties will have limited application if its cost is prohibitive. One of the reasons for the relatively wide selection of fired heater tube coil materials is that a particular material is often the most economical for a given application. Further, within an individual fired heater, it is not unusual to include two or more different tube materials in the most economical tube coil design. Below table presents a guide of process tube material.

Process	Internal scale	Typical metallurgy	Typical corrosion allowance	Typical failure causes
Crude	Coke	9 Cr	3.175mm	1, 3
Vacuum	Coke	9 Cr	3.175mm	1, 3
Coker	Coke	9 Cr	3.175mm	1, 3
Naphtha hydrotreating	Inorganic	1.25 Cr to 9 Cr, 347SS	3.175 to 4.763mm, 1.588mm	1, 2
Kerosene/Diesel hydrotreating	Inorganic	347SS	1.588mm	1, 2
Hydrocrackers	Inorganic	347SS	1.588mm	1, 2, 3, 4
Catalytic reformers	Clean	2.25 Cr to 9 Cr	1.588mm	1, 4
Reboilers	Clean	C.S.	3.175mm	1

Note) Typical failure causes: 1. Overheating caused by low flow or flame impingement 2. Hydrogen sulfide(H_2S) attack 3. Sulfur attack 4. Long term creep deformation

The use of a steel at high temperature levels will result in creep or permanent deformation even at stress levels well below the yield strength of the material. The tube will eventually fail by creep rupture, even when a corrosion or oxidation mechanism is not active. For steels at lower temperature levels, creep effect are negligible, indicating that under such circumstances the tube will last indefinitely unless corrosion or oxidation effects show themselves. There are, therefore, two different design considerations for heater tubes. At lower temperature, in the "elastic range", the design stress is based on the yield strength. At higher temperatures, in the "creep rupture range", the design stress is based on the rupture strength. For the temperature range where elastic and rupture stresses cross, the design tube wall thickness requirement must satisfy both conditions.

8.2.2 Tubing materials

Carbon steel, the most widely used material for heater tubing, is suitable where corrosion or oxidation is relatively mild. The widespread usage of this material reflects its relatively low cost, generally good service performance and good weldability.

The alloy steels (below Table) used for elevated temperature service generally contain either molybdenum, chromium or silicon. The molybdenum is added mainly to give higher strength; chromium is added to suppress graphitization and to yield improved oxidation resistance; and silicon is added to provide a further improvement in oxidation resistance.

Class	Material	Type or grade	Limiting design metal temperature (℃)	Characteristics
Base	Carbon steel	B	538	Most commonly used
Low Alloy	Carbon-0.5 Mo	T1 or P1	593	Higher temperature strength and improved resistance to hydrogen attack/hydrogen sulfide
	1.25 Cr-0.5 Mo	T11 or P11	593	
	2.25 Cr-1 Mo	T22 or P22	649	
	5 Cr-0.5 Mo	T5 or P5	649	
	7 Cr-0.5 Mo	T7 or P7	704	
	9 Cr-1 Mo	T9 or P9	704	
Austenitic Stainless Steel	18 Cr-8 Ni	304 or 304H	816	Excellent resistance to naphthenic acid/polythionic acid stress corrosion cracking
	16 Cr-12 Ni-2 Mo	316 or 316H	816	
	18 Cr-10 Ni-Ti	321 or 321H	816	
	18 Cr-10 Ni-Cb	347 or 347H	816	
Nickel Alloy	Ni-Fe-Cr	Alloy 800H	982	Excellent resistance to chloride stress corrosion cracking
Austenitic Stainless Steel	25 Cr-20 Ni	HK-40	1,010	Excellent high temperature strength

The austenitic stainless steels are essentially alloys of iron, chromium and nickel and, as a group, are used for handling many corrosive materials or for resisting severe oxidation. Type 304, the most popular of the austenitic stainless steels, has excellent resistance to corrosion and oxidation and has high creep strength. Types 321 and 347 are similar to Type 304 except that titanium and columbium, respectively, have been added. These additives combine with carbon and minimize intergranular corrosion that may occur in certain media after welding Type 316, which contains molybdenum, is used for high strength service up to about 816℃ and will resist oxidation up to about 899℃. For service above 871℃, Type 309 and 310, which contain about 25% chromium and 12 and 20% nickel, respectively, are used. These steels have excellent strength at these temperatures and, because of their chromium content, can be used in applications where extreme corrosion or oxidation is encountered. Alloy 800 (20% chromium and 32% nickel) has excellent strength at up to 982℃ and resists oxidation and carburization. It is used for tubes in pyrolysis heaters and steam super heaters, and for outlet pigtails and manifolds on steam hydrocarbon reformer heaters. Centrifugally cast materials such as HK-40 (25% chromium and 20% nickel), are widely used for tubing in steam hydrocarbon reformer heaters and in pyrolysis heaters.

Return bends. The least expensive method of connecting the tubes is to join them by means of 180° return bends. The return bends are welded to the tube ends in this arrangements, which represents the majority of modern heaters. As noted already, internal cleaning of all welded coils is performed by steam air decoking procedures, or by high velocity abrasive techniques (pigs). All welded design permits return bends to be positioned either in the path of the flue gases, where they function as heat absorbing surfaces, or in header boxes external to the firebox and the flue gas flow. Return bends can be of cast material.

Plug type headers. Many types of plug type headers using several closure designs have been developed. Compared to 180° return bends, these headers are somewhat more expensive and their use in new fired heater equipment is relatively rare. As noted above, plug type headers are used where mechanical cleaning of the tubes by turbining is anticipated and, sometimes where tube internal inspection is planned. Plug type headers cannot be placed in the firebox or in the path of the flue gases; they must be installed in header boxes external to the fire box and the flue gas flow. Plug type headers consist of cast material, and since they are external to the heat transfer zone, can be designed for a lower temperature than the heat absorbing tubes.

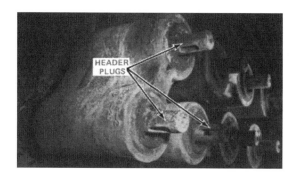

8.2.3 Extended surface improves convection

The surface area required in the convection section is controlled by film resistance on the flue gas side. As a means of increasing the convection transfer rate per lineal foot of tubing, extended surface devices have found almost universal acceptance. In today's designs, bare tube convection sections are generally reserved for those relatively rare applications where the combustion of extremely poor grades of fuel oil presents the risk of heavy ash deposition on the convection section tubes. Popular devices employed in heater convection sections are reviewed here. Note that extended surface is never installed in the radiant section.

Serrated fins. This arrangement uses a V-notched or serrated fin that is helically wrapped around and continuously welded to the tube(Fig. below). Fins can be supplied in many combinations of thickness, height and density (number of fins per unit length of tube). Typically, thickness ranges from 0.889 to 4.7625mm, height from 0.625 to 38.1mm, and density from 2 to 7 fins per 25.4mm.

Solid fins. This type is a noninterrupted fin that is helically wrapped around and continuously welded to the tube (Fig. below). These fins are available in the same ranges of thickness, height and density as serrated fins. Solid fins are mechanically stronger than serrated fins, but generally display a slightly lower heat transfer rate for the same fin configuration and flue gas mass flow.

Studs. Here, nominally cylindrical studs are flash-welded to the tube circumference(Fig. below). A stud diameter of 25.4mm is fairly standard for the industry, although 9.525mm studs are sometimes specified. Stud height ranges from 12.7 to 50.8mm.

Cylindrical studs are the only extended surface that can be effectively employed on tubes positioned normal to, as well as parallel to, the flow of flue gas.

Dimensions of extended surface devices used in gas fired heaters (12.7mm dia. studs or minimum 1.27mm thick fins) are somewhat smaller than those used in oil-fired units (12.7mm dia. studs or minimum 2.54mm thick fins). Fin dimensions should preferably be limited to a maximum height of 19.05mm and a maximum density of 3 fins per 25.4mm. Table below presents calculated maximum tip temperatures for various extended surface devices.

	Extended surface material	Tip temperature (℃)
Fins	Carbon steel	454
	5 Cr	593
	11-13 Cr	649
	18 Cr-8 Ni	816
Studs	Carbon steel	510
	5 Cr	593
	11-13 Cr	649
	18 Cr-8 Ni	816

8.2.4 Tube supports and guides

Proper mechanical design of a fired heater requires that the tube coil be adequately supported by tube hangers and tube sheets that are connected to the structural framework of the heater and not to the refractory.

Horizontal tubes. Horizontal coils having internal return bends are supported by intermediate tube hangers in the radiant section, and by tube sheets in the convection section. When the return bends (or plug type headers) are located external to the heater environment in header boxes, and tube sheets and intermediate tube supports are provided. A common arrangement features header boxes having external return bends for the convection section, and internal return bends for the radiant.

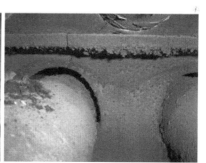

Radiant hanger   Convection tube sheet

The design of intermediate tube supports used in a horizontal radiant section should ensure that the supports can be removed without tube removal and with a minimum of refractory replacement. It is also desirable for intermediate radiant tube supports to have restraints, or "keepers" to prevent the tubes from lifting off the supports during operation. Intermediate tube sheet castings in the convection section should be sectionalized to minimize the amount of tube removal needed when a casting is replaced. Typically, the maximum unsupported length of horizontal tubes should not exceed 35 times the tube outside diameter, or 610cm, whichever is less.

Vertical tubes. Vertical coils can be supported either from the top or from the bottom. Top supported tubes are provided with bottom guides, bottom supported tubes with top guides. No intermediate supports are used. If necessary, intermediate guides are provided to restrain the vertical tubes from bowing inward toward the flame, or laterally toward adjacent tubes.

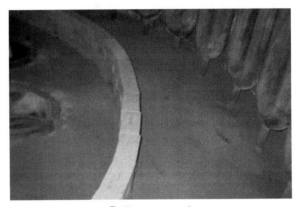

Bottom support

8.2.5 Materials used for tube supports

End tube sheets for tubes with external return bends or plug headers are normally of carbon steel plate, 12.7mm thick. If the tube sheets temperature exceeds 427℃, alloy material should be used. End tube sheets are insulated on the hot side with castable refractory- generally 76.2mm minimum in the convection section and 127mm minimum in the radiant section.

In the radiant section, intermediate supports for the horizontal tubes as well as top supports for vertical tubes are usually of cast 25% chromium-12% nickel, although many companies prefer the higher grade 25% chromium-20% nickel. The same materials are used for intermediate guides and bottom guides of vertical tubes, although the use of 18% chromium-8% nickel for bottom guides is fairly common. Bottom supports for vertical tubes normally consist of alloy cast iron and are shielded from flame radiation by the floor refractory.

When the firing of fuel oil containing more than 100ppm of vanadium is considered, radiant section supports and guides should be made of higher alloys such as 50% chromium-50% nickel, 60% chromium-40% nickel, or IN- 657, which is a proprietary 50% chromium-50% nickel alloy stabilized with columbium for additional high-temperature strength. These materials, however, are substantially more expensive than the conventional, lower grade alloys. Intermediate tube sheet castings exposed to hot gases in the convection section are normally fabricated from the same materials used for radiant section supports exposed to flame radiation. In the color flue gases, alloy cast iron is commonly employed. Table below lists a variety of tube support materials, and their usual temperature limits.

Material	Type or grade	Limiting design metal temperature (℃)
Carbon steel	A-283 Grc	427
5 Cr-0.5 Mo	GrC5	621
Alloy cast iron	A319 Class III Type C	649
18 Cr-8 Ni	Gr CF8	760
25 Cr-12 Ni	Type II	982
50 Cr-50 Ni		982
50 Cr-50 Ni-Cb	IN657	982
60 Cr-40 Ni		1,038
25 Cr-20 Ni	Gr HK40	1,093

8.3 Burners

The fundamental criteria for selecting a burner include (1) the ability to handle fuels having a reasonable variation in calorific value, (2) provision for safe ignition and easy maintenance, (3) a reasonable turndown ratio between maximum and minimum firing rates, and (4) predictable flame patterns for all fuels and firing rates.

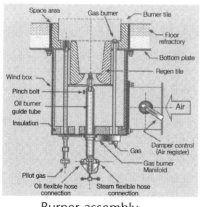

Burner assembly

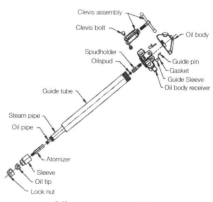

Oil gun assembly

8.3.1 Gas fired burners

Burners designed for gaseous fuel only are classified into two basic categories: premix inspiriting and raw gas burning.

Premix inspiriting. The premix burner relies on the kinetic energy made available by the expansion of the fuel gas through an orifice to aspirate and mix combustion air prior to ignition at the burner tip. Approximately 50 to 60% of the combustion air is inspirited as primary air into the burner ahead of the ignition point.

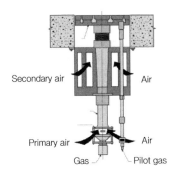

Some of the advantages of this type of burner are:

1. Operating flexibility is good over a range of conditions. The amount of air inspirited varies with the fuel gas pressure, and consequently requires only limited adjustment of secondary (non-inspirited) combustion air. Premix burners can operate at low excess air rates and are not significantly affected by changes in wind velocity and direction.

2. Flame length is short, and flame pattern sharply defined at high heat release rates.

3. Burner orifices or spuds are fairly large, and, since they are located in a cold zone, are less subject to plugging than the smaller openings on non-inspiriting gas burners.

Some of the disadvantages of inspiriting burners are:

1. Relatively high gas pressures must be available. Below a gas pressure of 10 psig at the burner, the percentage of inspirited air falls rapidly and flexibility is greatly reduced.

2. Flashback of the flame from the burner tip to the mixing orifice may occur at low gas pressures, or when the fraction of gases having high flame-propagation velocities, such as hydrogen becomes too high.

3. The noise level of premix inspiriting burners is higher than that of non-inspiriting types.

Raw gas burning. The nozzle mixing, raw gas burner receives fuel gas from the gas manifold without any premixing of combustion air. The gas is then burned at a tip equipped with a series of small ports as Fig. below.

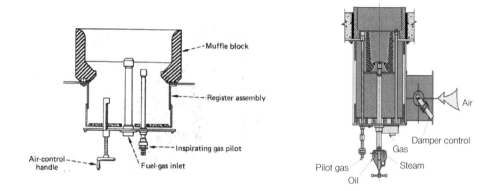

Some of the advantages of this type of burner are:

1. It has the greatest available turndown ratio for any given combustion condition.

2. It can operate at very low gas pressures on a wide variety of fuels and without flashback.

3. Noise level is reasonably low.

Some of the disadvantages of raw gas burners are:

1. Flexibility is limited over its wide turndown range. Because no primary air is inspirited, combustion air adjustments must be made over the full operating range of the burner.

2. The drilling of the burner ports is very sensitive, and any enlargement of the port opening will generally result in unsatisfactory flame conditions.

3. Flames tend to lengthen, and flame conditions become unsatisfactory as the burner is pushed beyond its design level.

4. The gas orifices or burner ports are exposed to the hot zone and are subject to plugging at low velocities and high temperatures.

8.3.2 Oil fired burners

Special measures must be provided for burning fuel oil, since mixing of fuel and combustion air occurs in the gaseous phase. To accomplish this, all liquid fuel burners use atomizing devices to break up the liquid mass into micron size droplets. This increases the surface to mass ratio, thereby allowing extremely rapid heating and vaporization of the oil mass.

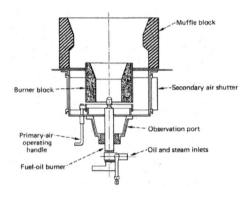

Oil burners in fired heaters almost always utilize steam as the atomizing medium. Such burners are designed with a double pipe in the feed tube to inject the steam and oil separately into a mixing chamber or atomizer immediately ahead of the burner tip. Steam pressure is slightly higher than oil pressure upon entering the atomizer, where the steam mixes with the oil due to the shearing action.

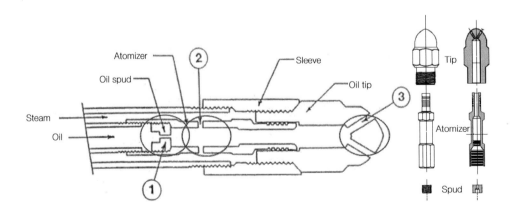

Furthermore, oil in contact with hot water vapor tends to foam or emulsify, thus contributing to the atomization process. The steam and the finely dispersed oil then issue through a series of orifices into the turbulent air stream.

For proper combustion of oil, the following important requirements should be met:

1. The oil must be heated high enough that its viscosity is not greater than *150-200 SSU (Seconds Saybolt Universal). * 32.1-38.85 centistokes

2. Oil pressure must be held constant, typically at about 5.27kg_f/cm^2.

3. The steam delivered at the burner must be absolutely dry. If available, a moderate superheat of approximately 10℃ is preferred. Typical steam pressure is about 7kg_f/cm^2.

Fig. above describes a typical oil burner. To maintain flame stability during the firing of high viscosity fuels, natural draft oil burners should be of the double block design, as illustrated. Atomizing steam needs fall in the range of about 1.5 to 3.5kg steam per kg of oil.

When volatile fuels such as naphtha or gasoline are burned, a safety interlock should be provided on each burner. The interlock sequentially shuts off the fuel, purges the oil gun with steam and shuts off the steam purge before the oil gun can be removed.

For those rare instances when oil must be burned and steam is not available, air atomization or mechanical atomization can be employed. The operating requirements of air atomized oil burners are similar to those of steam atomized ones, although a slightly higher oil temperature may be needed, to compensate for the cooling effect of the atomizing air. In addition, air (supplied by either a blower or a compressor) must be automatically modulated to adjust for changes in the firing rate. Uniformity of air supply is essential.

Mechanically atomized units take advantage of the oil's kinetic energy to atomize the fuel stream in the tip itself. Oil temperature and viscosity ranges are similar to those needed for steam atomization. Oil must be available at a supply pressure in the area of 21.1kg_f/cm^2. When wide turn down ratios are required, oil pressure may go as high as 70.3kg_f/cm^2.

8.3.3 Combination oil and gas burners

Combination burners are designed to burn all oil, all gas, or any combination of oil and gas simultaneously. Typically, these burners feature a double block design in which a single oil gun is arranged in the center of an array of gas nozzles. Gas firing in the newer combination burners is almost always accomplished with raw gas nozzles. Separate register adjustments allow independent control of primary air for oil combustion, and secondary air for fuel gas combustion.

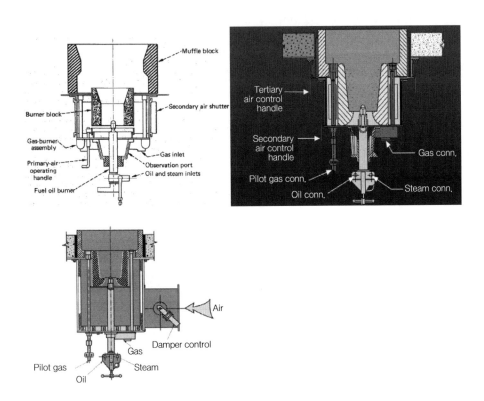

8.3.4 Radiant wall burners

The use of radiant wall burners is generally confined to high temperature specialty units such as steam hydrocarbon reformer heaters and pyrolysis heaters. Because the capacity of an individual burner is generally low, these heaters require many more burners than a conventional fired heater having the same total heat release. However, this feature affords the advantages of improved heat distribution and pinpoint firing control. Radiant wall burners generally burn only fuel gas, and are available as either inspiriting or as raw gas types.

The former type is designed to inspire 100% of the combustion air requirement. The flame from inspiriting burners is flat and radial, rather than projected into the firebox. Impingement of the flame on a refractory burner block heats the block to incandescence thus achieving the radiant wall effect.

The raw gas design is linear in shape, with the burners installed along two or three firing levels on the heater wall. The flames are thrown upward along the vertical or inclined firebox wall to create a radiant wall effect. This burner type has been adapted to burn light oils in a radiant wall mode of firing.

8.3.5 Low NO$_x$ burners

In low NO$_x$ burner, up to now, various technologies have been developed such as staged air(lean air), staged fuel(lean fuel), combination of both staged air and staged fuel, and flue gas recirculation(lean air and lean fuel). They are actually applied to commercial plant. Staged air was the original design. Today, staged fuel is the standard burner used throughout the world. Some refiners adopt the latest generation Ultra Low NO$_x$ burners to meet their NO$_x$ regulation.

Type	Technology involved	NO$_x$- fuel gas firing, *ppmv	NO$_x$- fuel oil firing, *ppmv
Conventional	Law gas or Premix	~80-150	~200-300
Low NO$_x$	Staged air (1984) Staged fuel gas (1986) or both	~50-100	~150-200
Ultra low NO$_x$	Staged air + Staged fuel gas + Flue gas recirculation (1992)	~15-30	-
Next generation	Staged fuel gas, Enhanced flue gas recirculation, Premix primary, etc	~10-20	-

* Corrected to 3% O$_2$ dry basis. No air preheat considered. Fuel oil with about 0.3wt% fuel bound nitrogen

Staged fuel burner. 30% of the fuel mixes with 100% air in the burner throat. This high excess air results in a low flame temperature and low NO$_x$. The remaining 70% of the fuel is injected into secondary gas tip. The burning is slow because the primary flame has already consumed 30% of the oxygen. This slow burning flame (i.e., longer combustion time) allows heat to be radiated to the tubes during combustion. The low peak flame temperature reduces NO$_x$ emissions to 40ppm as compared to 100ppm for a standard burner.

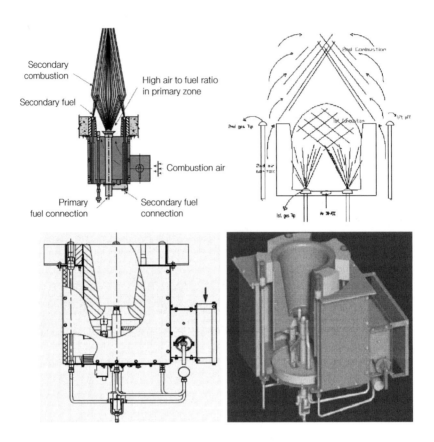

Ultra low NO$_x$ burner. Ultra low NO$_x$ burners use the energy from the burner tip to entrain flue gas into the flame (i.e., internal flue gas recirculation). The flue gas is cooler than the flame, so the peak flame temperature is lower. Each volume of primary fuel gas educts 5-7 parts of flue gas into the primary combustion zone. NO$_x$ is around 25-30ppm

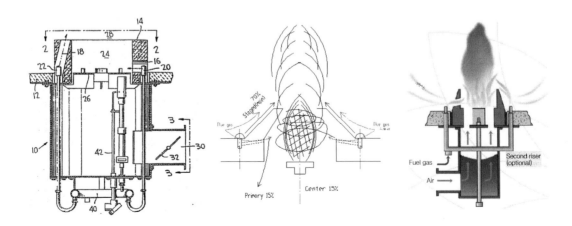

8.3.6 Pilot burners

These units are supplied as an integral part of the main burner. They are most often installed where it is desired to simplify burner light-off procedures, particularly with oil firing: where an extreme turndown to a fixed, minimum load is required: where intermittent on-off operation is required: and where extreme modulation of firing rate is needed.

The primary disadvantage of pilot burners is that they constitute a potential source of gas leakage into the firebox. The possibility always exists of a pilot being accidentally extinguished, permitting gas to be admitted to the heater during a shutdown. Also, because of their small port drillings, pilot burners clog easily and should be routinely inspected and cleaned. Pilot burners are almost always gas fired and are usually fueled from an outside source such as a propane or LPG drum which is not part of the process system. If the pilots are fueled from the main burner supply line, the gas offtake to the pilots must be upstream of the control and block valves.

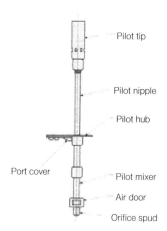

8.3.7 Burner noise- what to do about it

Burner noise results from the flow of fuel and air through the burner and from the combustion process itself. Although at present little can be done to reduce combustion noise, it is possible to minimize propagation of sound to the outside of the heater.

Noise emitted by premix inspiriting gas burners originates primarily at the premix venturi-inspirator, and secondarily at the outlet burner orifices. A method of reducing the inspirator noise is to replace the single orifice inspirator with a multi orifice inspirator. Further noise suppression can be achieved by fitting a silencer, or mute, on the primary air intake.

In raw gas burners, the noise is emitted by the fuel gas as it passes through the burner nozzles. Noise from liquid fuel burners is generated in the same way as in raw gas burners, except that the noise is lower in frequency. Combustion noise from these burners can be effectively suppressed by acoustically absorbent air intake plenums. These guidelines for plenum chamber design will ensure reasonable noise reduction:

(1) As much of the plenum interior as possible, including the heater wall, should be lined with acoustic material.

(2) Absorbent surfaces should be arranged so that prior to escaping, sound waves from the burner air intake will undergo several reflections. For greatest effectiveness, the acoustic material should have a density of about 96.2kg/m³ and its thickness should exceed 101.6mm.

(3) Steel plate with a minimum thickness of 3.175mm should be used for plenum fabrication. Plenums should enclose the burner registers and be undercoated with sound deadening material.

(4) Provision should be made for inspection and for draining oil leaks. Acoustic lining should be omitted where drips collect.

(5) All of the combustion air should be taken from the plenum chamber. These should be no line of sight from the burner or the plenum interior to the outside. Hand holes, and openings for external air register control, oil gun, etc. should be well sealed.

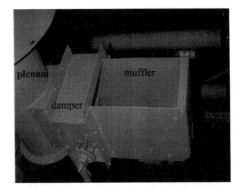

8.3.8 Dampers

The function of the stack damper is to control the heater's draft by maintaining a negative pressure of approximately -2.5mmH$_2$O in the region directly below the convection section. The damper adjusts as necessary to maintain this negative pressure. Stack dampers of single leaf construction are used for small diameter, and multi leaf construction for large diameter stacks.

In rectangular ductwork, which carries the flue gases between a heater and a separate stack, Louvre type dampers are normally used.

Stack dampers are usually manually operated from grade by means of cables. In the case of large dampers, it is now fairy common to maneuver them with pneumatic operators. Below table gives maximum flue gas temperatures for various damper materials.

Material	Gas temperature, ℃
Carbon steel	482
Cast iron	482
Alloy cast iron	538
11-13 Cr	649
18 Cr-8 Ni TP304	816
25 Cr-12 Ni TP309	982
25 Cr-20 Ni TP310	982

8.4 Cleaning the convection section

In order to maintain maximum thermal efficiency in a fired heater, it is necessary to keep convection heat transfer surfaces clean. Although the extended surface of the convection section improves heat transfer, its physical arrangement renders it susceptible to the accumulation of fuel ash deposits when oil fuels are fired. The major fouling constituents of an oil fuel are sulfur, vanadium, sodium and ash. The latter is a most important component, since a high ash content will significantly increase the deposition rate.

The viscosity range of the fuel also influences ash deposition. The larger ash particles that accompany the burning of high viscosity fuels carry over into the convection section to increase the fouling rate. Current methods used for onstream cleaning include manual lancing, soot blowing and water washing.

Manual lancing. This method requires a minimum of capital investment. Lance doors provided on the convection section sidewall permit workers to insert a steam or air lance and blow deposits off convection surfaces in the vicinity. The effectiveness of lancing is poor by comparison with that of soot blowing. In addition, lancing requires the participation of more than one operator to transport the equipment from grade to the convection section, which is normally located at a relatively high elevation. Consequently, the success of a lancing program is dependent on operator initiative. At the present time, manual lancing facilities are generally confined to retrofit installations on existing heaters- especially where a soot blower retrofit would involve substantial modifications to convection coils. Some operators have had reasonably good success in removing convection section deposits by sandblasting using a nozzle inserted manually through the lance doors. However, the erosive potential of sandblasting on the convection section refractory and on the heating surfaces must not be overlooked.

Radiant online chemical cleaning

Convection online chemical cleaning

Rotating element soot blowers. Also known as a fixed rotary type soot blower, this device is actually a multi-nozzle steam lance tube installed in the convection section. A mechanical valve and drive assembly external to the convection section rotates the element, and automatically opens and closes the steam supply valve. The effectiveness of this type of soot blower is limited by the physical strength of the ash. Some combinations of ash chemistry and flue gas temperature produce deposits that cannot be removed with such soot blowers. Further, highly corrosive fuels will materially reduce the life of the element or the lance tube.

Rotary soot blower

Retractable lance soot bowers. This apparatus differs from the rotating element type in that the lance tube remains retracted from the convection section when not in use. During soot blower operation, the lance is extended into the heater by means of an external drive. The lance is equipped with two cleaning nozzles, of larger diameter than those installed along the length of the multi nozzle rotating element design. The retractable lance blower cleans more effectively than does the fixed rotary type, because its two nozzles permit a greater concentration of the cleaning fluid stream. In addition, the retractable lance is exposed to high temperature flue gases only during the cleaning cycle.

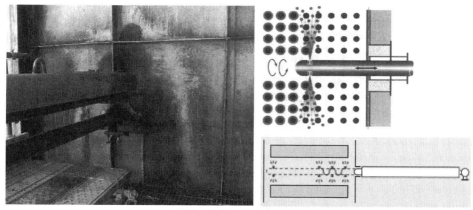

Retractable soot blower

8.4.1 Notes on soot blowers

The effective cleaning range of a soot blower is dependent on its orientation, the flue gas temperature, the arrangement of heating surface, and the blowing pressure. As a general guide, the vertical blowing capability of fixed rotary soot blowers will not exceed 3 rows up and 3 rows down, with a maximum horizontal blowing radius of 106.68cm. Corresponding limits for retractable lances are 4 rows up and 4 rows down, with a maximum horizontal blowing radius of 121.92cm.

The rate of steam consumption of a retractable lance soot blower will vary from 3.632 to 5.448 ton/hr depending upon the pressure at the nozzle. The cycle time required to operate a retractable lance is approximately 2.5min. The fixed rotary soot blower consumes steam at a rate of about 4.54 to 6.356 ton/hr with a blowing time of approximately 40sec. Soot blowers are normally operated once per 8hr shift.

Steam for soot blowing is normally taken from supply lines at the relatively low pressure of 10.5 to 14.1kg_f/cm^2. Below 1.5kg_f/cm^2, soot blowing is virtually ineffective. More effective cleaning can be achieved with pressures higher than 14.1kg_f/cm^2. When available, pressures of 28.1 to 42.2kg_f/cm^2 should be considered.

A completely automatic sequencing control system is recommended for multi blower installations. Upon activation by one pushbutton, each of the soot blowers will operate in a predetermined sequence until the entire blowing cycle is completed.

8.4.2 Water washing

Onstream water washing of convection sections represents a very effective method of removing deposits from the tubes. Permanently installed networks of high alloy headers and nozzles are positioned at selected elevations in the convection bank. Copious quantities of condensate or fresh water flow across the tube surfaces. Contact of the water with the hot tubes instantly vaporizes the water, thereby dislodging soot deposits.

Frequency of water washing depends on the nature and extent of the deposits. Typically, water washing is performed about once per week. Understandably, care must be exercised to avoid getting water on the refractory or on intermediate tube support castings. Water washing of stainless steel tube coils is not recommended in view of the danger of chloride stress corrosion.

8.5 Performance monitoring

Getting the optimum performance from a heater requires the close monitoring of key variables on both the process and combustion sides. Data can be taken that will reveal excess air, thermal efficiency, and heat absorption. In addition, certain data provides an indication as to how well the heater is being fired.

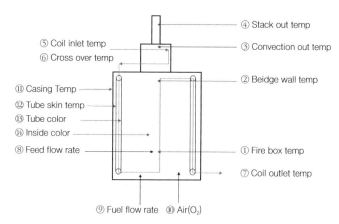

④ Stack out temp
⑤ Coil inlet temp
⑥ Cross over temp
③ Convection out temp
② Beidge wall temp
⑪ Casing Temp
⑫ Tube skin temp
⑬ Tube color
⑭ Inside color
⑧ Feed flow rate
① Fire box temp
⑦ Coil outlet temp
⑨ Fuel flow rate ⑩ Air(O₂)

Note) ④-⑤ is cold approach temp, ⑦-⑤ for entire coil △T, ⑥-⑤ for convection coil △T. And tube skin temp and tube color are indicators of heat distribution or balanced firing. Heater efficiency (%) = heat absorbed by the process/ heat fired x 100

Various monitoring variables

Process stream flow rate. In most applications, where provision is made for individual pass flow control, the flowrate to each parallel pass should be monitored. This is recommended for multi-pass heaters processing liquid hydrocarbons, where low flowrate in an individual pass can lead to excessive vaporization, increased pressure drop and further flow reduction- culminating in overheating and possible rupture of the tube.

Fuel firing rate. The rate of fuel input is normally controlled by the process fluid outlet temperature. Measuring the fuel firing rate permits direct determination of firebox heat release from the fuel's heating value.

Process stream temperatures. If individual pass flow control is planned, it is recommended that indicators be installed to show the fluid outlet temperature after each parallel pass. These temperatures can be used as a guide for adjusting the flowrates of each pass, as well as for determining the process side heat absorption.

Measuring the outlet fluid temperatures after each parallel pass in the radiant and convection sections enables one to determine the duty split on the process side between the radiant and convection sections.

Flue gas temperatures. Measurement of the flue gas temperature leaving the radiant section serves as an index to the firing balance in the firebox, and also as an indicator of over-firing conditions. This temperature measurement should be made at 15.24m intervals along the length of the firebox. Such measurements are often useful for establishing the maximum firing rate. Flue gas temperatures should also be monitored at the inlet to each convection coil, and at the outlet from the convection section. These temperatures provide an indication of heater efficiency and of convection tube fouling.

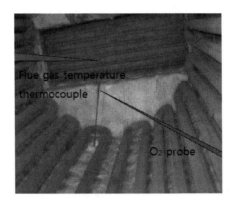

Flue gas draft profile. Draft (or buoyancy) is the force generated due to density difference over certain height: F = Δρ x g x h (F driving force, ρ density, g acceleration of gravity, h height). The pressure inside the furnace is negative because the hot gases are less dense than the outside air. Draft is usually measured in three places: at the firebox floor, below the convection section, and below the stack damper. The most important point is below the convection section because the negative pressure is smallest here. The small negative pressure is due to the tubes in the convection section which obstruct the flow of the upward moving gases. The resistance to flow can cause the pressure in the convection section to shift from slightly negative to slightly positive. When the pressure shifts positive, there is a loss of draft. If the loss is not, heat will be built up just under the furnace arch and roof which can damage the structure of the furnace. A loss of draft also means that no air is pulled into.

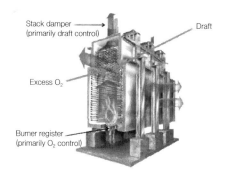

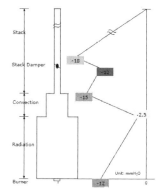

Typical natural draft profile in heater

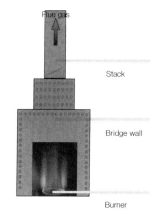

O_2 measuring point in heater

Tube skin thermocouple

Draft readings provide information on the pressure drop of combustion air and flue gas. The information is helpful for adjusting the burner registers and the stack damper. For example, opening the damper allows more flue gas to flow out the stack which, in turn, increases the draft throughout the entire furnace. Draft readings also indicate how close the heater is to its limiting operating conditions.

Flue gas sampling. Provisions for flue gas sampling are recommended at the exit of the radiant section and at the outlet of the convection section. Sampling to measure the oxygen in the flue gas at the first site provides an indication of the operator's firing technique. Measurement of combustibles is also recommended for this location. A determination of the oxygen content in the flue gas leaving the convection section is also needed for calculating the heater's combustion efficiency. If the oxygen content of the gases leaving the radiant section is known, the extent of air leakage into the convection section can be estimated.

Tube skin temperatures. Tube skin thermocouples are recommended, as a minimum, for the outlet tube of each pass and for one shock tube of each pass. Skin temperatures help set maximum firing rates. They also serve as indicators of local overheating.

9. How Combustion Conditions Influence Design and Operations

The process design for a fired heater must reconcile complex relationships between a number of variables. Among these are the physical properties and phase behavior of the process fluid, the heating value and combustion behavior of the fuel(s), the economic proportioning of heating duty between the convection and radiant sections, the pressure drops of the process fluid and the flue gas, and the economic stack height. The relative importance of these variables will be examined here.

9.1 Combustion basics

For those fuels containing hydrogen, two sets of heating values are reported. The gross or higher heating value, HHV, is determined by assuming that all of the water vapor produced in the combustion process is condensed and cooled to 16°C. (i.e., the water formed is considered as a liquid. In other words, the credit is taken for its heat of condensation). The net or lower heating value, LHV, assumes that the water vapor formed by combustion remains in the vapor phase (i.e., no credit is taken for the heat of condensation of water in the flue gas) and is numerically equal to the gross heating value less the latent heat of vaporization of the water.

In concept, the gross heating value can be considered as the actual heat release, whereas the net heating value can be considered as the useful portion of the actual heat release. For those fuels that do not contain hydrogen, for example, CO, only one heating value is reported.

Unlike the boiler industry, which works on the basis of the gross heating value, the fired heater industry almost always uses a net value. In this discussion, any reference to heat of combustion, heat input, or thermal efficiency will imply a net basis.

Although the chemistry of the combustion process is exceedingly complex, it can be readily simplified in terms of final reaction products. As with all chemical reactions, in order for the combustion process to go to completion within a reasonable time, an excess of one of the reactants must be present. Under no circumstances can an excess of fuel be tolerated. Combustion in a fired heater normally uses air as the source of oxygen. Consequently, there must be an excess of air above stoichiometric requirements to ensure complete combustion of the fuel. Reasonable excess air requirements for natural draft fired heaters are 20% for gas firing and 25% for oil firing. For forced draft heaters, in which a greater degree of air control is possible, reasonable design excess air values are 15% and 20% for gas and oil firing, respectively.

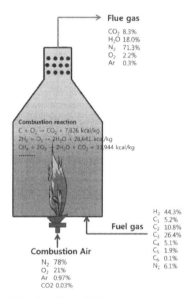

Fuel Combustion of One Case

Components	Formula	M.W.	Reaction chemistry	LHV (kcal/Nm³)
Hydrogen	H_2	2	$2H_2 + O_2 = 2H_2O$	2,576
Carbon monoxide	CO	28	$2CO + O_2 = 2CO_2$	3,017
Methane	CH_4	16	$CH_4 + 2O_2 = CO_2 + 2H_2O$	8,559
Ethane	C_2H_6	30	$2C_2H_6 + 7O_2 = 4CO_2 + 6H_2O$	15,232
Propane	C_3H_8	44	$C_3H_8 + 5O_2 = 3CO_2 + 4H_2O$	21,805
n-Butane	C_4H_{10}	58	$2C_4H_{10} + 13O_2 = 8CO_2 + 10H_2O$	28,433
n-Pentane	C_5H_{12}	72	$C_5H_{12} + 8O_2 = 5CO_2 + 6H_2O$	34,903
n-Hexane	C_6H_{14}	86	$2C_6H_{14} + 19O_2 = 12CO_2 + 14H_2O$	41,458
Benzene	C_6H_6	78	$2C_6H_6 + 15O_2 = 12CO_2 + 6H_2O$	33,808
Hydrogen sulfide	H_2S	34	$H_2S + 3O_2 = 2H_2O + 2SO_2$	
Carbon	C	12	$C + O_2 = CO_2$	
Sulfur	S	32	$S + O_2 = SO_2$	

Typical fuel compositions and reactions

Component	Mole%, wet	Mole%, dry
CO_2	9.25	11.0
H_2O	16.1	0.0
N_2	72.17	86.0
O_2	2.48	3.0
SO_2	0.001	0.0013

Sample product of combustion

	Gas fired	Oil fired
Natural draft	20%	25%
Forced/Balanced draft	15%	20%

Typical excess air

(Reference)

(1) NO_x (corrected to 3% O_2 dry basis): Raw NO_x x (21-3)/(21-O_2 measured)

(2) Excess air (%) = Excess O_2/(21- Excess O_2) x 100

9.1.1 Fuel gas

Table presents basic combustion constants for components of most industrial gaseous fuels.

Gas	Formula	Molecular weight	Heat of combustion, Gross (Btu/lb)	Heat of combustion, Net (Btu/lb)	Combustion air, Lb/combustible lb
Carbon monoxide	CO	28.01	4,347	4,347	2.462
Hydrogen	H_2	2.016	61,095	51,623	34.267
Methane	CH_4	16.042	23,875	21,495	17.195
Ethane	C_2H_6	30.068	22,323	20,418	15.899
Propane	C_3H_8	44.094	21,669	19,937	15.246
n-Butane	C_4H_{10}	58.12	21,321	19,678	14.984
n-Pentane	C_5H_{12}	72.146	21,095	19,507	15.323
n-Hexane	C_6H_{14}	86.172	20,966	19,415	15.238
Ethylene	C_2H_4	28.052	21,636	20,275	14.807
Propylene	C_3H_6	42.078	21,048	19,687	14.807
Butylene	C_4H_8	56.104	20,854	19,493	14.807
Benzene	C_6H_6	78.108	18,184	17,451	13.297
Toluene	C_7H_8	92.134	18,501	17,672	13.503
p-Xylene	C_8H_{10}	106.16	18,633	17,734	13.663
Acetylene	C_2H_2	26.036	21,502	20,769	13.297
Naphthalene	$C_{10}H_8$	128.164	17,303	16,708	12.932
Ammonia	NH_3	17.032	9,667	7,985	5.998
Hydrogen sulfide	H_2S	34.076	7,097	6,537	6.005

Properties used to predict heating values and air requirements for gaseous fuels

The following example illustrates the determination of heating value and stoichiometric air requirement for a multi-component gaseous fuel. (Basis: 100 lb moles of fuel)

Component	Molecular weight	Mole %	Lb	LHV (Btu/lb)	Btu	Lb air/lb fuel	Lb air
CH_4	16.42	86.0	1,379.6	21,500	29,564,500	17,195	23,720
C_2H_6	30068	8.6	258.6	20,420	5,280,100	15,899	4,110
C_3H_8	44.094	1.3	57.3	19,940	1,142,400	15,246	874
H_2	2.016	1.5	3.0	51,620	154,900	34,267	103
CO_2	44.010	2.6	114.4	-	-	-	-
Total	-	-	1,812.9	-	36,231,900	-	28,807

Average molecular weight = 1,812.9/100 = 18.13

Net heating value = 36,231,900/1,812.9 = 19,990 Btu/lb

Air requirement = 28,807/1,812.9 = 15.9 lb air/lb fuel

9.1.2 Fuel oil

Unlike gaseous fuels, the heating value of a fuel oil is normally not developed from a component analysis. Instead, the heating value of a liquid fuel can be expressed as a function of the specific gravity of the oil, with an accuracy sufficient for most engineering computations.

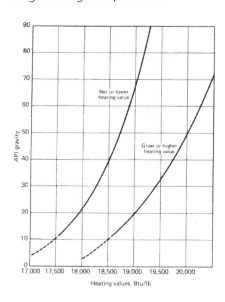

Left Fig. shows the relationship between the gross and net heating values of fuel oil as a function of API gravity. The API gravity, in turn, is related to the specific gravity, as follows:

$$API = (141.5-131.5)/Sp. \ Gr.$$

Fuel oil specific gravity	°API	Sulfur wt%	Inert wt%	C to H ratio
1.08	0	2.95	1.15	8.80
1.04	5	2.35	1.00	8.55
1.00	10	1.80	0.95	8.06
0.97	15	1.35	0.85	7.69
0.93	20	1.00	0.75	7.65
0.90	25	0.70	0.70	7.17
0.88	30	0.40	0.65	6.79
0.85	35	0.30	0.60	6.50

Fuel oil quality

If a fuel oil analysis is available, the stoichiometric air requirement can be approximated from the weight percentages of carbon, hydrogen, oxygen and sulfur, according to the following expression:

Lb air/lb fuel = (0.1159) %C + (0.3475) %H + (0.0435) %S – (0.0435) %O

For a fuel oil consisting by weight of carbon 84.6%, hydrogen 10.9%, sulfur 1.6%, and oxygen 2.9%, the stoichiometric air requirement is:

(0.1159)(84.6) + (0.3475)(10.9) + (0.0435)(1.6) – (0.0435)(2.9) = 13.54 lb air/lb fuel

Fuel oil No.2 (light distillate) and No.6 (heavy) are mostly used as below.

	HHV, Btu/lb	LHV, Btu/lb
Fuel oil #2 (light distillate)	19,000 - 19,500	18,250 - 18,500
Fuel oil #6 (heavy)	17,500 - 18,000	16,700 - 17,200

Fuel oil heating value

9.1.3 Combustion products

For the two types of fuel, the determination of flue gas quantity per unit quantity of fuel is performed simply by assuming 20% excess air for the fuel gas and 25% excess air for the fuel oil.

For the gas

   Total air = (1.20) (15.9) = 19.07 lb air/lb fuel

   Flue gas = 19.07 + 1 = 20.07 lb flue gas/lb fuel

For the oil

   Total air = (1.25) (13.54) = 16.93 lb air/lb fuel

   Flue gas = 16.93 + 1 = 17.93 lb flue gas/lb fuel

A generalized relationship that gives the weight ratio of flue gas to fuel is presented in Fig.1 below for several representative fuel gases and fuel oils.

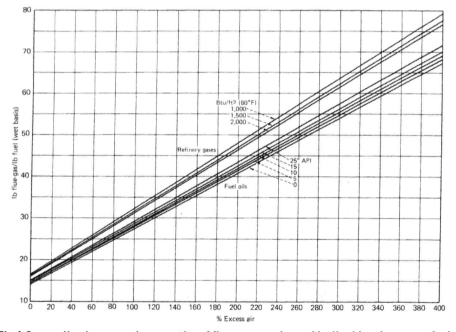

Fig.1 Generalized curves give quantity of flue gases released by liquid and gaseous fuels

A preferred fuel for ease of firing is one with a higher H/C (C/H) weight ratio of from 0.33 (3.0) to 0.25 (4.0) such as methane or ethane. A typical H/C (C/H) weight ratio of 0.12 (8.3) is characteristic of #6 oil.

9.1.4 Heating-coil arrangement

The normal flow of process fluid through a fired heater starts at the inlet to the convection section. The stream moves through the convection section counter current to the flow of flue gases, then into the shock bank. After leaving the shock bank, it passes into the radiant section, where the major portion of the heat is absorbed.

As described previously adjacent tubes are connected by means of 180° return bends or plug-type headers. Each bank of consecutive tubes in which the fluid travels from entry until exit is known as a pass or parallel stream. (This definition differs from that applied to unfired heat transfer equipment). In a two pass heater, the fluid is distributed into two streams at the inlet, with each stream flowing separately through its respective tube coil and then recombining after exiting from the heater. A single pass or series flow heater with 40 tubes for example, would be equivalent in area to a two pass heater having two coils of 20 tubes each.

The primary constraint affecting the selection of the number of passes and tube size is the allowable pressure drop. The fewer the number of passes, the less the potential for flow maldistribution.

Each heater application must be evaluated individually with regard to the number of passes and tube size. Such evaluation must examine not only the cost of the heater but also the cost of external distribution and collection manifolds, pass control valves, and other ancillaries. For normal heating applications a tube coil consisting of 4 in. IPS (iron pipe size) tubes often represents lowest heater investment (excluding externals). As tubes diameter increases or decreases from this size, the heater cost tends to become progressively more expensive.

Nomenclature	
A_c Convection surface area, ft^2	L_e Hydraulic length, ft
A_f Fin surface of extended surface tube, ft^2/ft	LMTD Log mean temperature difference, °F
A_i Internal tube surface, ft^2/ft	L_s Stack height, ft
A_o External tube surface, ft^2/ft	Δp Pressure drop, psi
A_R Radiant surface area, ft^2	p' Ambient pressure, psia
A_t Total surface of extended surface tube, ft^2/ft	p Design pressure, psig
c_p Specific heat, Btu/lb/°F	q Transfer rate, Btu/h/ft^2 of outside tube surface
d_i Tube inside dia., in.	q_i Transfer rate, Btu/h/ft^2 of inside tube surface
d_o Tube outside dia., in.	Q Heat absorbed, Btu/h

D Tube outside dia., ft	Q_c Convection section heat absorption, Btu/h
D' Stack dia., ft	Q_F Heat fired, Btu/h
e Thermal efficiency	Q_R Radiant section heat absorption, Btu/h
E Fin efficiency	R_i In-tube film resistance, $[Btu/h/(ft^2)(°F)]^{-1}$
f Fanning friction factor	R_o External film resistance, $[Btu/h/(ft^2)(°F)]^{-1}$
F_i In-tube fouling resistance, $[Btu/h/(ft^2)(°F)]^{-1}$	R_w Tube wall resistance, $[Btu/h/(ft^2)(°F)]^{-1}$
F_o External fouling resistance, $[Btu/h/(ft^2)(°F)]^{-1}$	R_t Total resistance, $[Btu/h/(ft^2)(°F)]^{-1}$
g Flue gas mass velocity, lb/s/ ft^2	S Design stress, psi
g' Fluid mass velocity, lb/s/ ft^2	t_c Thickness of coke/scale layer, in.
G Flue gas mass velocity, lb/h/ft^2	t_m Thickness of tube wall, in.
h_c Convection film coefficient, Btu/h/$(ft^2)(°F)$	T_a Ambient temperature, °R
h_i In-tube fluid film coefficient, Btu/h/$(ft^2)(°F)$	T_B Bulk fluid temperature, °F
h_o Total convection heat transfer coefficient, Btu/h/ $(ft^2)(°F)$	T_g Flue gas temperature, °F
h_{rg} Gas radiation coefficient, Btu/h/($ft^2)(°F)$	T_{ga} Flue gas temperature, °R
h_w Tube wall coefficient, Btu/h/$(ft^2)(°F)$	T_m Tube metal temperature, °F
H_A Heat available in flue gas, Btu/lb of fuel	u Bulk viscosity, lb/ft/h
H_F Net heating value of fuel, Btu/lb	u_w Viscosity at wall temperature, lb/ft/h
k Thermal conductivity, Btu/h/$(ft^2)(°F)$/ft	U Overall coefficient, Btu/(h)(ft^2)/°F
K_c Thermal conductivity of coke/scale layer, Btu/h/ $(ft^2)(°F)$/in.	V Specific volume, ft^3/lb
K_m Thermal conductivity of tube wall, Btu/h/(ft^2) $(°F)$/in.	W Flow rate, lb/h

9.2 Thermal efficiency

For each fuel, profiles can be developed at constant values of excess air to express the heat extracted from the products of combustion as a function of flue gas temperature. The percentage of the heat extracted from the flue gas will range from 0% at the flame temperature, to 100% at the datum flue gas temperature of 16℃. This heat includes the heat absorbed by the charge stock plus the heat lost from the heater casing. At any flue gas temperature, the differential between the percent heat extracted and 100% constitutes the percent heat loss up the stack.

Below two Figs illustrate for two typical fuels (a 19,700 Btu/lb refinery gas and a 15°API fuel oil), profiles of heat available in the flue gas as a function of temperature and excess air.

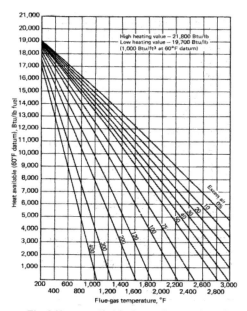

Fig.2 Heat available from the combustion of a 19,700 Btu/lb(LHV) refinery gas

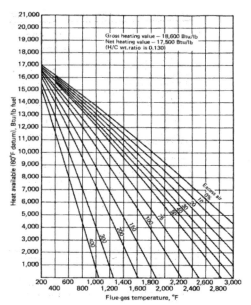

Fig.3 Heat available from the combustion of a 15° API fuel oil

Assuming that the fuel and combustion air are supplied at the datum temperature of 16℃, for a given combination of flue gas temperature and excess air, the% heat extracted from the flue gas is:

% Heat extracted = Heat available (Btu/lb fuel at flue gas temperature)/Net heating
value of fuel(Btu/lb) x 100
$$= H_A/H_F \text{ x } 100$$

The heat transmitted from the heater casing to the atmosphere via radiation and convection (the so called "radiation loss") is generally assessed at 1.5 to 2% of the heat fired for conventional installations, and at 2 to 2.5% for heaters with extensive hot duct runs and/or air preheaters.

The actual or calculated thermal efficiency of the heater is computed as the percent heat extracted minus the percent radiation loss. The heat fired is then determined from the relationship:

Thermal efficiency (e) = Heat absorbed (Btu/h)/Heat fired (Btu/h) x 100
$$= Q/Q_F \text{ x } 100$$

The quantity of fuel consumed is obtained as follows:

Fuel consumed (lb/h) = Heat fired (Btu/h)/Fuel heating value (Btu/lb)

$$= Q_F / H_F$$

The fuel gas mass flow rate is approximated by multiplying the fuel consumption rate by the appropriate weight ratio of flue gas to fuel, obtained from the curves in Fig.1.

The previously mentioned relationships define the interdependence of excess air, flue gas temperature and thermal efficiency. In the design of heaters having convection sections, thermal efficiency is determined by selecting the flue gas temperature. Specification of this temperature is normally dependent upon that of the process fluid at the inlet. A typical design basis would assume a temperature approach of about 65℃ between flue gas temperature and inlet fluid temperature.

A temperature differential of this magnitude represents a reasonable balance between thermal efficiency and capital investment. However, many criteria affect the temperature approach for a particular application. These include fuel value, project payback time, minimum stack height requirement, type and material of extended surface, auxiliaries such as soot blowers, etc. In the case of all radiant heaters without convection section, the designer does not have the degree of freedom to select the design flue gas temperature. For these heaters, the temperature is equal to the residual radiant section gas temperature.

9.2.1 Radiant section

In order to evaluate the split in heat absorption between radiant and convection sections, it is necessary to determine the radiant efficiency- the fraction of heat liberated that is absorbed by the heat transfer surface in the combustion chamber. For a given fuel at a given value of excess air, the radiant efficiency can be shown to be a function of the residual radiant section gas temperature..

In turn, for a particular heater configuration, this temperature can be graphically as a function of excess air, tube metal temperature, and radiant transfer rate.

The selection of the average radiant transfer rate is, essentially, the first step in the process design of a fired heater. By definition, the average radiant rate represents the heat transferred to the charge stock in the radiant section, divided by the total radiant section heat transfer surface, based on the tube O.D. The higher the design radiant rate, the less the amount of heat transfer surface, the smaller the heater, and the lower the investment cost.

Excessive high radiant rates, however, result in higher maintenance costs. Because the refractories and tube supports are exposed to higher temperatures, they have shorter service lives. Furthermore, high tube wall temperatures reduce tube life and raise the potential for coke deposition and product degradation. Actual radiant transfer rates will reflect the experience of both the user and the designer. A tabulation of typical rates for various commercial heating services is shown in below table.

Typical heat transfer rate across the radiant sections of heaters in various services

Service	Average radiant rate (Btu/hr ft^2)
Atmospheric crude heaters	10,000-14,000
Reduced crude vacuum heaters	8,000-10,000
Reboilers	10,000-12,000
Circulating oil heaters	8,000-11,000
Catalytic reformer charge and reheat heaters	7,500-12,000
Delayed coker heaters	10,000-11,000
Visbreaker heaters- heating section	9,000-10,000
Visbreaker heaters- soaking section	6,000-7,000
Propane deasphalting heaters	8,000-9,000
Lube vacuum heaters	7,500-8,500
Hydrotreater and hydrocracker charge heaters	10,000
Catalytic cracker feed heaters	10,000-11,000
Steam super heaters	9,000-13,000
Natural gasoline plant heaters	10,000-12,000

* **For tubes spaced at twice the nominal diameter, fired on one side and backed by refractory**

Distribution of the radiant heat transfer rate around a tube is not uniform. In fact, the rate varies substantially around the tube circumference, depending upon the ratio of tube spacing to tube diameter, and upon the firing mode- i.e., whether the tubes are fired from one side or from both sides.

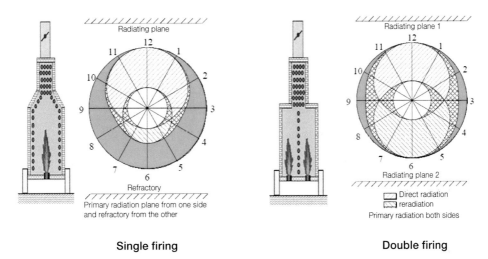

If the maximum radiant rate is assumed to occur at the front 60° of the fired tube face, the ratio of maximum radiant rate to average radiant rate can be obtained from below Fig.4 for various tube coil configurations and firing modes.

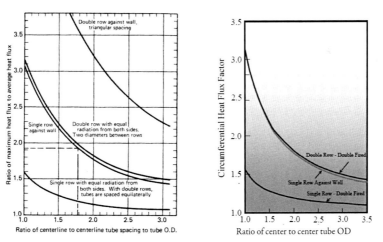

Fig.4 Distribution of radiant heat transfer rate around tube depends on coil arrangement and firing mode

As noted previously, the tube metal temperature is always hotter than the bulk fluid temperature at any given location. The magnitude of the metal temperature depends on the fluid temperature, the radiant transfer rate, the tube side film coefficient, the thermal conductivity of the tube material, and the thermal resistance of coke or scale deposits. Typical fouling factors for typical process feed are shown below.

Service	Fouling factor (hr.ft^2. $^\circ$F/Btu)
Crude oil	0.004-0.006
Gasoline	0.002
Naphtha, Kero, Light gas oil	0.002-0.003
Heavy gas oil	0.003-0.005
Atmospheric tower bottoms	0.007
Vacuum tower bottoms	0.01
Coker light/heavy gas oil	0.003-0.005
Catalytic reformer charge	0.0015
Recycle gas	0.001

The metal temperature can be expressed as the sum of the bulk fluid temperature plus the temperature differentials across the fluid film, the coke/scale layer, and the tube wall.

$$T_m = T_b + q_i/h_i + q_i t_c/K_c + q_i t_m/K_m$$

For vertical cylindrical heaters and horizon-tall-tube heaters, Fig.5 a and b estimate the residual radiant section gas temperature- the so called bridge wall temperature (BWT) as a function of the radiant transfer rate and the average metal temperature. The charts assume that the tube coil is fired from one side, and that the tubes are spaced at twice their nominal diameter. As an expedient, the metal temperature can be reasonably approximated for most conventional heating applications by adding 23°C to the average bulk fluid temperature in the radiant section.

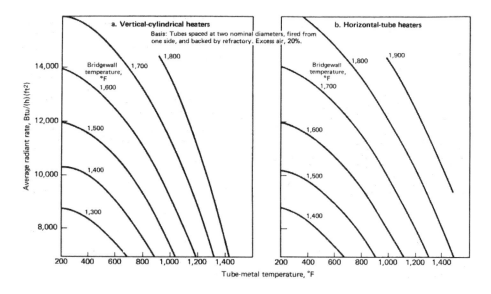

Fig.5 Bridge wall temperature enters into calculation of duty split between radiant and convection section

Once the bridge wall temperature and the flue gas temperature (FGT) are established, the split in duty between radiant and convection sections can be obtained as a function of the heat available in the products of combustion at these temperatures. The split is governed by the relationship that the ratio of the heat available at the bridge wall temperature to that at the flue gas temperature is equal to the ratio of the radiant duty to the total duty:

$$H_{A,BWT}/H_{A,FGT} = Q_R/Q$$

The radiant heat transfer surface area is obtained as the quotient of the radiant duty divided by the average radiant rate:

$$A_R = Q_R/q$$

9.2.2 Convection section

All combustion should be completed before the flue gas reaches the convection section. Although convective heat transfer is the primary mode of heat transmission in this section, radiative effects also contribute. The first two or three rows of tubes adjacent to the point at which the flue gas enters the convection section are known as the shield bank. Because these tubes are generally oriented at a staggered triangular pitch normal to the flue gas flow, they will usually screen and absorb most of the residual radiative component. When the shield bank can see the firebox, as in most modern installations, industry practice is to consider the surface area of one row of shield tubes to be radiant transfer area. Consequently, this surface is usually included in the radiant section surface requirement. To estimate a film coefficient based on pure convection for flue gas flowing normal to a bank of bare tubes, a method was developed by Monrad and was subsequently revised to the following form:

$$h_c = 2.14 g^{0.6} T_{ga}^{0.28} / d_o^{0.4}$$

Where g=flue gas mass velocity, lb/s/ft^2 at minimum cross section; T_{ga}= average flue gas temperature, °R; and d_o= tube outside diameter, in.

This equation does not take into account radiation from the hot gases flowing across the tubes, or re-radiation from the walls of the convection section. As an approximation, the radiation coefficient of the hot gas may be obtained from the following equation:

$$h_{rg} = 0.0025\, T_g - 0.5$$

Where T_g = average flue gas temperature, °F

Re-radiation from the walls of the convection section usually ranges from 6 to 15% of the sum of the pure convection and the hot gas radiation coefficients. A value of 10% represents a typical average. Based on this value, the total heat transfer coefficient for the bare tube convection section can be computed as:

$$h_o = (1.1)(h_c + h_{rg})$$

Convection section surface area requirements are controlled by the film resistance on the flue gas side. In order to increase transfer rates per unit length of tubing, convection sections in modern heater designs are almost always equipped with extended surface. Various types of extended surface employed in heater convection sections have been already described. With all of this extended surface type, radiant transfer to the convection tubes is so small that it can be neglected. For example, the heat transfer coefficient of a serrated fin surface, h_o can be obtained from the above equation as a function of the Reynolds number:

$$h_o = Jc_pG/(uc_p/k)^{2/3}$$

Where G = flue gas mass velocity, lb/h/ft^2; c_p = flue gas specific heat, Btu/lb/°F; u = flue gas viscosity, lb/h/ft; and k = flue gas thermal conductivity, Btu/h/(ft^2)(°F)/ft

The effective outside heat transfer coefficient is calculated from:

Effective $h_o = h_o(EA_f + A_o)/A_t$

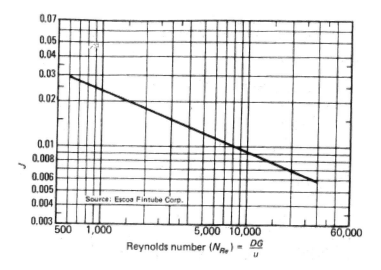

Fig.6 Heat transfer coefficient on the flue gas side of serrated fins

The fin efficiency E, takes into account the variability of fin effectiveness as a function of fin configuration, the thermal conductivity of the extended surface, and the convection film coefficient. For serrated fins, values of E may be approximated from Fig. below.

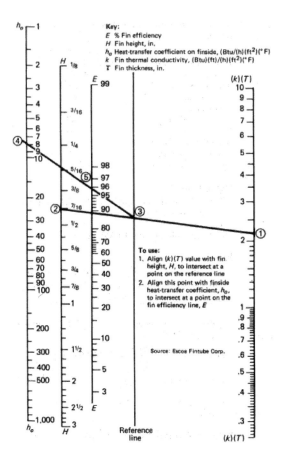

Fin efficiency varies with design and thermal conductivity, and with convection film coefficient.

The overall coefficient, U, is obtained from several relationships:

$$R_o = (1/\text{Effective } h_o) + F_o$$
$$R_w = A_t/h_w A_i$$
$$R_i = A_t(1/h_i + F_i)/A_i$$
$$R_t = R_o + R_w + R_i$$
$$U = 1/R_t$$

The convection surface requirement is calculated as:

$$A_c = Q_c/\text{LMTD}(U)$$

The highest convection transfer rate normally occurs at the lowest extended surface row. In many designs, the density of the extended surface is reduced for the lowest row or two in order to keep the convection rate from becoming excessive. A reasonable estimated value for the maximum convection heat transfer rate (expressed on an equivalent bare tube basis) for conventional heating applications is twice the average radiant transfer rate.

The maximum tube metal temperature in the convection section is estimated in the same manner as for the radiant section but with bulk fluid temperature, in-tube film coefficient and transfer rate appropriate for the convection zone.

9.3 Fluid pressure drop

The optimum combination of tube size and number of passes is dependent on an accurate evaluation of the fluid pressure drop. Historically, for heaters processing heavy oils, an important criterion for choosing the tube size and the passing arrangement has been the cold oil velocity. However, with the advent on numerous all vapor applications, a more meaningful basis has proved to be the fluid mass velocity which governs choice of tube size and number of passes. Table below presents a listing of typical fluid mass velocities for the more frequently encountered heating applications.

Service	Mass velocity, lb/s/ ft^2
Atmospheric crude heaters	175-250
Reduced crude vacuum heaters (outlet tube)	60-100
Reboilers	150-250
Circulation oil heaters	350-450
Catalytic reformer charge and reheat heaters	45-70
Delayed coker heaters	350-450
Hydrotreater and hydrocracker charge heaters	150-200
Steam superheaters	30-75
Steam generators (forced circulation)	100-150
Catalytic cracker feed heaters	300-400

The pressure drop for a single phase fluid, either all vapor or all liquid can be predicted with reasonable confidence using well established hydraulic principle. The following relation allows pressure drop computations to be made with an accuracy sufficient for most engineering purposes:

$$\Delta p - (0.00517) f(g')2(V)(L_e)/d_i$$

Where Δp = pressure drop, lb/in^2; f = Fanning friction factor; g' = fluid mass velocity, lb/s/ft^2; V = average specific volume, ft^3/lb; L_e = hydraulic length, ft.; and d_i = tube inside diameter, in.

The total hydraulic length, L_e, is based on the sum of the actual tube lengths plus the equivalent lengths of return fittings and elbows. These equivalent lengths can be approximated for various fittings as a multiple of the tube inside diameter. Typical equivalent lengths are 50 diameters for 180° return bends, 30 diameters for 90° elbows, and 100 diameters for plug type headers.

The pressure drop for a mixed phase flow regime can be approximated using the foregoing relation- although with a lesser degree of confidence than for single phase computations. In the case of mixed phase flow, the specific volume is very sensitive to the quantity of vapor present at a given location. Consequently, a more accurate estimate of overall pressure drop can be obtained by raising the number of incremental pressure drop determinations over the length of the coil.

In most vaporizing hydrocarbon services, vapor formation is not linear along the tube length; instead, it increases near the heater outlet. If an arithmetic average of the specific volume at the inlet and outlet of a mixed phase zone is used, the resultant pressure drop value will be overly conservative. To compensate for this, it is suggested that the log mean average of the inlet and outlet specific volumes be used for each mixed phase zone under consideration.

In view of the vast quantity of literature, both theoretical and empirical, concerning mixed phase pressure drop, it would seem simplistic to apply a single friction factor to a mixed phase flow system. Yet for many years, a single friction factor has been used with great success to design countless mixed phase fired heaters. For example, in conventional hydrocarbon vapor liquid systems, a factor of 0.0045 has proven to be of sufficient engineering accuracy for pressure drop determinations.

Because it is unusual for a charge stock to enter a fired heater at its bubble point, vaporization normally begins during through the tube coil. Under this condition, individual pressure drop determinations should be made for an all liquid, single phase regime, and for a separate mixed phase regime. The point of initial vaporization is arrived at via trial and error, mixed phase pressure drop calculations. Below represents some cases of typical fluid flow regimes.

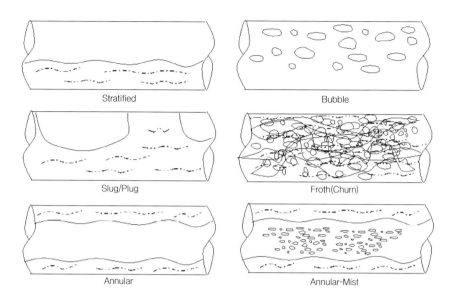

Many heaters heat two phase streams or vaporize liquid to create two phase streams. The orientation of vapor and liquid in a tube (flow regime) depends on the relative flow rates of the two streams. High velocity and pressure drop are required to produce flow regimes with fully wetted tube walls. Liquid cools tube walls better than vapor because liquid heat transfer coefficients are typically ten times that of vapor. Two phase heaters are designed to operate in flow regimes with fully wetted walls.

In heaters having high percentages of vaporization, it is not unusual for a temperature peaking condition to occur. As the mixed phase fluid flows through the coil, it undergoes a substantial drop in pressure per unit length of flow. This causes the rate of vaporization to surpass the rate at which latent and sensible heat can be supplied by the products of combustion. Thus the bulk fluid temperature drops even though the enthalpy increases. It may well be that at some intermediate location along the coil, the bulk fluid temperature will rise above the outlet temperature.

Although this temperature peaking effect is a perfectly normal condition, it is necessary upon the designer to properly assess its magnitude and the consequences, particularly in high temperature services.

9.4 Tube wall thickness

As mentioned previously, there are two design criteria that have been applied to heater tubes. At lower temperatures in the elastic range, the design stress is based on the yield strength. At higher temperatures in the creep rupture range, the design stress is based on the rupture strength. For both conditions, the recommended tube wall thickness is determined from the mean diameter formula:

$$T_m = Pd_o/(2S + P)$$

In the elastic range, the corrosion allowance is added to the wall thickness as determined above. The design pressure, P, corresponds to the maximum pressure excursion during an upset condition. The design stress, S, is based on two thirds of the yield strength for ferritic steels or 90% of the yield strength for austenitic steels.

In the creep rupture range, a fraction of the corrosion allowance is added to the wall thickness, as determined by the equation above. The design pressure, P, corresponds to the operating pressure, and stress, S, is based on 100% of the minimum rupture strength for a specified service life. Typical design service lives range from 60,000 to 100,000 hours.

A more detailed presentation of the foregoing, as well as design stress values and a procedure for evaluating the corrosion fraction is to be found in the 2nd ed. of the API's Recommended Practice 530, "Calculation of Heater Tube Thickness in Petroleum Refineries".

When the fluid being heated is water or steam, the requirements of the ASME Boiler and Pressure Vessel Code normally prevail. The tube wall thickness should be calculated in accordance with Section I, using Code formulas and stresses. Design stresses are very sensitive to temperature, particularly in the higher regions. Consequently, the thickness calculation requires an accurate determination of the tube wall temperature. It should be noted that the curves in Fig.4, which show the ratio of the maximum radiant transfer rate to the average rate do not consider convection heat transfer to the radiant tubes, circumferential heat transfer by conduction along the tube wall, or variations in the transfer rate at different zones of the combustion chamber.

The highest transfer rate in horizontal tube cabin heaters generally occurs at the wall, 1.52 to 3.05m above the floor, or at the shield bank, where convection effects are most pronounced. The transfer rate in the lower third of cylindrical heaters may approach 1.1 to 1.5 times the average radiant transfer rate, depending on the ratio of effective tube length to tube circle diameter. The magnitude of this local variation is compensated somewhat by the combined conductive/convective effect, which tends to reduce the difference between maximum and average transfer rates.

9.5 Stack design

The main functions of a stack are to induce the flow of combustion air into the heater, and to produce draft sufficient to overcome all obstructions to the flow of flue gas while maintaining a negative pressure system throughout. Never should a heater be operated with greater than atmospheric pressure at any point within the structure. Positive pressure within a heater setting creates a driving force for the outward movement of hot gases, which can lead to serious overheating and corrosion of the steel structure.

The draft produced by a column of hot flue gas depends on the density difference between the hot gas and the ambient air. The draft, in inches of water, can be expressed as:

$$\text{Draft} = (0.52)(L_s)(p')(1/T_a - 1/T_{ga})$$

Where L_s = stack height, ft; p' = atmospheric pressure, psia; T_a = ambient temperature, °R; and T_{ga} = flue gas temperature, °R

Because of heat loss through the stack casing, the flue gas temperature at the top of stack is substantially lower than at the inlet at the stack base. The magnitude of the differential depends on several factors, including the stack dimensions and the effectiveness of the stack insulation. For most applications, the average flue gas temperature in the stack may be conservatively estimated at 24°C less than the inlet gas temperature.

The relationship between stack diameter and stack height is also affected by cost optimization. Assuming that there are no requirements dictating a minimum stack height, a reasonable basis for selecting the diameter would result in a flue gas mass velocity in the range of 0.75 to 1.0 lb/s/ft^2. As noted above, a major objective in stack design is to ensure negative pressure throughout the heater. The most critical location in this respect occurs where the flue gas enters the convection section. Positive pressure presents itself first in this zone. It is recommended that the stack design be based on a negative pressure of 1.27 mmH$_2$O at this entry point.

The use of a design pressure less negative is not realistic considering the confidence level of the design computations. On the other hand, the use of still lower pressure (i.e., more negative) would tend to create too great a driving force for the leakage of air through various cracks and seams in the structures. The frictional loss of the flue gas flowing through the stack, in in. H$_2$O is:

Loss per foot of stack height $= (g)^2 (T_{ga})/(211{,}000)(D')$

Where g = mass velocity in stack, lb/s/ft^2; and D' = stack diameter, ft.

The reminder of the flue gas losses can be expressed in terms of the velocity head, based on the flue gas mass velocity at the location under consideration:

Velocity head (in. H_2O) $= (0.0030)(g^2)(V_g)$

The losses can be estimated as the product of the velocity head at each location and a factor. For bare convection tubes, this factor is 0.2 (per row); for finned tubes it is 1.0 (per row); for the stack entrance it is 0.5; for the damper it is 1.5; and for the stack exit it is 1.0. It should be noted that the convection section adds a stack effect due to its physical height and serves to reduce the overall draft requirement. In order to assure good flue gas distribution throughout the convection section, it is usual for a flue gas withdrawal opening to be provided at every 40 ft of convection section length.

At best, the determination of flue gas flow is only an approximation. Operation at excess air levels greater than design produces increased quantities of flue gas. Furthermore, efficiency decrease due to solids deposition on heat transfer surfaces also results in greater flue gas quantities. Consequently, a stack design based only on the design flue gas flow affords virtually no overcapacity in terms of the firing rate, and it severely restricts increases in throughput. To ensure flexibility for nominal capacity increments, the designer will often add an overcapacity of 25% to the flue gas design flow when estimating the various flue gas frictional losses.

10. Sonic Velocity for Piping Design

Sonic velocity is also known as speed of sound (340m/sec). In chemical engineering, sonic velocity is defined as the sound velocity of the internal fluid. This term, sonic velocity is important to verify if the internal fluid across the valve achieves choked condition or critical condition. When the fluid achieves that condition, it is described as choked flow. Choked flow is an undesired condition to the process flow system. It is a condition in which the mass flow rate will not increase with further decrease in the outlet pressure and fixed inlet pressure. When choked flow is achieved, the Mach number will be equal to 1 (Ma=1). The formula of Sonic Velocity is as defined below: (Rule of thumb) [Source: Branan, C.R., The Process Engineer's Pocket Handbook, Vol.1, Gulf Publishing Co., 1976]

Sonic Velocity, V_s= square root $(K \times g \times R \times T/M_w)$

Where,

V_s= sonic velocity with unit of ft/sec

K= C_p(constant pressure heat capacity)/C_v(constant volume heat capacity)

g(acceleration of gravity) = 32.2 ft/sec^2

R(gas constant) = 1,544 ft lb_f/°R lb-mol

M_w= molecular weight

T= absolute temperature, °R

It is possible to ensure if the flow achieves choked flow throughout the simulation analysis. In order to monitor the progress of the flow condition (chocked flow or not), another flow speed is inserted next to the flow sonic flow. The flow speed can be described as V= (actual volume flow)/(internal piping cross sectional area).

Sonic velocity and flow velocity are plotted in a strip chart. As the integrator is activated, the calculated result of sonic velocity and flow velocity will be plotted in the strip chart. From the strip chart, it can determine if the flow velocity must always be lower than the sonic velocity. Some people do not allow the velocity in the downstream piping to ever achieve a flow greater than 75% to 80% of sonic.

This sonic velocity is important because in heater design it is very closely related to the pressure drop of process fluid, the number of tube passes, flow regime & mass velocity. As an example, below shows the results of 2 phase sonic velocity for vacuum tower heater in a vacuum residue hydrocracker.

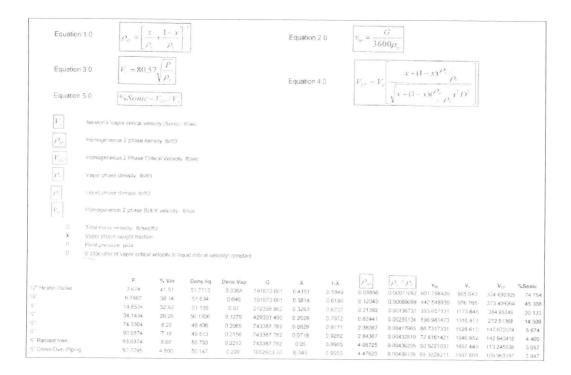

Equation 1.0
$$\rho_{tp} = \left[\frac{x}{\rho_v} + \frac{1-x}{\rho_l}\right]^{-1}$$

Equation 2.0
$$V_{sp} = \frac{G}{3600\rho_{tp}}$$

Equation 3.0
$$V = 80.57\sqrt{\frac{P}{\rho_v}}$$

Equation 4.0
$$V_{tp} = V_v \sqrt{\frac{x + (1-x)\dfrac{\rho_v}{\rho_l}}{x + (1-x)\dfrac{\rho_v}{\rho_l}\dfrac{V'}{D^2}}}$$

Equation 5.0
$$\%Sonic = V_{sp}/V_v$$

V_v	Newton's Vapor critical velocity (Sonic), ft/sec
ρ_{tp}	Homogeneous 2 phase density, lb/ft3
V_{tp}	Homogeneous 2 Phase Critical Velocity, ft/sec
ρ_v	Vapor phase density, lb/ft3
ρ_l	Liquid phase density, lb/ft3
V_{sp}	Homogeneous 2 phase BULK velocity, ft/sec
G	Total mass velocity, lb/sec/ft2
X	Vapor phase weight fraction
P	Point pressure, psia
D	0.333(ratio of vapor critical velocity to liquid critical velocity) constant

	P	% Vin	Dens.liq	Dens Vap	G	X	1-X	ρ_{tp}	ρ_v/ρ_{tp}	V_{sp}	V_v	V_{tp}	%Sonic
12" Heater Outlet	3.674	41.51	51.7713	0.0368	191873.001	0.4151	0.5849	0.08856	0.00071082	601.798426	805.043	334.692309	74.754
10"	6.7607	38.14	51.634	0.046	191873.001	0.3814	0.6186	0.12043	0.00089089	442.548039	976.765	373.409964	45.308
8"	14.8534	32.63	51.195	0.07	272358.902	0.3263	0.6737	0.21392	0.00136732	353.657131	1173.646	384.85346	30.133
9"	34.1434	20.28	50.1306	0.1279	429301.490	0.2028	0.7972	0.82441	0.00255134	190.981473	1316.411	272.91368	14.508
5"	74.3304	8.29	49.406	0.2085	743387.762	0.0829	0.9171	2.36087	0.00417065	86.7317331	1528.610	147.072074	5.674
6"	90.0874	7.18	49.513	0.2156	743387.762	0.0718	0.9282	2.84367	0.00432819	72.6181421	1646.952	142.943418	4.409
6" Radiant Inlet	93.0874	5.00	50.733	0.2213	743387.762	0.05	0.9500	4.08725	0.00436205	50.5221037	1652.449	113.245638	3.057
5" Cross-Over Piping	97.7795	4.500	50.147	0.220	1052693.30	0.045	0.9550	4.47620	0.00439109	65.3228211	1692.808	109.963397	3.847

The followings represent the some formulas in furnace design.

Heat Transfer Coefficient for Two-Phase Flow

$$h_i = 0.022 \left[\frac{K_L}{D_i}\right]\left[\frac{D_i G}{2.42\mu_L}\right]^{0.8}\left[\frac{2.42 C_{p_L}\mu_L}{K_L}\right]^{0.4}\left[\frac{\mu_L}{\mu_w}\right]^{0.167}$$

Btu/Hr-Ft^2-°F (Film Coefficient Based on Inside Area)

Inside Film Temperature at Tube Wall

$$T_f = \left[\frac{Q_p}{h_i}\right]\left[\frac{D_o}{D_i}\right] + T_b , °F$$

Critical Velocity

$$\hat{V}_{TP} = \hat{V}_v \left[\frac{X + (1-x)\rho_v/\rho_L}{\sqrt{x + (1-x)(\rho_v/\rho_L)^2 D^2}}\right] \quad Ft/Sec. \ (Critical \ Velocity)$$

$$\hat{V}_v = 80.57\sqrt{P/\rho_v} , \ Ft/Sec. \quad (Newton's \ Formula \ for \ Vapor \ Critical \ Velocity)$$

$$V_{TP} = \frac{G\left[\dfrac{x}{\rho_v} + \dfrac{1-x}{\rho_L}\right]}{3600} \quad Ft/Sec. \ (Bulk \ Velocity)$$

Where

C_{p_L}	=	Liquid Specific Heat at Bulk Temperature, Btu/Lb-°F
D	=	=0.333 (Ratio of Vapor Critical Velocity to Liquid Critical Velocity)
D_i	=	Inside Tube Diameter, Ft.
D_o	=	Outside Tube Diameter, Ft.
Q_p	=	Peak Radiant Section Heat Flux Based on Outside Area, Btu/Hr-Ft^2
G	=	Total Mass Flow Rate, Lb/Ft^2-Hr
K_L	=	Liquid Thermal Conductivity at Bulk Temperature, Btu/Ft-Hr-°F
P	=	Absolute Pressure, psia
T_b	=	Bulk Fluid Temperature, °F
x	=	Wt Fraction Vapor
ρ_L	=	Liquid Density at Bulk Temperature, Lb/Ft^3
ρ_v	=	Vapor Density at Bulk Temperature, Lb/Ft^3
μ_L	=	Liquid Viscosity at Bulk Temperature, Centipoise
μ_w	=	Liquid Viscosity at Wall Temperature, Centipoise

11. Sample of Hand Calculation

The following sequence of calculations illustrates the principal steps in the process design of a fired heater.

(Sample: process requirements)

Fluid	Dowtherm A
Heat absorbed, Btu/h	270,390,000
Flow rate, lb/h	550,000
Inlet temperature, °F	490
Outlet temperature, °F	580
Inlet vaporization	Nil
Outlet vaporization	Nil
Outlet pressure, psig	150
Allowable pressure drop, psi	25
Design pressure, psig	250
Fuel gas, Btu/lb (LHV)	19,700

[Tip 1] W= AUρ= Qρ= AG: W mass flow rate (mass velocity), Q volumetric flow rate, G mass flux, V velocity, A cross sectional area ($\pi D^2/4$ or πr^2)

(1) Design basis (operation requirement)

Vertical cylindrical heater with horizontal tube convection section
Excess air, 20%
[Tip 2] Excess Air (%)= Excess O_2/(21- Excess O_2) x 100

Average radiant rate, 10,000 Btu/h/ft^2

(2) Efficiency

Take flue gas temperature (FGT)= 490 +150= 640°F
Heat available at FGT= 16,600 Btu/lb (Fig.2)

[Tip 3] Flue gas temp.= Coil Inlet Temp. + Cold End Approach Temp.

% Heat extracted= (100)(16,600)/19,700= 84.3 %
Calculated efficiency= 84.3- 2.0(radiative loss)= 82.3% @LHV
Heat fired= 27,390,000/0.823= 33,280,000 Btu/h
Fuel consumed= 33,280,000/19,700= 1,689.3 lb/h
Flue gas flow= (1,689.3)(19.9)= 33,620 lb/h (Fig @ p113)

(3) Radiant/convection duty split

Select tube coil having 4 passes of 4 in. IPS, Sched.40 (4.5 in. O.D. x 0.237 in. avg. wall)
Fluid mass velocity= W/A= 550,000/(3,600)(4)(0.0884)= 432 lb/s/ft^2
Assume fluid temperature at radiant inlet= 520°F
Radiant section average fluid temperature= (520 + 580)/2= 550°F
Take radiant section average tube metal temperature= 550 + 75= 625°F
Bridge wall temp (BWT)= 1,470°F (Fig.5a)
$H_{A,BWT}/H_{A,FGT}$= 11,700/16,600= Q_R/27,390,000

[Tip 4] Available heat at BWT(1,470°F)/FGT(640°F) respectively in the Fig.

Q_R= radiant duty= 19,310,000 Btu/h
Q_C= convection duty= 8,080,000 Btu/h
A_R= 19,310,000/10,000= 1,931 ft^2

[Tip 5] Radiant surface area= radiant duty/avg. radiant heat flux

Assume 52 radiant tubes on 8 in. centers
Tube circle dia.(TCD)= (52)(8)/(12)(π)= 11.03 ft

[Tip 6] Circle= Dia. x π (so, dia.= circle/π)

Take effective tube length (ETL) of shield bank= 10 ft
With 8 tubes per row on 8 in. equilateral centers:
Shield bank free area= (8)(10.0)(8.0-4.5)/(12)= 23.33 ft^2

[Tip 7] 4 in. IPS Sched. 40 (4.5 in. O.D. x 0.237 in. avg. wall)

g= Flue gas mass velocity= 33,620/(3,600)(23.33)= 0.400 1b/s/ft^2
Consider surface equivalent to one shield row as radiant surface
Surface of one shield row= (8)(10.0)(1.178)= 94 ft^2

[Tip 8] 4.5 in. O.D. x 2.54cm/in. x π x ft/30.48cm= 1.178

Vertical tube radiant surface= 1,931-94= 1,837 ft^2
Vertical tube ETL= 1,837/(52)(1.178)= 30 ft
ETL/TCD ratio= 30.0/11.03= 2.72

(4) Shield bank

Shield bank heat absorption is approximated via a trial and error process.
Use three low shield bank.
Take average fluid temperature in shield bank= 515°F
Fluid temperature increase in shield bank is small and can be neglected.
Assume flue gas temperature drop across shield= 210°F.
LMTD calculation:
BWT(shield bank in) 1,470°F and shield bank out 1,470-210= 1,260°F
Shield bank average fluid temperature 515°F
Δt_1= 1,470 - 515= 955°F
Δt_2= 1,260 - 515= 745°F
LMTD= (955-745)/ln(955/745)= 846°F
h_c= (2.14)(0.400)$^{0.6}$(1,825)$^{0.28}$/(4.5)$^{0.4}$= 5.54
h_{rg}= (0.0025)(1,325)- 0.5= 2.91
h_o= (1.1)(5.54+2.91)= 9.29
Assume clean tube design and no internal or external fouling resistances.
h_i= (0.027)k/D[(DG)/(u)]$^{0.8}$[($c_p u$)/(k)]$^{0.333}$[(u)/(u_w)]$^{0.14}$
In view of the small variation in viscosity between bulk and wall temperatures, the value of $(u)/(u_w)^{0.14}$ may be taken to be 1.0.
hi=(0.027)(0.064)/(0.375) [(0.375)(1,555,200)/(0.605)]$^{0.8}$ x [(0.543)(0.605)/(0.064)]$^{0.333}$
(1.0)= 487
R_i= $A_o/h_i A_i$= 1.178/(487)(1.054)= 0.002295
h_w= K_m/t_m= 324/0.237= 1,367
R_w= $A_o/h_w A_i$= 1.178/(1,367)(1.054)= 0.000818
R_o= 1/h_o= 1/9.29= 0.1076
R_t= R_i+R_w+ R_o= 0.11076
U= 1/R_t= 9.03
Shield bank transfer rate= (U)(LMTD)= (9.03)(846)= 7,639 Btu/h/ft^2
[Tip 9] Q= U A ΔT (so, Q/A= U ΔT)

Three rows shield bank surface= (3)(8)(10.0)(1.178)= 283 ft^2
Shield bank heat absorption= (283)(7,639)= 2,162,000 Btu/h

Check of flue gas temperature above shield bank:

$H_{A,BWT}/H_{A,shield}$= 11,700/$H_{A,shield}$= 19,310,000/21,472,000

[Tip 10] Radiant duty + shield bank duty= 19,310,000 + 2,162,000= 21,472,000 Btu/h

$H_{A,shield}$=13,010 Btu/lb of fuel

Flue gas temperature above shield bank= 1,260°F (as assumed)

(5) Convection bank

Finbank heat absorption= 8.08 MM Btu/h- 2.162 MM Btu/h= 5,918,000 Btu/h

[Tip 11] Convection duty - shield bank duty= fin bank duty

Fluid temperature leaving fin bank= 510°F

LMTD calculation:

Fin bank in(shield bank out) 1,260°F and Fin bank out= 490°F + 150°F= 640°F

Fin bank out (shield bank in) 510°F and Fin bank in (COT) 490°F

Δt_1= 1,260 - 510= 750°F

Δt_2= 640 - 490= 150°F

LMTD= (750-150)/ln(750/150)= 373°F

Try 3 fins/in., with each fin 3/4 in. height x 0.05 in. thick. (A_t= 7.33 ft²/ft)

Fin bank free area= (8)(10.0) x [(8.0-4.5)/(12)-(2)(0.05)(0.75)(3)/(12)]= 21.83 ft²

G= W/A= 33,620/21.83= 1,540 lb/h/ft²

Reynolds number= DG/μ= (4.5)(1,540)/(12)(0.084)= 6,875

J(heat transfer coefficient on the flue gas side of serrated fins)= 0.011 (Fig.6)

h_o= (0.011)(0.28)(1,540)/[(0.084)(0.28)/(0.030)]$^{2/3}$= 5.58

Fin efficiency (E)= 84% (Fig. @ p126)

Effective h_o= $h_o(EA_f + A_o)/A_t$= (5.58)[(0.84)(6.152)+(1.178)]/(7.33)= 4.83

R_i= (7.33)/(487)(1.054)= 0.01428

R_w= (7.33)/(1,367)(1.054)= 0.00509

R_o= 1/h_o= 1/4.83= 0.20701

R_t= R_i+R_w+ R_o= 0.22638

U= 1/R_t= 4.42

A_c= convection surface= 5,918,000/(373)(4.42)= 3,590 ft²

[Tip 12] Q= UAΔT (so, A= Q/UΔT)

Surface area per convection row= (8)(10.0)(7.33)= 586 ft²

Number of finned rows= 3,590/586= 6.1(use 6)

Convection surface= (6)(586)= 3,516 ft²

Summary of results

	Total	Radiant section	Shield bank	Convention section
	Vertical cylindrical	52 tubes	3 bare rows	6 fin rows
Duty(Q), Btu/h	Q_T= 27,390,000	Q_R= 19,310,000	Q_S= 2,162,000	Q_C= 5,918000
	100%	71%	8%	22%
Area(A), ft^2	A_T= 5,730	A_R= 1,931	A_S= 283	A_C= 1,683
	100%	34%	5%	61%
Flue gas temp, °F	-	BWT= 1,470	Shield out= 1,260	Convention out= 640
Fluid temp, °F	-	515 to COT 580	510 to 515	CIT 490 to 510

(6) Pressure drop

Reynolds number= (4.026)(1,555,200)/(12)(0.581)= 898,000
Fanning friction factor= 0.0038
$L_{e, convection+shield}$= (2)(13.0)+(16)(11.5)+(17)(50)(4.026)/(12)= 495 ft
$L_{e, crossover}$= 20+(3)(30)(4.026)/(12)= 50 ft
$L_{e, radiant}$= (2)(33.0)+(11)(29.0)+(12)(50)(4.026)/(12)= 586 ft
Total L_e= 1,131 ft
Average specific volume= 0.0194 ft^3/1b
Δp= (0.00517)(0.0038)(432)^2(0.0194)(1,131)/(4.026)= 20.0 psi
(vs. 25 psi allowed)

(7) Tube wall thickness

Average radiant rate= 10,000 Btu/h/ft^2
Factor (Fig.4) max. to avg. flux ratio= 1.93

[Tip 13] Ratio of centerline to centerline tube spacing to tube O.D.= 8in./4.5 in.= 1.78 vs. 1.93 ratio of max. heat flux to avg. heat flux on the curve of single row against wall

Factor for local flux variation= 1.25 (assumed)
Factor for conductive/convective effects= 0.85 (assumed)
Maximum local radiant rate= (10,000)(1.93)(1.25)(0.85)= 20,500 Btu/h/ft^2
At outlet fluid temperature, (h_i)= 522
$T_m= T_B+q_i/h_i+ q_i t_c/K_c+ q_i t_m/K_m$
    = 580+(20,500)(4.50)/(522)(4.026)+(20,500)(4.5)(0.237)/(315)(4.263)
    = 640 °F

To allow for fouling effects and to provide safety margin, use design tube metal temp.= 750°F.

With carbon steel tubes, assume 1/16 in. corrosion allowance.

t_m= Tube wall thickness= $Pd_o/(2S+P)$= (250)(4.5)/[(2)(15,500)+250]+0.063= 0.099 in. min. wall

[Tip 14] Corrosion allowance 1/16 in. (= 0.063 in.) plus

As the least wall thickness, use Schedule 40 (0.237 in. average wall).

(8) Stack design

Size stack for mass velocity of 0.8 lb/s/ft^2 at 125% of design gas flow

Cross sectional area= (1.25)(33,620)/(3,600)(0.8)= 14.59 ft^2

[Tip 15] W= AUρ= Qρ= AG (so, A= W/G)

Stack diameter(D′)= 4 ft 4 in.

[Tip 16] πr^2= 14.59, r^2= 4.644, r= (4.644)$^{1/2}$= 2.15, D= 2xr= 2x2.15= 4.3ft

Assume average gas temperature in stack= 640-75= 565°F.

Assume stack exit gas temperature= 490°F.

① Draft under arch= 0.050 in.

② Shield bank loss= (3)(0.2)(0.0030)(1.25x0.400)^2(46.3)= 0.021 in.

[Tip 17] Velocity head (in. H^{2}O)= (0.0030)(g)^2(V$_g$) for 3 bare-tube rows at shield bank

③ Fin bank loss= (6)(1.0)(0.0030)(1.25x0.428)^2(35.6)= 0.183 in.

[Tip 18] Velocity head (in. H$_2$O)= (0.0030)(g)^2(V$_g$) for 6 fin-tube rows at fin bank

④ Stack entrance loss= (0.5)(0.0030)(0.8)^2(27.8) = 0.027 in.

⑤ Damper loss= (1.5)(0.0030)(0.8)^2(27.8)= 0.080 in.

⑥ Stack exit loss= (1.0)(0.0030)(0.8)^2(24.0)= 0.046 in.

Subtotal= ①+②+③+④+⑤+⑥= 0.407 in.

⑦ Convection section draft gain= (0.52)(8.5)(14.69)[(1/540)-(1/1,515)]= 0.077 in.

Required stack draft= 0.407 - 0.077= 0.330 in.

[Tip 19] Draft= (0.52)(L$_s$)(p′)(1/T$_a$- 1/T$_{ga}$): L$_s$stack height(ft), p′atmospheric pressure(psia), T$_a$ ambient temp.(°R)
T$_{ga}$ flue gas temp. (°R)

⑧ Stack draft gain/ft= (0.52)(1.0)(14.69)[(1/540)-(1/1,025)]= 0.00669 in.

⑨ Stack frictional loss/ft= (0.8)^2(1,025)/(211,000)(4.33)= 0.00072 in.

[Tip 20] Loss per foot of stack height= (g)^2(T$_{ga}$)/(211,000)(D′): g flue gas mass velocity(lb/s/ft^2),

T$_{ga}$ flue gas temp.(°R), D′stack diameter(ft)

Net stack effect/ft = 0.00669-0.00072= 0.00597 in.

Stack height required= 0.330/0.00597= 55 ft 3 in.

12. Example of NHT Charge Heater Design

This example demonstrates the design of a naphtha hydrotreating charge heater. Design a vertical cylindrical NHT charge heater.

Process conditions

Flow rate, BPSD (lb/h)	20,256 (632,624)
Process inlet, °F (°C)	566 (297)
Process outlet, °F (°C)	694 (368)
Process duty, MM Btu/h	20.31

Gas only natural draft burners are required.

(1) Fired heater specification

This chart is possibly the most important item in the document. It summarizes the design practice for new heater designs. NHT heaters are usually vertical cylindrical designs. Evaluate corrosion rate at 100°F(38°C) above process outlet (clean conditions) and 300°F(149°C) above process outlet (fouled conditions).

Process	Radiant Flux Btu/h-ft^2	Mass Velocity, lb/sec-ft^2	Process ΔP.psi	Coil Metallurgy	Type of Heater
Crude	10,000	200 min	150	5 Cr or 9 Cr	Vertical Tube Box Cabin or V.C.
Vacuum	8,000	250 min	75	5 Cr or 9 Cr	Vertical Tube Box Cabin or V.C.
Visbreaking	7,000-8,000	350 min (5-6 fps cold oil)	250-300 clean	9 Cr	Cabin with brick centerwall
Coking	9,000	400 min (6 fps cold oil)	300 clean	9 Cr	Cabin with brick centerwall or double fired horizontal tube
Semi-Regn. Platformer	10,000	40-50	7	2-1/4 Cr	V.C. , cabin, wicket
CCR Platformer	10,000 SFU 15,000 DFU	23-35	3 per cell	9 Cr	Single Fired U-tube Double Fired U-tube
Reboiler	10,000	150-300	50	C.S.	V.C.
Hot Oil	10,000	400-600	65	C.S.	V.C.
Naphtha Hydrotreater (max. 15 mils/yr)	10,000	70-180	50	9 Cr	V.C.
Naphtha Hydrotreater (with cracked stock)	10,000	70-180	50	347	V.C.
Kerosene Hydrotreater	10,000	150-250	50	347H SS	V.C.
Diesel Hydrotreater	10,000	150-250	50	347H SS	V.C. or double fired vertical tube
Hydrocracker Gas Only	10,000 - 13,000	50-80	30-50	347H SS	double fired vertical tube
Hydrocracker Gas&Liquid	15,000	180-300	50-80	347H SS	double fired vertical tube

(2) NHT heat and weight balance

This is a typical heat and weight balance for a naphtha hydrotreater. It shows the typical split of 80% duty in the combined feed exchanger and 20% duty in the fired heater. Fouling often occurs in the CFE. A loss of 10% surface area in the CFE will result in a 40% increase in fired heater duty. Most NHT heaters operate at 300-350 psig pressure for virgin feed to remove sulfur. At this pressure level, only 50% of nitrogen compounds are removed. High nitrogen feeds require a higher operating pressure, 600-800 psig. Recycle gas ratio should be at least 400 SCFB H_2/Bbl.

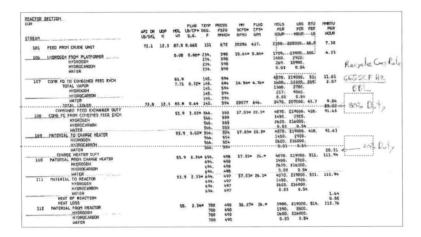

(3) NHT charge heater

This charge was designed for operating pressure of 300-350 psig. For higher operating pressure, scale the pressure drop proportional to operating pressure.

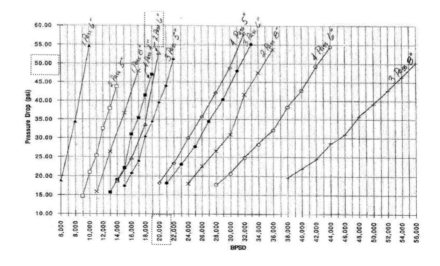

(4) Select radiant flux rate

Use the flux curve to determine the BWT. A conservative design is 10,000 Btu/ft^2 h. The maximum flux rate for new designs is 12,000 Btu/ ft^2 h.

Design new units for conservative flux rates.

Consider future process capability.

Example, new NHT Charge heater flux= 10,000 Btu/ft^2 h

BWT= 1,440°F (782°C)

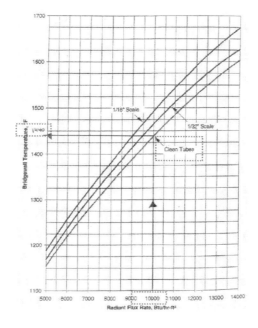

(5) Run fuels program

A cold end approach of 75°F (24°C) is chosen to obtain a stack temperature. The stack temperature is used to calculate a fuel efficiency (assuming 2% radiation losses). The fuels program is used to determine the flame temperature. The BWT and stack temperature was input to determine the duty split between the radiant and convection section. In this example, 70.5% of the process duty absorbed is in the radiant section and 29.5% is in the convection section.

· Stack= process inlet + 75°F

· BWT from charts

· Example:

NHT heater BWT= 1,440°F @10,000 Btu/ft^2 h

Stack= 566+75= 644°F

· Fuels program output:

Efficiency= 81.5%

Fuel fired= 20.31/0.819= 24.80 MM Btu/h

Radiant convection split (57.7%/81.9%)= 70.5% radiant

Radiant duty= 20.31x0.705= 14.32 MM Btu/h

Convection duty= 20.31-14.32= 5.99 MM Btu/h

Amount of flue gas= (24,800,000/18,299)x17.6= 23,852 lb/h

				FLUE GAS INFO	Btu/lb
adiabatic flame temp, °F	3188				
LHV of fuel, BTU/lb	18299		20%xs air	bwt	enthalpy
LHV, Btu/SCF	899.15			1400	383.38
lbs of wet air/lb fuel	17.60			1500	414.72
(Flowing @ Amb Air) Ft3 of wet air/lb fuel	236.18		BW temp, °F=	1440	396
air preheat temp, °F	60		% Radiant Efficiency=	57.57%	
air preheat content, btu/lb	0.00		FG out stack, °F=	641	157.67
Corrected heat content fuel, btu/lb	18299		Heat Loss=	2.00%	
Corrected heat content fuel, Btu/SCF	899.2				

Note: ⁻P air side burner ~ (Mass Flow)^2 * Specific Volume (ft3/lb)

% Total Efficiency=   81.90%

(6) Pipe factors for heaters

Most heaters are designed with 6" tubes. Use the target mass velocity to determine the number of heater passes. Do a trial and error calculation to determine the tube length that results in an even number of tubes per pass and a tube length/tube circle diameter (L/D)= 2.0. Minimum tube length is 20ft and maximum tube length is 60ft. Determine the number of burners and the heat release per burner. The flame length should be 1/3 to 1/2 of the radiant box height. Since the crude heater is a potential coking service, A heater maker prefers a burner to tube clearance that is 1 feet (12 inch) additional off the tubes. Remember the flame radiation is proportional to the square of the distance.

Size I.P.S.	O.D. Inches	I.D. Inches	Cold Oil Vel.Ft/Sec 1000 H/D @ 60°F	O.D. Sq. Ft. per ft.	t In.	Vol.,Ft³/Ft. I.D. Cross Sec-tional Area Sq. Ft.	(I.D.)⁵ Inches	(I.D.)^{0.2} Inches	Eq. Lg. - 180° bend 50 d	75 d	100 d	RBL ft
2"												
Sch. 40	2.375	2.067	2.79	.622	.154	.0233	37.72	1.156	8.6	12.9	17.2	
Sch. 80	2.375	1.939	3.17	.622	.218	.0205	27.41	1.142	8.1	13.1	16.2	0.524
Sch. 160	2.375	1.689	4.17	.622	.343	.0156	13.74	1.110	7.0	10.6	14.1	
2-1/2" 58λ mm												
Sch. 40	2.875	2.469	1.96	.753	.203	.0332	91.8	1.198	10.3	15.4	20.6	
Sch. 80	2.875	2.323	2.21	.753	.276	.0294	67.6	1.184	9.7	14.5	19.4	0.655
Sch. 160	2.875	2.125	2.64	.753	.375	.0246	43.3	1.163	8.9	13.3	17.7	
3"												
Sch. 40	3.500	3.068	1.27	.916	.216	.0513	271.8	1.251	12.8	19.2	25.6	
Sch. 80	3.500	2.900	1.42	.916	.300	.0459	205.1	1.237	12.1	18.1	24.2	0.785
Sch. 160	3.500	2.624	1.73	.916	.438	.0376	124.4	1.213	10.9	16.4	21.9	
3-1/2"												
Sch. 40	4.000	3.548	0.95	1.047	.226	.0687	562	1.288	14.8	22.2	29.6	
Sch. 80	4.000	3.364	1.05	1.047	.318	.0617	431	1.275	14.0	21.0	28.0	0.916
4" 114.3												
Sch. 40	4.500	4.026	0.74	1.178	.237	.0884	1,058	1.321	16.8	25.2	33.6	
Sch. 80	4.500	3.826	0.81	1.178	.337	.0799	820	1.308	15.9	23.9	31.9	1.047
Sch. 120	4.500	3.626	0.91	1.178	.438	.0716	625	1.294	15.1	22.7	30.2	
Sch. 160	4.500	3.438	1.01	1.178	.531	.0645	480	1.280	14.3	21.5	28.7	
5"												
Sch. 40	5.563	5.047	0.47	1.456	.258	.1390	3,275	1.382	21.0	31.5	42.1	
Sch. 80	5.563	4.813	0.52	1.456	.375	.1263	2,583	1.369	20.1	30.1	40.1	1.309
Sch. 120	5.563	4.563	0.57	1.456	.500	.1136	1,978	1.355	19.0	28.5	38.0	
Sch. 160	5.563	4.313	0.64	1.456	.625	.1015	1,492	1.340	18.0	27.0	35.9	
6"												
Sch. 40	6.625	6.065	0.32	1.734	.280	.2006	8,206	1.434	25.3	37.9	50.5	1.571
Sch. 80	6.625	5.761	0.36	1.734	.432	.1810	6,346	1.419	24.0	36.0	48.0	
Sch. 120	6.625	5.501	0.39	1.734	.562	.1650	5,037	1.406	22.9	34.4	48.8	
Sch. 160	6.625	5.189	0.44	1.734	.718	.1469	3,762	1.390	21.6	32.4	43.2	
8"												
Sch. 40	8.625	7.981	0.19	2.258	.322	.3474	32,400	1.515	33.3	49.9	66.5	2.094
Sch. 80	8.625	7.625	0.20	2.258	.500	.3171	25,800	1.501	31.8	47.7	63.5	
Sch. 100	8.625	7.439	0.215	2.258	.593	.3018	22,800	1.494	31.0	46.5	62.0	
Sch. 120	8.625	7.189	0.23	2.258	.718	.2819	19,200	1.484	30.0	44.9	59.9	
Sch. 160	8.625	6.813	0.257	2.258	.906	.2532	14,700	1.468	28.4	42.6	56.8	
10"												
Sch. 40	10.750	10.020	0.119	2.81	.365	.5475	101,000	1.586	41.8	62.6	83.5	
Sch. 80	10.750	9.564	0.130	2.81	.593	.4989	80,000	1.571	39.9	59.8	79.7	
Sch. 100	10.750	9.314	0.137	2.81	.718	.4732	69,400	1.563	38.8	58.2	77.6	
Sch. 120	10.750	9.064	0.145	2.81	.843	.4481	61,200	1.554	37.8	56.6	75.5	
Sch. 160	10.750	8.500	0.165	2.81	1.125	.3941	44,400	1.534	35.4	53.1	70.8	

(7) Layout radiation section

Most of the heaters in the refinery were designed by hand calculations with a slide rule. Many good designers do the heater design by simple hand calculations and enter the design into a computer program to check the results.

· Number of passes= [(219,000 lb/h)/(3,600 sec/h)]/[150MV x 0.1810 ft^2/pass]

= 2.2 (use 2 passes)

[Tip 1] W= AUρ= Qρ= AG(lb/ft^2 sec) & 150 Target Mass Velocity (lb/ft^2 sec)
& 219,000 lb/h (feed + recycle gas)

· 20.31 MM Btu/h x (0.705 radiant)/10,000 radiant flux= 1,432 ft^2 radiant surface

· 1,432/1.734 ft^2/ft= 826 feet of 6" pipe required

· Try 40 tubes:

Length= 826/40= 20.65'-1.6' return bend= 19' (use 20' as shortest)

Tube circle diameter= (12 center to center)x(40 tubes)/(12)(π)= 40/π= 12.7'

L/D= 20/12.7= 1.6

· Try 3 burners:

Fuel fired= (24.8 MM Btu/h)/3= 8.3 MM Btu/burner

Flame length= 1 feet x 8.3= 8.3 feet

[Tip 2] 1ft from L/D=1.6 & 8.3 from heat release/burner

% tube coverage= 8.3/20= 42%

(8) Burner to tube clearance

Many tube failures can be traced to flame impingement and burners placed too close to the tubes. The burner to tube spacing in API 560 is not adequate, especially for oil firing. A heater maker adds 1'-0"(1 feet i.e., 12") to these charts for coking (black oil) or fouling services (hydrotreating).

Gas Firing		
Maximum Heat Release per Burner, MM Btu/h	UOP Spacing horizontal to centerline wall tube from burner centerline, inches	API Spacing horizontal to centerline wall tube from burner centerline, inches
4	39	30
6	39	36
8	45	42
10	48	48
12	54	54
14	60	60

Oil Firing

Maximum Heat Release per Burner, MM Btu/h	UOP Spacing horizontal to centerline wall tube from burner centerline, inches	API Spacing horizontal to centerline wall tube from burner centerline, inches
4	48	33
6	48	39
8	54	45
10	57	51

Oil Firing- Staged Air Burner

Maximum Heat Release per Burner, MM Btu/h	UOP Spacing horizontal to centerline wall tube from burner centerline, inches	API Spacing horizontal to centerline wall tube from burner centerline, inches
4	54	33
6	54	39
8	60	45
10	63	51

(9) Layout radiant section

Many problems can occur if the burners are spaced too close together. CFD (Computation Fluid Dynamics) programs are used to optimize burner spacings on Ultra-Low NO_x designs and retrofits.

Try 3 burners at the burner to tube spacing of 48"(4'):

Burner circle diameter= 12.7'-2(4')= 4.7'

Space between burner= {4.7' x 12 x π - 3(24")}/3= 35"

[Tip 3] 24"(2') from a burner diameter

Burner to burner= 24+35= 59"= 2.5 tile OD

Increase burner to tube clearance since a standard burner only requires 4" spacing between burners.

Low NO_x burners require 1.5 tile OD spacings between burners.

Ultra low NO_x burners require 2.0 tile OD spacings between burners.

(10) Layout convection section

The convection tube length is set at the radiant section TCD. The number of tubes per row is set to obtain a flue gas mass velocity of 0.3 to 0.4 lb/ft^2 sec. The number of tube rows is set to obtain a 75°F using charts or tables.

Set tube length to cover radiant section (TCD)= 12.7'.

Set number of tubes per row to obtain flue gas mass velocity= 0.3-0.4 lb/ft^2 sec.

Set number of tube rows to obtain 75°F cold end approach.

Use fins for extended surface:

Use maximum fin length= 1".

Use maximum fin density= 5 fins/inch.

Check convection flue gas pressure drop, target= 0.2-0.3" H$_2$O.

(11) Select convection

This table gives cold end approach temperatures and flue gas pressure drops.

· If 0.3MV= 11 rows & If 0.4MV= 12 rows

· 4 tubes/row= 0.34 MV & 6 tubes/row= 0.23 MV

· Use 4 tubes per row at 12 tube rows to obtain a 71°F cold end approach with a 0.22 "ΔP.

Number Tube Rows	4" Tube Cold End Approach, °F	Flue Gas ΔP, Inch H$_2$O	6" Tube Cold End Approach, °F	Flue Gas ΔP, Inch H$_2$O
Convection Section Layout — Gas Fired, 450°F Process Inlet — 5 fins/inch, 0.75"high(4"),1.0"high(6"),0.05" thick				
0.2 lb/sec/ft^2 Mass Velocity				
6	219	0.032	214	0.032
7	149	0.040	147	0.041
8	102	0.046	100	0.047
9	69	0.053	68	0.055
10	44	0.059	47	0.061
11	29	0.065	32	0.068
12	18	0.071	21	0.075
0.3 lb/sec/ft^2 Mass Velocity				
6	298	0.072	298	0.073
7	218	0.089	220	0.092
8	160	0.104	163	0.108
9	117	0.119	121	0.123
10	83	0.132	89	0.139
11	60	0.147	66	0.154
12	43	0.159	48	0.167
0.4 lb/sec/ft^2 Mass Velocity				
6	357	0.127	363	0.130
7	274	0.157	281	0.163
8	210	0.183	218	0.192
9	161	0.211	169	0.221
10	120	0.235	131	0.248
11	93	0.260	101	0.275
12	71	0.284	79	0.300

(12) Design summary

Designer designs for flexibility.

> · Standard burner designs are simple.
> · Low NO_x designs require large burner to burner clearance and larger burner to tube clearance.
> · Energy efficiency is controlled by the number of tube rows.
> · Many vendors use a design that does not have enough stack height.

(13) Heater tube thickness

This is the current formula from API. The corrosion allowance is typically 1/8" (3mm). Austenitic stainless steel typically has 1/16" (1.6mm) corrosion allowance. The elastic allowable stress is two thirds the yield stress at temperature for ferritic steels and 90% of the yield strength at temperature for austenitic steels. Stainless tubes have less safety margin. Heater tubes can be purchased to either a piping (i.e. A-335 P9) or a tubing specification (i.e. A-200 T9). The piping specification mill tolerance is ±12.5%, while the tubing specification has a mill tolerance of -0% to +28%. The minimum thickness is used in the formula below. If inspection thicknesses are available, the lowest recorded value is used. Many of the older tubing specifications such as A-200 have lower allowable stress. Most of the heater tubes today are purchased to piping standards such as A-335.

> · API 530 heater tube thickness
> · Check existing tube thickness.
> · Determine retirement thickness.
> · Creep-High temperature-Permanent damage
>
> For tube metal temperatures in the elastic (low temperature) zone
> $t_m = t_s + CA$
> $t_s = P_E D_o /(2S_E + P_E)$
> Where,
> t_m = minimum tube wall thickness required, in
> t_s = stress thickness required, in
> P_E = design pressure, psig
> D_o = tube outside diameter, in
> S_E = elastic allowable stress at maximum tube wall temperature, psi

In the low temperature elastic range there is no permanent damage to the tube. In the high temperature creep range there is permanent damage to the tube. The time and temperature relationship is defined by the series of stress curves to the right side of the curves. The elastic allowable stress is two thirds the yield stress at temperature for ferritic steels and 90% of the yield strength at temperature for austenitic steels. This means there is less safety margin on a stainless steel heater tube.

· API 530 heater tube thickness
· Elastic-Low temperature-No permanent damage
· Creep-High temperature-Permanent damage

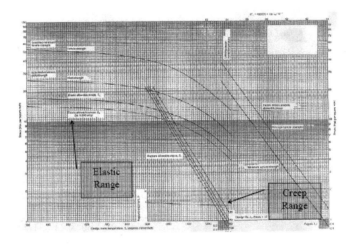

13. Controlling Film Temperature in Fired Heaters

Direct-fired heaters have been widely used in the oil refining and chemical process industries to heat the crude oil contained in tubular coils by the combustion of fuel within an internally insulated enclosure. A successful fired heater design relies on many factors. Film temperature control is one of the key factors that play a crucial role in fired heater design, particularly for units processing heavy feedstocks that are thermally unstable, such as Canadian oil sands-based feedstocks.

Film temperature determines the susceptibility of a process fluid towards coking. Bulk oil temperature plus a temperature rise across the oil film sets the film temperature. In most applications, it is the oil film temperature, not the bulk oil temperature that limits the heater duty and the oil life. Film temperature is an important factor in fired heater design for many reasons. Firstly, oil degradation starts in the fluid film, since this is the hottest place for the bulk oil. Fluid life is shortened because of degradation, which can lead to a costly result. Secondly, if the film temperature exceeds the limitation, the stationary fluid film on the inside tube surfaces is subject to thermal decomposition, which results in coke deposition at that location. Coke deposits increase resistance to heat transfer and raise the tube metal's temperature. Once the tube wall temperature reaches the design temperature, the heater must be shut down for decoking to avoid coil damage. Thirdly, overheating of the fluid film accelerates the fouling rate. Fouling requires more heat input and a hotter tube metal temperature to maintain the same heater outlet temperature. These factors cause heaters to shut down much more frequently and eventually reduce the whole plant's profitability.

Due to the importance of the film temperature, its control has become a hot topic for fired heater designs, especially for crude heaters, vacuum heaters and coker heaters. So, some feasible methods of controlling the film temperature for fired heater design are introduced as follows.

13.1 Steam injection

Steam injection is one of the best options for lowering the film temperature, as long as there are no unintended consequences to downstream of the units. Steam reduces the oil residence time by increasing the fluid velocity. High fluid velocity improves heat transfer in the film layer, which lowers the differential temperature between the tube wall and the bulk fluid. Fig.1 shows the effect of steam injection on film temperature for one crude heater.

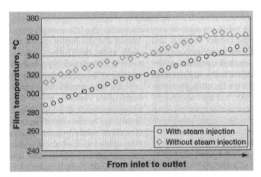

Fig.1 Effect of steam injection on film temperature

The horizontal axis represents the radiant coil growth from inlet to outlet. It can be seen that the film temperature drops around 20 ℃ after injecting 1 wt% of steam into the fluid. The results shown in Fig.1 arise from the simulation of a crude heater at a design duty of 54.2 MM Kcal/hr with diluted bitumen as the process feed.

Application	Average Heat Flux Btu/hr(ft^2)	Maximum Film Temperature °F	Allowance for Fouling, °F
Crude Atmospheric	12,500	850	50
Crude Vacuum	10,000	950	100
Hydrocracker Reactor[1]	15,000	900	—
Resid Hydrotreater[1]	10,000	850	—
Gas Oil Hydrotreater[1]	15,000	900	—
Light Lube Hydrocracker[1] Light Lube Hydrofinisher[1]	15,000	900	—
Heavy Lube Hydrocracker	10,000	800	—
Heavy Lube Hydrofinisher[1]	12,000	750	—
Reboiler, Light Oil	12,000	—	—
Reboiler, Heavier Oil	10,000	900	—
Delayed Coker	9,000	1,100	100
Steam Reformer Hydrogen[1]	15,000	1,700	—
Catalytic (Naphtha) Reformer	11,000	—	—
Hot Oil System Heater	8,000 to 10,000	—	50
Fire tube, heating water	10,000	—	—
Fire tube, heating oil	8,000	—	—
Fire tube, heating amine or glycol	7,500	—	—

Heat flux limits

Selecting the correct location and amount of steam injection is critical. It must be injected to the upstream of the heater tubes with the highest film temperature, yet far enough downstream in the radiant section to minimize incremental pressure drop to ensure that charge pump capacity is not exceeded.

Steam injection can also change the flow regime for two phase flow. The problem associated with slug flow can be mitigated by steam injection. But caution is advised in selecting a suitable steam condition to match the process fluid condition, to make sure that no condensation occurs after steam injection into the process fluid.

13.2 Reducing the size

Depending on the allowable pressure drop, fired heater coils are usually divided into multiple passes to accommodate the total flow and to meet pressure drop requirements. For a given flow rate and the number of flow passes, oil mass velocity increases with reducing tube size.

A higher oil mass velocity reduces the oil residence time and increases the film heat transfer coefficient, therefore lowering the film temperature. Fig.2 shows the effect of tube size on film temperatures.

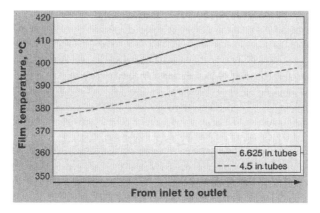

Fig.2 Effect of tube size on film temperature

The horizontal axis represents the heater coil growth from radiant inlet to radiant outlet. The results come from the simulation of a small heater with a design duty of 12.9 MM Kcal/hr. The heater is a vertical cylindrical type with four pass flow. To demonstrate the effect of tube size on the film temperature, the radiant tube size is changed from 6.625in. to 4.5in., with no change in the size of the convection tubes. More radiant tubes have been added to the radiant section to compensate for surface loss caused by the reduction in tube size, to keep the same heat flux. Case 1 is for the heater with all tubes sized at 6.625in., while case 2 is for the same heater with tube sizes changed from 6.625in. to 4.5in., for the radiant coils only. It can be seen that the film temperature in the radiant section has been decreased because of the reduction in tube size.

Reducing the tube size increases the heater pressure drop, which requires a much higher pump head upstream of the heater. Since film temperature control is basically intended to control the peak film temperature, it is advisable to reduce tube sizes for tubes with peak film temperature only, to minimize the increase in pressure drop caused by reducing tube sizes.

13.3 Double fired vs. single fired

A fired heater can be single or double fired. The heat flux on the tube's circumferential surface is not uniform because of the shading of radiant heat. The single fired heater receives radiant heat on one side of the process tubes (directly from the burner flame), while the other side of the tubes, facing the heater wall, gains radiant heat from the refractory. The portion of the tube facing the burners has a higher local heat flux, while the side facing the refractory is much lower. For a given fired heater with nominal two diameter tube spacing and a very uniform longitudinal heat flux distribution, the local peak heat flux (Q_m) is approximately 1.8 times the average heat flux (Q_a) for single fired heating. In contrast, the double fired heater has radiant heat on both sides of the tubes, which greatly reduces the peak flux to about 1.2 times the average heat flux. The correlations mentioned above for single and double fired can be simply represented in the following equation:

$$q_m = xq_a \quad (1)$$

Where X represents the time factor, which is approximately equal to 1.8 and 1.2 for single and double fired, respectively. The local film temperature can be calculated by the following equations:

$T_f = T_b + \Delta T_f$ (2)

$\Delta T_f = q_m / K_f (D_o / D_i)$ (3)

Where T_f and T_b are film temperature and oil bulk temperature, respectively. ΔT_f is the film temperature rise and K_f is a film heat transfer coefficient.

From the equations, it can be seen that it is the localized heat flux, not the average heat flux, that directly governs the film temperature. For a heater with a given average heat flux, a double fired heater has a lower localized heat flux distribution than a single fired heater. A lower localized heat flux reduces the film temperature at that location. Fig.3 shows a comparison of the film temperature between single and double firing for the same heater with the same average heat flux. The results are from the simulation of a vacuum heater with a design duty of 43 MM Kcal/hr. It can be seen that using double fired can greatly reduce the film temperature of the radiant coil for the heater.

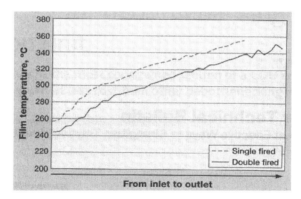

Fig.3 Comparison of the film temperature between single and double firing

13.4 Lowering the average heat flux

The first step in designing a fired heater is to set up the allowable average radiant heat flux. For a given heater, either single or double fired, it is helpful to control the film temperature by lowering the average heat flux. From Eq.(1), the localized heat flux reduces with lowering average heat flux, no matter whether it is single or double fired. Oil film temperature depends on the heat flux and oil mass velocity. Decreasing the heat flux reduces the oil film temperature at a fixed mass velocity.

The average radiant section heat flux is defined as the total radiant section absorbed duty divided by the total radiant section tube surface area. For a given radiant duty of a fired heater, the only way to lower the average heat flux is to increase the radiant section's surface area. It may be reasonable to assume a relatively low average heat flux to design a fired heater with very tight film temperature control. However, it is also noted that oil residence time increases as the surface area increases, which may partially counteract the benefit of decreasing the film temperature by lowering the average heat flux. A lower average heat flux means more capital cost in the heater coils, which is another drawback in controlling the film temperature by reducing the average heat flux. Thus, the effect on heater design of lowering the average heat flux should be carefully evaluated before a decision is made.

13.5 Other design considerations

There are other design considerations that should not be neglected when designing a fired heater with better film temperature control. The radiant section heat flux at any point in the heater is controlled by the temperature difference between the hot flue gas and oil in the tube. The heat transfer rate increases with the temperature difference between the hot flue gas and the cold oil. In a vertical up fired heater, it is not rare to see that heat flux is low at the floor and gradually increases along the length of the flame. It is highest at the point where maximum combustion takes place in the flame, then reduces at the top of the fire box. Thus, real fired heaters may encounter more or less significant heat flux imbalances. This heat flux imbalance can cause high film temperatures and high rates of fouling formation. Efforts should be made to optimize the design parameters to minimize the heat flux imbalance. These design parameters include radiant section height to width ratio, burner to tube distance, number of burners, flame shape and dimensions, and radiant section tube layout.

Flame impingement can cause extremely high localized heat flux, which results in a higher film temperature and rapid coke formation. Flame impingement occurs when a flame actually touches or engulfs the tubes. Some precautions need to be considered in the design of fired heaters to prevent flame impingement occurring; for instance, an adequate fire box to contain the flame, more and equally spaced burners, the correct type of burners, and improved distribution of combustion air flow.

14. How to Reduce Your Fuel Bill

Demands on the greater energy conservation have had direct impact on the design and operation of fired heaters. Fuel savings have been realized with a variety of measures, ranging from the inexpensive- such as the fine tuning of operating procedures and the upgrading of maintenance techniques- to the capital intensive- such as the installation of complex heat recovery facilities at substantial initial outlay. The following will focus on methods that have proved successful in raising the thermal efficiency of fired heaters.

14.1 Reduction of excess air

In existing installations, excess air is the most important combustion variable affecting the thermal efficiency of a fired heater. Although heater operation is easier to control at high excess air levels, it is very costly. The higher the excess air, the greater the fuel consumption for a given heat absorption. The extra fuel is consumed in heating the excess air volume from ambient temperature of the exiting flue gases.

In order to exercise greater control over excess air, the O_2 content of the flue gas should be monitored above the combustion zone. Often, excess air in the combustion section may run as low as 10 to 15%, but stack gas analysis reveals an O_2 content equivalent to as much as 100% excess air. This differential results from air leakage into the heater that occurs between the combustion zone and the stack. Such leakage cannot be corrected by burner adjustments.

Air leakage into a heater can occur at many locations. One route of entry is through the seams of the steel casing, between adjacent plates and stiffening members. Air can also enter through distorted or poorly gasketed header boxes. The terminal tubes in the tube coil, where they enter and leave the heater casing, can also be a source of air leakage. Only through a rigorous, continuing maintenance effort can air leakage into heater settings be controlled effectively. There are various schemes for controlling excess air via the monitoring of flue gas O_2 content. Four such monitoring schemes are listed here, in order of increasing sophistication and cost:

(1) The least costly scheme requires only periodic checking of O_2 content using a

portable O_2 analyzer and a portable draft gage. With these readings as a guide, the operator can make the adjustments necessary to operate at minimum excess air levels.

(2) One of the most common monitoring systems employs a continuous oxygen analyzer equipped with a local readout device and a permanently mounted draft gage. Continuous readings enable the operator to adjust heater operation whenever necessary.

(3) Bringing the O_2 and draft readouts into the control room and adding a remotely controlled, pneumatic damper positioner is the next step toward improving excess air control.

(4) Although generally justifiable only for large heaters, further sophistication in excess air control can be achieved with automatic stack damper control. In this control scheme the damper is positioned automatically according to a draft signal received from a probe located below the convection section. The controller changes the damper position so as to hold a set point draft corresponding to the targeted excess air. The draft set point is normally changed only to achieve variations in heater duty.

14.2 Further recovery of convection heat

The potential for additional heat recovery in the convection section exists when the heater operates at a relatively high flue gas temperature. First consideration should be given to increasing the convection section surface area by adding several rows of convection tubes in the same heating service. Many current installations have anticipated such expanded service, and have been designed to accommodate the future installation of two rows of convection tubes.

The installation of supplementary heat exchange units to reclaim additional convection heat often provides a quick economic return. Falling into this category are a number of units that recover process heat. Examples are feed preheaters for pyrolysis and steam hydrocarbon reformer heaters, reboilers operated in conjunction with catalytic reformers, and superheaters providing process steam for petroleum refinery distillation units.

Very often the supplementary unit recovers convection heat for use in a steam generating facility rather than in a process stream heating application. Steam generating units driven by convection heat are routinely coupled with catalytic reformer heaters and steam hydrocarbon reformer heaters. Fired heaters in such installations have convection sections equipped with several independent coils so that the same convection section provides heat for boiler feed water heating, steam generation, and steam superheating. Coils installed in the convection sections of steam generating fired heaters are almost always of the forced circulation type.

Occasionally, flue gases from several fired heaters are routed to a central waste heat recovery facility. Typically, the recovered heat is utilized for steam generation. Waste heat boilers of this type are normally designed with a flue gas bypass around the boiler, which allows the heaters to operate even when the boiler is taken out of service. Where several heaters are involved, provisions for positive isolation at the flue gas ducts from each heater will enable the user to take an individual heater out of service while the remaining heaters and the waste heat boiler stay on line. Before any add-on heat recovery device is retrofitted to a heater, the structural integrity of the existing steel work and foundation must b examined to assess the effects of the increased loads.

Furthermore, it should be noted that the additional convection section heat recovery will result in a lower stack gas temperature, thereby reducing the stack draft. In additions, the installation of more surface area will increase the flue gas pressure drop. Therefore, it is imperative that the effect of any alterations on the stack draft be analyzed beforehand in order to determine whether additional stack height will be required.

14.3 Replace bare tubes with extended surface

Many older generation fired heaters are equipped with bare tube convection and operate with high flue gas temperatures at 65 to 70% thermal efficiency. By replacing the bare tubes with extended surface tubes, efficiency improvement in the neighborhood of 10% may be realized in some installations. This corresponds to a stack gas temperature reduction of about 149℃.

The most economical conversion from bare to extended surface tubes can be made when the reduction in tube size is such that the tip-to-tip diameter across the extended surface tube is the same as the outside diameter of the original bare tube.

On this basis, it is very likely that the existing convection section tube sheets can be retained. However, the reduction in tube I.D. will result in higher fluid pressure drops unless an increase in the number of convection section parallel passes can be tolerated. Conversely, if the same tube I.D. is maintained, the conversion from bare to extended surface will necessitate the replacement of the convection section tube sheets - with accompanying down-time and expense.

When conversion is considered for a heater firing liquid fuel, it is recommended that soot blowers be installed, in view of the greater fouling tendency of extended surface devices. The cavities required to accommodate the soot blower lances can usually be created simply by omitting the installation of a row or two of tubes at selected locations in the convection section. Again, it is mandatory that the stack draft be assessed to confirm whether the existing stack is adequate.

14.4 Preheat the combustion air

Fuel consumption in a fired heater can be reduced significantly by preheating the combustion air. In the preheater, heat is transferred from the flue gas to the combustion air, reducing the exit temperature of the flue gas and raising the thermal efficiency. With air preheat systems, exit flue gas temperatures often range in the 149 to 177°C and efficiency levels commonly reach 90 to 92% (LHV). With such systems, the attainable thermal efficiency is no longer controlled by the approach between the flue gas and inlet fluid temperatures. The significant features of the more popular air preheaters are noted here.

14.4.1 Recuperative type air preheaters

These devices, typically of tubular construction, transfer heat by convection from flue gas to combustion air. Customarily, the air flows inside the tubes, whereas the flue gas flows across the tube bundle. The preheater can be installed in the fired heater above the process convection section or, as is more usually the case, at grade alongside the heater.

Materials selected for heat transfer surfaces vary from designer to designer. If the flue gas is well above the acid dewpoint, a manufacturer may select cast iron tubes having internal and external fins. In the zone approaching the dewpoint, fins are provided on the gas side only, in order to keep tube metal temperatures are high as possible. Below the dewpoint, plain tubes of borosilicate glass are used in order to minimize acid corrosion.

Tubular type air preheaters are essentially tight between air and gas, with no leakage of air into the flue gas side. Fig.1 shows a bank of tubular cast iron heating elements equipped with internal and external fins.

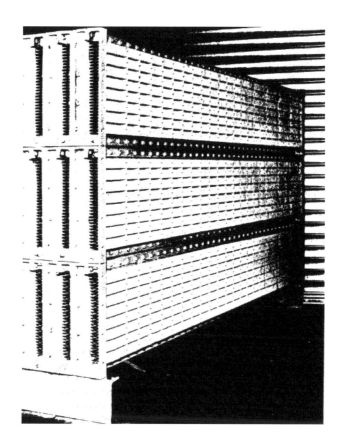

**Fig.1 Recuperative preheater warms combustion air.
Unit is equipped with finned, cast iron tubes**

A schematic arrangement for a typical recuperative air preheater installation is shown in Fig.2. (The arrangement shown is equally applicable to regenerative preheater systems, described below). The system employs a force draft fan to supply combustion air, and an induced draft fan to maintain a negative pressure and draw the flue gas through the system to the stack. A cold air bypass enables the operator to route a portion of the incoming combustion air around the preheater when ambient temperature is very low. The preheater surface area can thus be maintained above the acid dewpoint, at a nominal sacrifice in thermal efficiency.

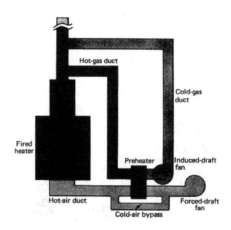

Fig.2 Typical process scheme used for both recuperative and regenerative air preheaters

14.4.2 Regenerative type air preheaters

This apparatus consists of heat transfer elements housed in a subdivided cylinder, which rotates inside a casing. Hot flue gases pass though one side of the cylinder, cold air through the other side. As the cylinder slowly rotates, the elements continuously absorb heat from the flue gas and release it to the incoming air stream.

The cylinder is subdivided by baffles which, like the seals between the cylinder and the casing, help to minimize leakage of air into the flue-gas stream. This leakage, which results from the pressure differential that exists between the air side and the flue-gas side, is normally 10 to 15% of the total air flow. The preheater system, particularly the forced draft and induced draft fan, must be designed to accommodate this leakage.

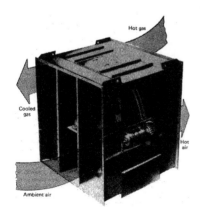

Fig.3 Regenerative preheater. Rotating cylinder absorbs heat from flue gases, releases it to incoming air

The heating elements of the regenerative air preheater are installed in two or three layers. Corrosion of the elements is usually confined to the final portion of the cold end layer, where cold air enters and cooled flue gas leaves. If such corrosion doses occur, the elements of the cold end layer can be removed and reversed to extend their service life. For those applications where the elements are exposed to very corrosive atmospheres, porcelain enameled heating surfaces are available. Fig.3 illustrates the basic construction of a regenerative air preheater. The main components of the overall process arrangement are shown in Fig.2.

14.4.3 Heat medium air preheaters

Instead of direct heat exchange between air and flue gas, these units employ an intermediate fluid to transfer heat from the flue gas to the incoming combustion air. The heat medium is contained in a closed loop that includes a reheat coil located in the flue gas flow downstream of the process convection coil, and a preheat coil positioned in the air stream. The circulating fluid extracts heat from the flue gas, lowering the gas temperature and raising the fluid temperate. In turn, the hot fluid releases its heat to the incoming air.

The process scheme for a heat medium preheat system (Fig.4) includes, in addition to the preheat and heat coils, a fluid surge vessel and a circulating pump. Depending on the design of the heater and the available draft, the system may be provided with forced draft/induced draft, forced draft only, or induced draft only.

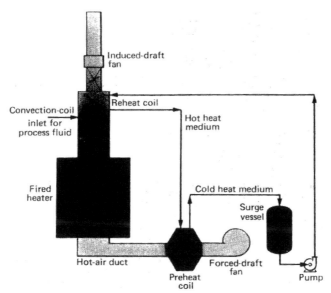

**Fig.4 Heat medium preheater transfers heat from flue
gases to air via intermediate, circulating fluid**

14.4.4 Process fluid air preheaters

These systems take off a portion of the process stream entering the convection section and send it to a preheat coil mounted in the air stream, which warms the air. The subcooled process fluid is then returned to a reheat coil located in the flue gas flow downstream of the main process convection coil, which reheats the fluid to its original temperature level and returns it to the main process stream ahead of the convection coil. Only a portion of the process fluid stream is drawn off for preheat service. The mass flows and specific heats of the combustion air and process fluid must be carefully balanced in order to achieve the desired temperature changes in each stream.

A schematic arrangement of a process fluid air preheater is shown in Fig.5. As with heat medium preheaters, these systems can be supplied with forced draft/induced draft, forced draft only, or induced draft only.

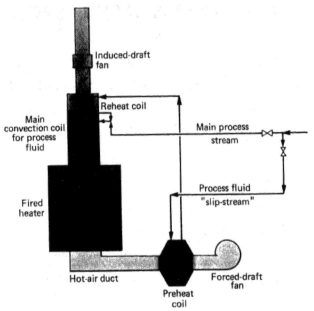

Fig.5 Typical arrangement for air preheater that uses process fluid as heat exchange medium

14.5 Process effects of air preheating

Fired heater operation with preheated combustion air results in higher adiabatic flame temperatures than with ambient air. From the combustion kinetics, then, one might expect heaters using preheated air to generate higher amounts of NO_x than those using ambient air. However, N_2 fixation from the atmosphere is usually a relatively minor factor in NO_x production. The N_2 which most readily converts to NO_x is that which is chemically bound up in the fuel. The quantity of oxides of N_2 formed from this source is not significantly affected by elevated combustion temperature. In fact, with an air preheater, the reduction in the amount of fuel consumed corresponds directly to a reduction in the amount of NO_x formed from fuel bound N_2, and is quite likely to offset the additional atmospheric fixation due to the higher flame temperature.

Another process effect that shows itself when an air preheat system is retrofitted to an existing heater is the shift in duty split between the radiant and convection sections. The higher combustion temperature due to preheating, coupled with the lower convection transfer rate at reduced flue gas flow rates, causes a greater proportion of the total heat absorption to take place in the radiant section. Consequently, operation with preheated combustion air results in higher radiant duties than with ambient air. The magnitude of the shift in duty split is, of course, dependent upon the degree of preheat applied to the combustion air.

14.6 Gas turbine exhaust used as combustion air

The exhaust gas from gas turbines, widely used as drivers for compressors, pumps, and electrical generators, usually ranges from 427 to 482℃ and contains from 17 to 18% O_2. These gases are well suited as a source of preheated combustion air for fired heater, and can be used to cut heater fuel consumption.

In most systems utilizing gas turbine exhausts, an auxiliary source of combustion air permits independent operation of the heater and the gas turbine. Typically, the overall system is designed so that the gas turbine exhaust can be vented to atmosphere whenever the heater is operated on ambient air.

14.7 Conversion from gas to liquid firing

In the face of cutbacks of natural gas deliveries, many industries are seeking alternative fuel sources, primarily liquid fuels. The following areas of concern should be examined closely before any conversion from gas to liquid firing is undertaken.

14.7.1 Impact on radiant section

The major factor affecting the combustion chamber is the size of the oil flame compared to the size of the gas flame. At the same level of heat release, an oil flame is generally longer and wider than a gas flame. For this reason, the distance from burner to tube should be examined to assess the potential for flame impingement on the tube coil. Similarly, vertical clearance dimensions should be reviewed, since the longer oil flame may impinge on the shield tubes, the reradiating cone, or the baffle provided at the top of some older vertical cylindrical heaters.

If the contemplated liquid fuel contains substantial vanadium and sodium concentrations, consideration should be given to protecting exposed tube supports from the corrosive combustion environment. Also, if the heater contains a reradiating cone or baffle, the heater manufacturer should be consulted regarding the possibility of removing such vulnerable equipment.

14.7.2 Impact on convection section

Convection section extended surface in gas fired heater is very often of the high density, finned tube type. Under heavy liquid fuel firing, such tubing is difficult to keep clean using conventional onstream cleaning techniques. Before conversion, therefore, the designer should consider replacing high density finned tubes with either heavy low density tubes, or studded tubes. Since replacement will reduce the total effective surface area, additional rows of convection tubes will be required to maintain thermal efficiency.

Facilities for onstream convection cleaning, such as soot blowers, should be installed as part of the conversion. The addition of such equipments, as well as the necessary ladder and platform access, requires that the structural integrity of the steelwork and foundation be assessed.

14.7.3 Adjustments to draft

Many heaters designed for gas firing operate virtually at their draft limit, usually because they are pushed well beyond their original design capacity. Because oil firing requires higher excess air levels to achieve acceptable combustion, draft limited gas fired heaters will suffer a capacity decrease when converted to liquid firing unless additional stack height is provided.

14.7.4 Alterations to burners

Burner replacement in a gas to liquid fuel conversion can, in most instances, be made on a one-for-one basis. However, for those conventional gas fired heaters that operate with relatively small burner heat releases in the range of 2 to 3 MM Btu/h, substitution of oil burners on a one-for-one basis may result in unstable combustion. In these cases, a totally new firing arrangement should be specified as part of the conversion. The final burner arrangement should take into account such parameters as the combustion chamber configuration, burner turndown requirements and fuel oil combustion behavior.

From a practical standpoint, an important consideration that must not be overlooked in any conversion study is the heater's mechanical condition and its remaining useful life. Investment costs for gas to oil conversions are expensive, particularly when convection sections must be revamped. Furthermore, the downtime necessary to effect the conversion is costly, due to lost revenues. Consequently, if conversion is contemplated for a heater that is in poor mechanical condition, an evaluation of maintenance costs, the conversion investment and revenue losses may well show replacement to be economically more attractive.

A good part of conversion cost lies in the fuel handling and distribution system. Conversion of only a few large units, particularly if they are located in the same general plot area, will reduce this cost.

14.8 Various configuration of heat recovery

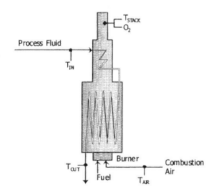 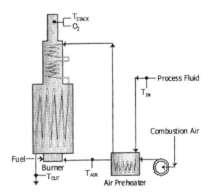

Fig.1 No Air Preheat: Stand-Alone Single Furnace

Fig.2 Air Preheated by Process Fluid

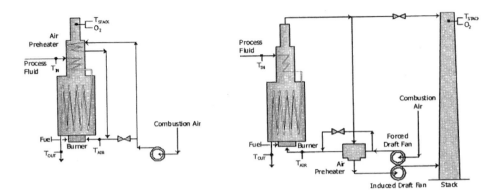

Fig.3 Air Preheated by Stack Gas

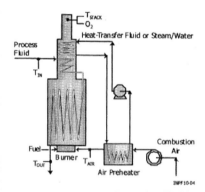

Fig.4 Air Preheated by Transfer Fluid

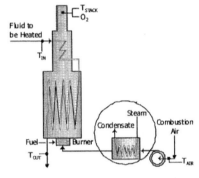

Fig.5 Air Preheated by Steam or Other Waste Heat

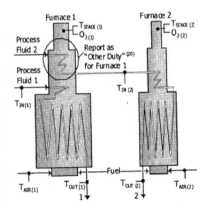

Fig.6 Convection Duty to a Different Process Stream

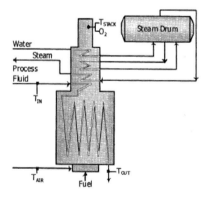

Fig.7 Convection Duty to Steam Generation and/or Superheat

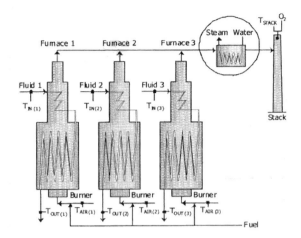

Fig.8 Shared Convection Duty

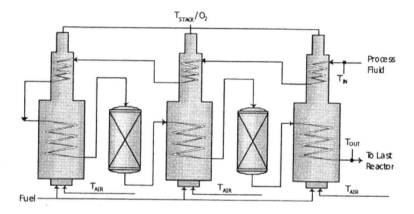

Fig.9 Naphtha Reforming Furnaces Reported as One Fired Heater

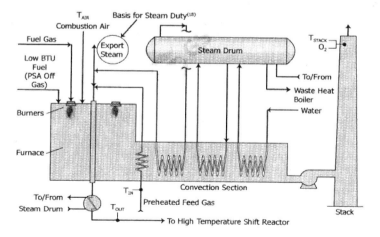

Fig.10 Hydrogen Plant Reformer

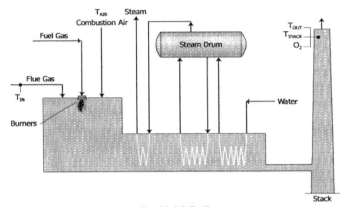

Fig.11 CO Boiler

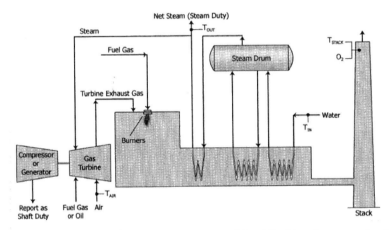

Fig.12 Integrated Gas Turbine and Boiler (Fired Turbine Cogeneration)

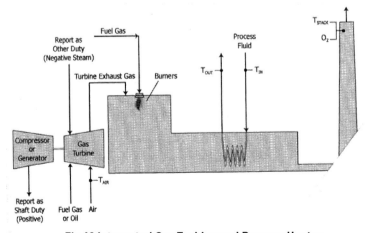

Fig.13 Integrated Gas Turbine and Process Heater

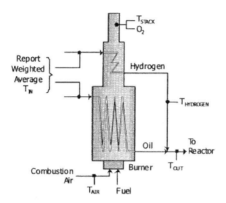

Fig.14 Hydrotreater Reactor Charge Heater

Legend: T_{AIR}= combustion air temperature, T_{STACK}= stack gas outlet temperature, T_{IN}= furnace inlet temperature, T_{OUT}= furnace outlet temperature, O_2= stack oxygen content

15. Generalized Method Predicts Fired Heater Performance

Purchasers and users frequently must make their own ratings of fired heaters in order to ① estimate heater sizes, fuel consumption, and waste heat recovery, during the conceptual design stage. ② evaluate vendor's offerings with regard to reasonableness of designs and consistency between competing proposals. ③ predict the effects of changes in feed rates, feed properties and similar operating variables. ④ anticipate the effects of proposed modifications to an existing heater. Final design is usually done by the heater manufacturer, following the purchaser's duty specification. In general, each manufacturer has its own rating method for its particular design. Such a method can be expected to be more accurate for its range of application than any general procedure.

15.1 Suggested method validated by experience

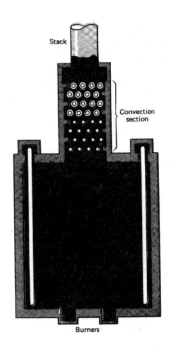

Fig.1 Typical fired heater is shown in cross section

The method proposed for computing the performance of both the radiant and convection sections of fired heaters is based on fundamental correlations for heat transfer by radiation and convection, adjusted on the basis of operating experience with industrial heaters. Experience has shown the method to be well suited for the type of calculations outlined.

Fig.1 shows a cross section of a typical process plant fired heater. It consists of a firebox or radiant section, a convection section, and a stack system to dispose of the combustion products and provide draft. The radiant section, where the fuel is burned, contains heat absorbing tubes, which remove a portion of the heat from the combustion products before they flow to the convection section. In this type of furnace, the tubes are located around the outside of the firebox, combustion taking place in the open central space.

There are also designs in which the tubes are suspended in the center of the furnace, with combustion space on both sides. In these furnaces, the burners may be arranged to fire against refractory walls so as to enhance radiant heat transfer to the tubes. In any case, the tubes must be arranged for uniform and efficient radiant heat absorptions, and there must be adequate volume for complete combustion without local over heating or flame impingement on the tubes.

The convection section recovers additional heat from the flue gas at a lower temperature than in the radiant section. Here, since the primary heat transfer mechanism is convection, the tubes are arranged to create high mass velocities and turbulence in the gas. Tubes having fins and other types of extended surface are frequently installed to improve convective heat transfer.

The first rows of tubes in the convection section, which are exposed to radiation from the hot gas and refractory in the radiant section, are generally called shield or shock tubes. They are usually a structural part of the convection section and are rated with it. However, special allowance must be made for the radiant heat added to these tubes. The stack system collects and disposes of the flue gas. In natural draft furnaces, the stack height must provide adequate draft to draw the gas through the firebox and convection section. In induced draft furnaces, stack height is usually set by gas dispersion requirements. Because the governing heat transfer mechanism is different in the radiant and convection sections, the two sections are rated by different methods.

15.2 Radiant section design

Radiant heat transfer between solid surfaces in various arrangements, and between hot gases and solids, has been extensively investigated. Applying basic radiation concepts to process type heater design, Lobo and Evans developed a generally applicable rating method. Almost all methods published since have followed their basic approach, as does the method described here (with simplification made by eliminating minor variables and by including general correlations).

Radiant heat transfer is basically described by the Stefan-Boltzman equation. A black body at absolute temperature T radiates energy at a rate W_B:

$$W_B = T^4 \quad (1)$$

For radiant heat transfer between two real surfaces at temperatures T_a and T_b, the relation becomes:

$$q_r = \sigma AF\,(T_a^4 - T_b^4) \quad (2)$$

Here, A is the area of one of the surfaces, and F is an exchange factor that depends on the relative area and arrangement of the surfaces, and on the emissivity and absorptivity of each. For transfer inside a furnace, it is generally best to use the heat receiving, or cold, surface as the basis for calculation.

15.3 Simplify with equivalent cold plane surface

The usual heat absorbing system consists of a number of parallel cylindrical tubes. Part of the radiation from the hot gas strikes the tubes directly and is absorbed, and the remainder passes through. If the tubes are in front of a refractory wall, the energy that passes through is radiated back into the furnace, where part of it is absorbed by the tubes, and the remainder passes through.

Nomenclature
A Area, m^2
A_{cp} Cold plane area, m^2
A_e Total furnace envelope area, m^2
A_t Tube surface area, m^2
F Radiant exchange factor
G Mass velocity at minimum cross section, kg/(m^2)(s)
h Film heat transfer coefficient, kcal/(h)(m^2)(°C)
L Mean length of radiant beam, m
P Partial pressure of radiating components, atm
p_v Velocity head, mmH$_2$O
q Heat rate, kcal/h
T Temperature, °K
V Volume, m^3
W_B Radiant emission from a black body, kcal/(h)(m^2)
α Factor of comparison between a tube bank and a plane
ρ Density, kg/m^3
σ Stefan-Boltsmann constant, 4.92x 10^{-8} kcal/(h)(m^2)(°K^4)

Subscripts
a Combustion air
c Convective heat transfer
f Fuel
g Flue gas
L Losses
n Net heat of combustion
R Radiant section
r Radiant heat transfer
S Shield- and convection- tube radiant transfer
t Tube surface

This complicated situation is handled by expressing the tube area as an equivalent plane surface, A_{cp}. It equals the number of tubes, times their exposed length, times the center-to-center spacing. Because the tube bank does not absorb all the energy radiated to the cold plane area, an absorption efficiency factor, α, must be applied. Values for α as a function of tube arrangement and spacing have been developed and published by Hottel (in Chemical Engineers' Handbook), whose curves for the most common tube arrangements in process furnaces are shown in Fig. 2.

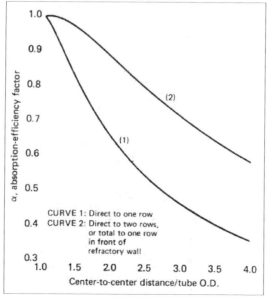

Fig.2 Heat absorption varies with tube arrangement

The product αA_{cp}, called the equivalent cold plane area, is the area of a plane having the same absorbing capacity as the actual tube bank. In the type of heater shown in Fig.1, the convection section forms part of the firebox enclosure. Because the convection section is a number of rows deep, it eventually absorbs all the radiation coming to it from the firebox. Therefore, α for the convection section equals unity. The working equation for calculating radiant heat transfer to the tubes thus becomes:

$$q_{Rr} = \sigma \alpha \, A_{cp} F(T_a^{\,4} - T_b^{\,4}) \quad (3)$$

15.4 Determining the exchange factor

The remaining term to be evaluated in Eq. (3) is the exchange factor, F. The gas in the firebox is a poor radiator, because the only constituents normally in flue gas that contribute significantly to the radiant emission are CO_2 and H_2O. The amount of radiating components can be expressed by a single term, the partial pressure of CO_2 plus H_2O multiplied by the mean beam length. Fig.3 shows the partial pressure, P, as a function of excess air for the usual hydrocarbon fuels. For the less usual fuels, such as hydrogen, P can readily be calculated by simple stoichiometry. The mean beam length, L, is calculated by the equation:

$$L = 3.6 V/A_e \quad (4)$$

Here, V is the total firebox volume inside the centerline of tubes, and A_e is the total firebox envelope area.

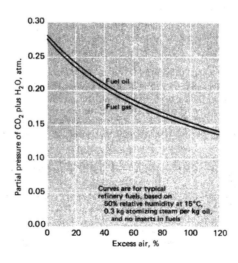

Fig.3 Excess air sets CO$_2$, H$_2$O pressure in flue gas

Gas emissivity is also a function of the temperatures of the gas and the absorbing surface. However, because the tube wall temperature effect has been found to be minor, gas emissivity can be correlated as a function of the product of PL, and of the gas temperature (Fig.4).

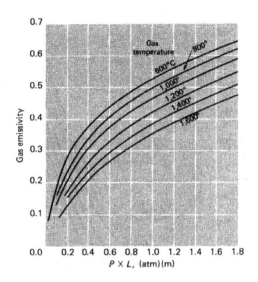

Fig.4 Flue gas emissivity depends on absorbing surface

The exchange factor also depends on the amount of radiation reflected from exposed refractory. (Energy striking refractory is reflected back toward the tubes, where it has a second chance to be absorbed). Thus a furnace having a large amount of exposed refractory will transfer more heat per unit of tube surface than one whose walls are covered by tubes.

Lobo and Evans correlated this effect on the basis of the ratio of exposed refractory area to total equivalent cold plane area. Fig.5 is based on the same correlation, except that the ratio is of cold plane to total firebox envelope area, which simplifies the calculations.

Fig.5 also takes into account that the tubes themselves are not perfect absorbers. The curves are based on a tube surface absorptivity of 0.9, which is typical for oxidized metal surfaces.

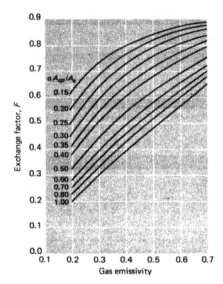

Fig.5 Overall exchange factor depends on reflected energy

15.5 Estimating convection transfer

Although radiation accounts for most of the heat transfer in the radiant section, convection cannot be neglected. The relationship for convective heat transfer is:

$$q_c = h_c A_t (T_g - T_t) \quad (5)$$

Because convective heat transfer is not the major contributor and cannot be calculated precisely, some simplifying approximations are made. For the usual furnace, h_c in the radiant section is about 10 kcal/(h)(m^2)(°K), A_t is about two times αA_{cp}, and F is about 0.57. This allows putting Eq. (5) into a form similar to Eq. (3):

$$q_{Rc} = (10)(2\alpha A_{cp})(F/0.57)(T_g - T_t)$$

$$= 35\alpha A_{cp} F(T_g - T_t) \quad (6)$$

The total heat transfer rate in the radiant section is the sum of the radiant and convective heat transfer:

$$q_R = \sigma\alpha A_{cp} F(T_a^4 - T_b^4) + 35\alpha A_{cp} F(T_g - T_t) \quad (7)$$

$$q_R / \alpha A_{cp} F = \sigma(T_a^4 - T_b^4) + 35(T_g - T_t) \quad (8)$$

Thus, the ratio $q_R/\alpha A_{cp}F$ is a function of gas and tube wall temperatures only (Fig.6).

Eq. (7) actually applies only to the tubes in the radiant coil and not to radiation to the convection section. Convective transfer to the convection section tubes, which is calculated separately, does not occur until after the flue gas leaves the firebox, so radiation to the convection section should be calculated by Eq.(3). Computer calculations can be easily programmed to maintain this distinction. For hand calculation, however, it is simpler to use Fig.6 for all tubes, the error introduced usually being negligible.

15.6 Tubes mounted centrally

For the foregoing thing, the furnace is assumed to be of the type shown in Fig.1, with refractory backed tubes around the periphery of the firebox. Certain adjustments are necessary when the tubes are arranged as in Fig.7, with a single or double row of centrally mounted tubes. In such cases, advantage is taken of the fact that a plane of symmetry in the firebox can be replaced, for computational purposes, by a refractory wall (indicated by dashed lines in Fig.7).

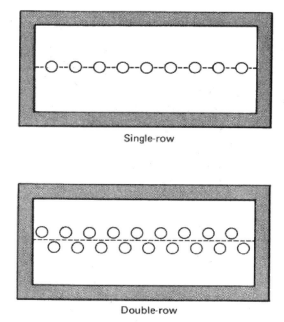

Single-row

Double-row

Fig.7 Dashed lines indicate the refractory wall assumed for calculations

Therefore, the cold plane area, A_{cp}, for the tube bank is twice the projected cold plane area, because each side is figured separately. On this basis, α for a single row is obtained from Curve 1, and for a double row from Curve 2, in Fig.2. Similarly, the mean beam length, L, must be calculated on the basis of half a furnace.

15.7 Temperature complexities

Fig.6 gives the rate of heat transfer between a mass of gas at one uniform temperature and a tube surface at another uniform temperature. In most actual furnaces, however, such is not the case. Average effective temperatures must be selected in order to use Fig. 6.

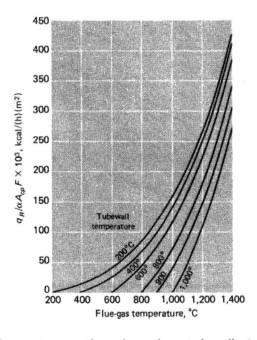

Fig.6 Temperatures set heat absorption rate in radiant section

Tube wall temperature depends on the temperature of the fluid and its transfer coefficient inside the tube, the thermal resistance of the tube wall, and the total heat flux. At tube wall temperatures below 500℃, when the radiant absorption rate is fairly insensitive to the receiving temperature, it is usually acceptable to use the average fluid temperature plus 50℃. At higher temperatures, more detailed calculations are often necessary. In installations, such as pyrolysis furnaces, having extremely high tube wall temperatures, it is advisable to divide the receiving area into zones of different average temperature, and to calculate each zone's heat absorption rate separately.

Because about 70% of the radiation to the convection section is received by the front row of tubes, the average tube wall temperature for these tubes should be used to calculate convection section radiation. Depending on the type of furnace, there may also be considerable variation in flue gas temperature within the firebox. However, because of the large amount of turbulent mixing that occurs, the transparency of the radiating gas, and the effect of secondary radiation from exposed refractory, it is very difficult to quantitatively allow for these temperature variations when making the heat flux calculation. In the absence of actual physical barriers between zones, it appears to be better to base all heat transfer calculations on a single effective radiating temperature throughout the firebox.

15.8 Heat balance provides firing rate

The previously mentioned procedure enables calculating the firebox temperature necessary to transfer a specific amount of heat into a specified radiant section coil. The next step is to determine the firing rate necessary to maintain that temperature. This is done through a heat balance around the firebox. Heat is put into the radiant section from three primary sources: (1) the net heat of combustion, q_n; (2) the sensible heat of the combustion air, q_a; and (3) the sensible heat of the fuel and any atomizing steam, q_f.

Heat is removed via absorption by the radiant section tubes (q_R), radiation to the shield and convection tubes (q_S), casing losses (q_L), and sensible heat of the exiting flue gas (q_{g2}):

$$q_n + q_a + q_f = q_R + q_S + q_L + q_{g2} \quad (9)$$

The terms q_a and q_f, being generally proportional to the amount of fuel burned, can be expressed as ratios to q_n. Similarly, the loss q_L is usually taken as 1 to 3 % of the net heat release, depending on the furnace design and experience. Finally, the fraction of net heat release remaining in the flue gas is a function of fuel composition, flue gas temperature, and excess air. For the common hydrocarbon fuels, the ratio q_g/q_n can be correlated on a single set of curves, as in Fig.8.

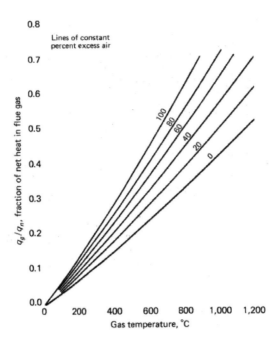

Fig.8 Flue gas heat content for common hydrocarbon fuels

On this basis, Eq. (9) can be rearranged to allow direct calculation of q_n:

$$q_n = (q_R + q_S)/\{(1) + (q_d/q_n) + (q_f/q_n) - (q_L/q_n) - (q_{g2}/q_n)\} \quad (10)$$

The last function required to close the computational loop around the radiant section is the relationship between the effective gas radiating temperature, T_g, and the exit flue gas temperature, T_{g2}. For box type heaters having an approximately square cross section, and no areas of refractory with direct flame impingement, the two temperatures may be assumed equal.

In the opposite extreme of a high temperature furnace having a tall, narrow firebox with wall mounted radiant burners, T_g may be 100 to 150°C higher than T_{g2}. Other types, such as narrow, bottom fired vertical cylindrical heaters, will fall somewhere in between. The magnitude of the difference must be determined empirically from experience with similar designs.

15.9 Convection section design

The relative importance of the convection section in fired heater design has increased markedly. One reason is the higher cost of fuel, which has resulted in setting furnace efficiency targets far higher than could formerly be economically justified. Another major factor is the development of chemical conversion processes that demand extremely high heat input fluxes, and correspondingly high firebox temperatures.

The result is a smaller fraction of the total heat released being removed in the radiant section, and a correspondingly greater convection section loading.

Whereas typical heat distribution formerly was roughly 50% radiation section, 20% convection section, and 30% losses, it now may be, respectively, 40%, 50% and 10%. Thus, there has been a trend toward larger convection sections and more use of extended surface. Also, there are more multiservice convection section that - via steam generation, feed water heating and similar extraneous services - take up heat that cannot be absorbed in the primary process. The result has been to make convection section design more difficult and more critical.

15.10 Transfer coefficients depend on tube type

As in the radiant section, heat is transferred in the convection section by both radiation and convection. The classical basis for calculating convection section heat transfer was developed by Monrad, who took into account direct convection, radiation from the gas, and radiation from the refractory walls. Monrad's basic method has been modified in light of later experimental results and adapted to extended surface tubes by Schweppe and Torrijos.

Because of the previously mentioned trend toward increased convection section heat recovery, most present day heater designs incorporate extended surface tubes. Because of the many types and sizes of extended surface available, it is not practical to present a specific procedure for calculating coefficients.

In any case, the amount of flue gas flowing through the convection tube bank must be known. As with the flue gas heat content, the flue gas quantity can be calculated from stoichiometric relationships involving the fuel consumption, the fuel heating value, and the amount of excess air. For normal hydrocarbon fuels, the quantity can be estimated closely from the net heat release and the excess air, as in Fig. 9.

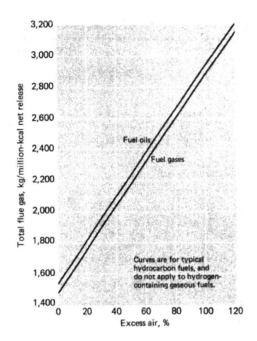

Fig.9 Flue gas flow estimated through convection tubes

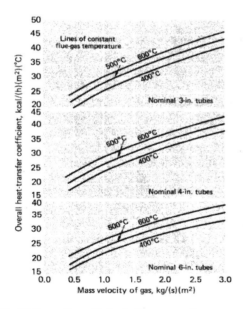

Fig.10 Overall coefficients for convection section tubes

For bare tubes, the inside tube heat transfer resistance is usually low enough that a reasonable estimate is satisfactory, at least for preliminary design. Fig.10 presents calculated overall heat transfer coefficients for typical sizes of bare tubes. These curves were calculated from the data of Schweppe and Torrijos, assuming usual staggered tube arrangements, fluid temperatures, and inside tube film coefficients. The average gas temperature used in this Fig. is the arithmetic average inside tube fluid temperature plus the log mean temperature difference from flue gas to fluid. The mass velocity is that through the minimum cross-section of the tube bank.

Having established the duty and overall heat transfer coefficients for each service in the convection section, one calculates the required surface via conventional heat balance and heat exchange procedures. For normal fuels, Fig.8 can be used to establish flue gas temperatures throughout the section.

15.11 Radiation from the firebox

In rating the radiant section, a quantity, q_s, representing the heat radiated directly from the firebox to the convection section, was calculated. This quantity of heat must be added to that derived by the convection section calculations, to determine the total heat put into the lower tubes in the convection section.

Because almost all the radiant heat is absorbed in the first two rows of the convection section, it is usually not necessary to make any detailed breakdown of the distribution.

15.12 Actual furnace performance

Some caution must be exercised when using published heat transfer coefficients for industrial heater design. Experimental coefficients are usually measured under ideal conditions, with uniform flow distribution and no bypassing. Actual furnace deviates from the ideal in some respects as follows:

In tall convection section, there can be considerable bypassing of hot flue gas through the tube header boxes. Particularly with extended surface tubes, there is often significant leakage area through the tube sheet holes. Under the driving force of the pressure drop through the convection section, hot flue gas flows out from the lower part into the header box, up through the box, and back into the upper convection section. This bypassing seriously undermines the overall performance of the convection section. Careful sealing of tube sheet holes, and in some cases partitioning of the header boxes, is necessary to avoid this difficulty.

Flow distribution may be far from uniform, particularly with long convection section tubes. Flow is most commonly excessive in the center, near the stack opening, and is nearly stagnant in area near the ends. Ample hood area above the convection tubes, and careful attention to flow patterns entering the section, are helpful in avoiding this problem. Because of these and similar effects, some safety factor should be applied when heat transfer coefficients are calculated from laboratory data. This is particularly true for extended surface designs. More reliance can usually be placed on good test data taken from actual furnaces.

15.13 Stack design

The stack system must create sufficient draft to draw the flue gases through the radiant and convection sections, and must discharge the gases at a suitable height. This is usually done in process furnaces by natural or induced draft.

15.14 Radiant section draft

A pressure drop across the burners is usually required to draw in the combustion air. Its magnitude depends on the type of burner and the fuel, and is typically about 6 mm of water (i.e., 6mmH$_2$O).

All sections of the furnace generally must also be below atmospheric pressure, so that flows through peepholes, tube openings and the like will be outside air moving in, rather than hot gas going out. A normal design point is about 2 mm negative pressures (i.e., -2 mmH$_2$O) at the inlet to the convection section. In most modern heaters that have tall radiant sections, this provides more than enough draft for the burners.

15.15 Velocity head in the convection section

Pressure drop calculations through the convection section and stack system are conveniently made in terms of velocity head. The velocity head in millimeters of water is given by:

$$p_v = 0.051G^2/\rho_g \quad (11)$$

Here, G is the mass velocity in kg/(m^2)(s), and ρ_g is the gas density in kg/m^3. If the flue gas composition is known, its density can be readily calculated from the usual gas laws. Flue gas density is relatively insensitive to fuel composition and amount of excess air. For the usual situation at sea level, the following equation is applicable:

$$\rho_g = 342/T_g \quad (12)$$

Here, T_g is the gas temperature in °K.

For banks of bare tubes, the frictional pressure drop is about one half of the velocity head per row. For extended surface tubes, it is best to rely on the manufacturer's data, or correlations developed specifically for the particular configuration. In the absence of such information, the generalized relation of Schweppe and Torrijos is serviceable.

15.16 Stack and damper head losses

Remaining losses may be estimated as follows:

Source	Velocity head loss
Stack entrance	0.5
Damper	1.5
Stack and ducts	1.0/50 dia.
Stack exit	1.0

Other losses may be similarly calculated from the usual published loss coefficients.

15.17 Stack draft

Draft depends on the difference in density between the hot flue gas and the surrounding air. The molecular weight of the flue gas is quite insensitive to fuel composition, and is about 28.5 for the usual hydrocarbon fuels. On this basis, draft per 100m of height is shown in Fig.11 Note that tall convection sections have their own draft, for which allowance should be taken in the computations.

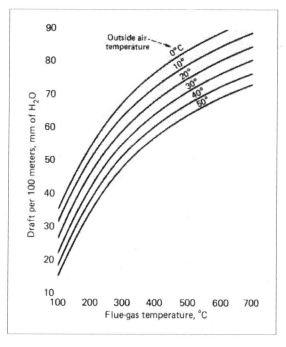

Fig.11 Stack draft is indicated per 100m of height

When calculating available draft, some allowance should be made for lower stack temperatures due to heat loss, air leaking in, and the like. The amount depends on the flue gas temperature and the degree of insulation of the stack and ductwork. For an unlined stack, and flue gas at about 400℃, a reasonable allowance would be 50℃.

Most process plants are close to sea level. However, for installations above 300m elevation, decreased atmospheric pressure must be taken into account. Lower pressure increases the volume of gas, and therefore the pressure drop, through the furnace and stack system. At the same time, it decreases the draft because of the smaller density differences. Corrections for gas density and stack draft are given in the below table.

Altitude correction for gas density and stack draft

Height above sea level, m	Correction factor
0	1.000
500	0.942
1,000	0.887
1,500	0.835
2,000	0.785
2,500	0.738
3,000	0.694

Note: Multiply sea level density, Eq.(12), and stack draft, Fig.11, by the correction factor

15.18 Overall furnace rating

Design of an actual furnace requires combining the various heat exchange and flue gas heat content relationships with the specified flows, temperatures and heat duties for the process fluids. This must be done for each stream and for each section of the heater. In almost all cases, too many factors are involved to permit vigorous simultaneous solution of all the equations. Instead, an iterative procedure with successive approximations must be followed.

The approach to any specific problem depends on the type of heater under consideration, and on the purpose of the calculations. For example, the procedure for rating a vendor's offering will not necessarily be the same as that for checking an initial design. Still another approach would be necessary to estimate the effect of modifications to an existing heater.

The preceding correlations and procedures have been presented on the basis of performing the calculations by hand. This is to permit the occasional user to obtain valid results with a reasonable expenditure of time and effort. The many repetitive calculations, however, suggest the advantages of computerization, particularly for anyone who handles a significant number of heater design problems.

In addition to speeding up the iterative calculations, computerization allows substituting exact calculations fitted to the specific situation for some of the generalized correlations. For example, Fig.3, 8, and 9 are based on the most common liquid and gaseous fuels. In a computerized operation, it is quite practical to calculate the flue gas composition, amount, and heat content for the specific fuel and the specific combustion conditions. In this way, abnormal fuels or other unusual conditions can be accommodated.

When developing such a program, procedures should be included so that a heater can be designed to meet specified performance criteria, or so that the performance of a fixed design can be estimated for various operating conditions. Beyond that, there are a number of options that might be included, depending on individual needs. Typical requirements are the ability to handle multiple fuels, gas turbine exhaust, combustion air preheat, and extraneous heat recovery streams in the convection section. The degree to which these variations can be accommodated is a major factor in determining the complexity of the program.

Computer programs of this type are valuable for handling the numerous repetitive calculations involved in optimizing a new plant design, for confirming the validity of vendors' proposals and for predicting the effect of different feedstocks or operating conditions on the performance of existing heaters.

There are, of course, many factors in addition to the rating calculation that must enter into the basic design of a heater. Typical ones to consider are allowable heat flux rates, peak to average flux ratios, coil arrangement, combustion volume, and burner size and placement. What has been presented here is a relatively simple procedure that can be used to calculate heat transfer rates and efficiency for a wide variety of fired heaters. Experience has shown that the results are accurate enough to meet the needs of most engineers.

16. Guide to Economics of Fired Heater Design

Economic design of a fired heater requires that three criteria be met: The heater must meet the required duty at the lowest capital investment; it should perform its function at the lowest possible level of fuel consumption; and, it should assure the lowest maintenance cost at a maximum reliable onstream performance.

A designer must consider the factors listed above, and relate them to the user's specific process requirements. For example, an economical design for a preheater on a catalytic cracking unit targeted for a three year run could be quite different from that of a low temperature, carbon steel reboiler with a minimal process liability.

In turn, an economical design for a 101 MM Kcal/hr crude heater must place a substantially greater emphasis on fuel efficiency than a similar design on a 1 MM Kcal/hr hot belt heater. These two examples are obviously extremes, but they do emphasize the point that materials of construction and layout should never be the only consideration for which a good designer assumes responsibility. Any approach to a sound and economical engineering selection must consider heater configuration, optimization of design, materials of construction, and installation cost.

16.1 Heater configuration

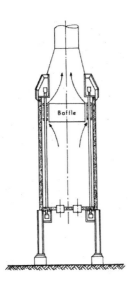

There has been a tendency in the chemical process industries to categorize generic heater design types as either economical or expensive by configuration or layout.

A typical example is the long held belief that vertical cylindrical heaters are always less expensive than horizontal tube types. This conviction originated at a time when the vertical integral radiant convection tube heater, with extended surface (Fig.1), came into use in competition with conventional horizontal tube heaters.

Plug type cleanout fittings were in general use at this time, and very few heaters were equipped with extended surface

Fig. 1 Cylindrical fired heater

tubes. The great competitive advantage of the new vertical tube unit was largely the result of the great reduction in the amount of bare convection surface required, and the attendant reduction in the number of expensive plug fittings needed, rather than the position of the tubes.

Within a few years, however, most horizontal tube heaters were also being equipped with extended surface tubes; further, with the advent of steam/air decoking resulting in the widespread elimination of plug fittings, the economic balance between vertical and horizontal tube heaters was largely restored.

Today, no categorical statements should be made regarding lower cost of one design approach vs. another purely on the basis of heater configuration. Each application must be studied, and a decision made on its own merits.

16.2 Basic design criteria

Specific process considerations and design requirements will definitely favor particular heater configurations. In full realization of the risk inherent in general statements, the following criteria are nevertheless useful in making basic design decisions:

A requirement for tall stacks 45.72m or more high tends to favor the use of vertical tube heaters when single units are involved. In the case of multiple heater installations, a single common stack with collecting ductwork generally is the most economical approach and also leaves the designer greater freedom in selecting the optimum layout for each individual service.

Heavy oil firing has a substantial effect on heater design economics. This is particularly true with fuels that contain significant amounts of sulfur, metallic salts, and vanadium.

When no basic operating prejudice exists against up-firing of heavy fuels, the floor fired cylindrical or box heater with vertically hung tubes presents the most economical approach. Both capital investment and maintenance cost are reduced since, in most instances, high alloy tube supports can be completely eliminated in the radiant section. By limiting the convection tube length, convection supports can often be eliminated also.

Where horizontal fuel oil firing is required by customer specifications, a multi lane, horizontally fired vertical tube unit can satisfy both the requirement for horizontal firing and the elimination of internal tube supports in the firebox by the use of hung tubes, with the return bends placed in header boxes (Fig.2).

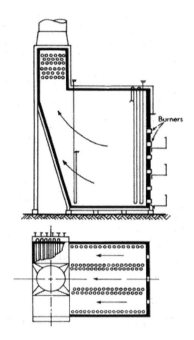

Fig. 2 Horizontally fired heater

Cost savings are most substantial in those instances where fuel composition requires 50-50 or 60-40 chrome-nickel tube supports to resist vanadium attack. These special alloys cost approximately two to three times as much per pound as the conventional 25-20 or 25-12 supports. In addition, however, the lower allowable stresses of the vanadium resistant materials require the use of heavier supports. In many cases, the added weight can drive the overall cost of intermediate tube supports up by a factor of five or six over a conventional design.

In the case of horizontal tube heaters, similar heavy fuel firing requirements generally justify the use of larger tube diameters, because of the greater permissible span between intermediate supports and the attendant reduction in the number, and consequently the cost, of these tube supports. Since, however, the use of larger tube sizes results in an increase in the cost of the heater coils, particularly in the case of alloy materials, the designer must check the economic balance between cost of tubes and cost of supports.

Apart from affecting heater configuration, the requirement for heavy fuel oil firing also generally involves a modification of the convection section extended surface. Many refiners consider the use of cylindrical stud surfaces mandatory when firing such fuels. This in turn has a profound effect on heater cost. The approximate relative investment in pressure parts for carbon steel studs vs. fins for a typical convection section in both carbon steel and 5% chrome tubing for a 25 MM Kcal/hr. Heater is illustrated below.

	Fins, $	Studs, $	Increases, %
Carbon steel	12,750	32,700	156
5% Chrome	30,700	57,400	87

16.3 Combustion air preheat

Combustion air preheat as a method for achieving maximum fuel efficiency is receiving ever increasing interest for new heater installations. Based on present day fuel costs and specified payout periods, air preheat can generally be justified on a majority of the new heater installations absorbing in the range of 25 MM Kcal/hr, or more. Unfortunately, the theoretical aspects of return on investment utilizing combustion air preheat frequently do not conform to stringent budget limitations that are often imposed on new projects.

For combustion air preheat to receive serious consideration, the designer must consider all possible methods for reducing the needed capital investment.

The vertical tube, floor fired heater presents the most attractive layout for use of air preheat. This type of design generally requires fewer burners than a horizontal tube heater. In addition, the entire flue gas and combustion air ductwork and burner plenum systems tend to be compact, simple, and as a result more economical.

It is also important to determine the optimum breakpoint between the use of heater convection surface and combustion air preheat surface. In the case of a heater with carbon steel tube coils, substantial use of convection surface in conjunction with a small air preheater will probably show the greatest return. On the other hand, should the heater convection surface consist of alloy materials (or expensive stud surfaces) a greater emphasis on air preheater surfaces will in all probability result in a more economical approach.

16.4 Shop fabrication and field erection

The continuing and rapid increase in field erection costs has resulted in an ever greater emphasis being placed on shop fabrication of heaters and heater components. This trend tends to favor horizontal tube heater designs, since they can be supplied in substantially greater capacities on a shop fabricated basis than is possible with vertical tube designs. This is primarily because the overall width of a horizontal tube heater can generally be held within shipping limitations- a restriction compensated for by increasing the heater length and height.

Some limitation on this approach must be observed with regard to the heater height to width relationship. Excessive height will result in a substantial maldistribution of heat input along the vertical dimension of the heater, and a consequent increase in the radiant heat transfer rates to the tubes in the lower portion of the unit. This in turn can result in overheating of the lower tubes and potentially require a limitation on throughput to assure adequate run lengths. Under such circumstances, the economics of maximum shop assembly could be completely negated by a reduction in furnace capability.

The type of return bend specified has a pronounced effect on the selection of an economical heater design. If welding "U" bends are called for, the cost of the fittings generally does not represent a sufficiently high percentage of the total materials cost to affect the selection of heater layout, either horizontal or vertical. Other considerations will tend to dominate the choice, particularly if the coil material is carbon steel.

If, on the other hand, cleanout type plug fittings are being considered, it becomes essential from an economics standpoint to maximize tube lengths in order to minimize the number of fittings required. The vertical cylindrical heater, with the usual cross flow convection section, is at a definite disadvantage because of the relatively short length of the convection tubes and the resultant large number of fittings required. In contrast, the horizontal tube heater, using convection tubes equal in length to the radiant tubes, requires far fewer fittings and generally presents a clear economic advantage. This relationship is illustrated in Fig.3, which provides the material cost distribution for a 25.2 MM kcal/hr heater in a vertical as well as a horizontal layout for both welding "U" bends and plug fittings.

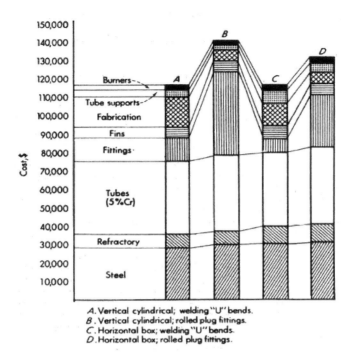

Fig. 3 Material cost distribution for typical 25.2 MM kcal/hr heaters having differing configurations and methods of fabrication

The same applies to any situation where the return bends are a high cost item, as in the case of alloy tube heaters operating at high temperatures and pressures. Consequently, reactor heaters in hydrocracking service, for example, are seldom designed as vertical cylindrical units if hydrocarbon convection surface is to be part of the design.

The cost of fittings relative to the overall cost of the heater increases on the smaller units. Economical design, therefore, requires a particularly careful review of alternative heater layouts when high cost fittings are combined with small heaters. For such services, an attempt to achieve high efficiency by use of convection surfaces can be prohibitively expensive.

For many applications of this type, helical coil heaters have considerable merit, since return bends are eliminated. Efficiency must be sacrificed for economy, however, because the helical design does not lend itself readily to the use of convection surfaces.

16.5 Auxiliary equipment

Auxiliary equipment, particularly soot blowers, can also have a marked effect on the economics of layout and design. The cost of soot blowers is largely a function of the number used, rather than the length of the soot blower lance. The horizontal tube heater, with its long, narrow convection section, requires far more soot blowers than the vertical heater with its relatively short (but wide) convection section. In practically all cases, the cost of the extra length of soot blower lance needed for the vertical unit is negligible when compared to the cost of the extra blowers and attendant control equipment needed on the horizontal unit. Foreshortening of the convection section on the horizontal heater can partially overcome this situation, provided that process objections do not preclude such an approach.

It must be pointed out, however, that the very steps that reduce the cost of a soot blower installation are exactly those that increase the cost of the convection surface by increasing the number of fittings and the cost of fabrication.

It is quite possible that a heater constructed with carbon steel or low alloy return bends, and employing soot blowers would be most economical as a vertical cylindrical unit: but if built with stainless steel pressure parts and fittings, it would be cheaper as a horizontal tube heater (despite an increase in the number of blowers).

The foregoing covers some of the factors having a substantial effect on the selection of a basic heater design. It is apparent, however, that the several major considerations are not at all complementary and that, depending on the importance attached to each, the heater configuration can take a variety of forms.

The difficulty of determining the most economical basic configuration for any given set of conditions should become quite apparent. A correct decision leading to selection of an inherently economical heater layout is only possible by a careful review of all the factors and their relationship to each other.

16.6 Optimization of design

Once a basic heater layout has been selected, the next step is to optimize the design. Primary consideration must be given to the pressure parts since they provide substantial design options and thereby present the best opportunity for optimization and lowering of heater cost.

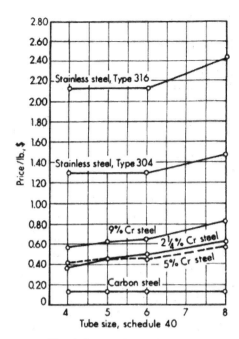

Fig. 4 Comparison of tube price against tube size

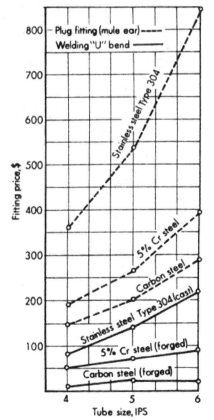

Fig.5 Fittings cost compared with Tube Size

One of the first determinations the designer must make is to select the number of parallel flow paths, or passes, to be used. He may choose a design based on using a minimum number of passes and larger tubes or, alternatively, more passes with smaller tubes. As a general rule, the cost of the pressure parts is directly related to the tube size. The larger sizes result in higher material costs, based on the need for greater wall thicknesses as the tube O.D. increases, and a higher per-pound cost for the tube metal in the case of alloys. The cost per pound of tube material in relation to tube size is given in Fig.4, which clearly emphasizes the importance of this variable. For example, changing from 4 in. IPS (iron pipe size) to 6 in. IPS in 2 and ¼ % Cr material results in a 20% increase in the cost per pound of the tube metal (making no allowance for any increased wall thickness requirements).

The cost relationship for return bends with changes in material and tube size is provided by Fig.5. A move to the larger tube sizes, although increasing the unit cost of the fittings, is compensated for (in part) by a reduction in the number of fittings required for any given layout.

The composite effect on the overall cost of a tube coil with changes in tube size or material is shown in Fig.6 for a typical, horizontal tube heater. Variations in unit costs, the total weight of the tube coil, and the number of fittings needed have all been taken into consideration.

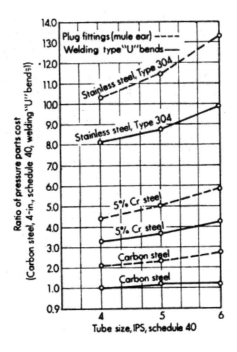

Fig. 6 Tube coil cost for horizontal box heater

Practical use can be made of this information in conjunction with Fig.7 and 8. The data on Fig.7 show the cost of the major heater components for a 4 in. IPS horizontal tube heater as a percentage of total heaters cost over a wide range of heater sizes.

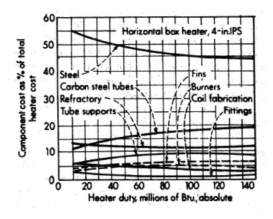

Fig. 7 Component costs making up total heater cost

The curves of Fig.8 provide the cost in dollars per million-Btu-absorbed for carbon steel, high efficiency, horizontal tube heaters over the same sizes as Fig.7.

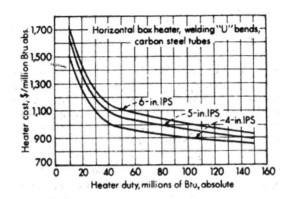

Fig. 8 Heater cost vs. duty for horizontal box heater

Conversion of the information on Fig.8 to other coil materials can be accomplished readily. First obtain from Fig.8 the cost in dollars of the heater size being considered and multiply this number by the combined percentage of tubes, bends and fabrication as shown on Fig.7. This will provide the cost of the carbon steel pressure parts. Then apply against this the appropriate ratio factor for the desired alloy or tube size from Fig.6, to obtain a new cost for the revised pressure parts. Adding this total back again to the balance of the heater cost obtained from Fig.8 will give the approximate overall cost of the heater for the new conditions.

The foregoing indicates that increasing tube diameters will generally raise heater costs. For a wide range of refinery applications, the amount of this increase will be materially influenced by the criteria used for establishing tube wall thicknesses. For most refinery applications other than catalytic reforming and hydrocracking, the methods in general use for determining tube wall thickness do not result in a controlling criterion. In most instances, calculated values for wall thickness based on temperature and pressure are so low that the tubes would not be acceptable from the structural standpoint, i.e. their ability to withstand handling, thermal stresses, and buckling at points of supports. As a result, the tube walls are increased by adding a retirement thickness or a corrosion allowance, or possibly some other arbitrary method is used for establishing minimum allowable wall thicknesses.

If IPS schedule thicknesses are called for, any increase in tube diameter will result in a substantial increase in coil metal weight. If, on the other hand, the design criteria are based on API RP-530, it is quite probable that tube wall thickness for a 6-in. IPS coil would be no greater than that for a 4-in. IPS coil. This would tend to favor the use of larger pipe sizes and a reduction in the number of passes if the considerable savings in external mani-folding, flow controls, and instrumentation were properly taken into account.

16.7 Heater radiant surfaces

The heater radiant surface is usually well defined, since most heater specifications generally are quite clear in stating maximum allowable, average, radiant transfer rates. Considerable flexibility is nevertheless possible, based on varying tube center to center spacing, or possibly a decision to subject the tubes to direct radiation from both sides. Maximum allowable transfer rates normally refer to tubes on a nominal two diameter spacing backed by a refractory wall and which receives direct radiant heating from one side only.

An increase in specified, average, radiant transfer rates is generally permissible if this can be accomplished without exceeding the maximum peak front facing rate corresponding to the overall average rate. Fig.9 provides a convenient method for determining how much the average heat transfer rate can be increased by varying the tube center to center spacing or by firing on both sides of the radiant tubes (or both).

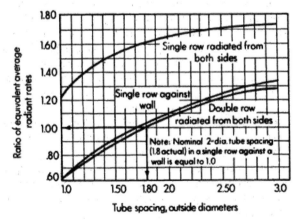

Fig. 9 Heat transfer rates vs. tube arrangement

Based on a single row of tubes on two nominal diameters, and fired from one side only, being equated to 1, the ratios given provide a direct multiplier for establishing the equivalent average rate for any other desired arrangement.

At first glance, modification of tube spacing or firing arrangement would appear to be an obvious method for reducing the amount and cost of radiant surface, and thereby provide a more economical heater design. However, any increase in tube spacing, or use of double firing, results in an attendant increase in the size, and hence cost, of the radiant enclosure of the heater. It again becomes a matter of establishing the economic balance, this time between the reduction in cost of the tubes, fittings, and supports versus the increased cost of the refractory and steel enclosure. Unfortunately, the breakpoint is not a clearly defined one, and from a practical standpoint its determination generally requires a relatively detailed layout and estimate for each alternative approach.

As an approximate guide, however, the use of double firing will generally not show any savings in alloy materials up to, and including, 5% Cr in schedule 40. The Schedule-80 wall thicknesses will probably present a different picture, particularly in the larger tube sizes. In most of the stainless alloys, the savings in tube material will substantially outweigh the extra cost of the heater enclosure for either wide tube spacing or double firing. The only probable exception is in the case of smaller heaters where the size range is such that a conventional single fired approach permits complete shop fabrication, and the alternative double firing results in a field erected design.

Double firing will also result in a lower in-tube pressure drop as a consequence of the reduction in radiant surface and attendant flow length.

16.8 Convection section optimization

In the case of high efficiency heaters, the single most important area for design optimization and the associated material savings rests in the convection section layout. The preponderant percentage of fired heaters built today utilizes some form of extended surface to increase the heat transfer on the external surface of the convection tubes. The wide range of variations possible in the use of extended surface with respect to type, material, spacing, height and thickness presents a larger number of variables to be considered. As a result, true optimization becomes a laborious and uncertain task. In addition, the position of the tubes with respect to horizontal and diagonal spacing has a significant effect on convection section effectiveness.

The optimization of convection sections is an ideal application for the computer. It can be of tremendous assistance in permitting a rapid run through of the involved iterative calculations essential to finding an economical design. A typical printout is illustrated in Fig.10, for one row of a row-by-row calculation. This program checks a variety of arrangements, printing out complete data for each row, and at the end of its run gives the cost of all the convection section materials, including tubes, fittings, extended surface, supports and the convection enclosure.

```
ROW NUMBER  5
0.562 INCH FINS,            4.00 FINS/INCH
   88.7 SQUARE FEET (TOTAL) ON THIS ROW
FLUE GAS MASS VELOCITY = 0.409

TEMPERATURES                  IN          OUT
FLUE GAS                     946.8       813.2
PROCESS FLUID                436.1       448.2
FILM                         460.1       472.3
TUBE METAL                   468.6       480.8
FIN TIP                      533.3       543.8

GENERAL INFORMATION
CUMULATIVE SURFACE (SQ FT)              295.9
OUTSIDE FLUX (MAX)                     2814.2
BARE TUBE FLUX (MAX)                  16728.8
FIN EFFICIENCY                           0.893
H(OUTSIDE)                               7.619
FLUE GAS DELTA P  (IN H2O/ROW)           0.0076
CUMULATIVE FLUE GAS DELTA P (IN H2O)     0.0378
DUTY THIS ROW (MM BTU/HR)                0.249
CUMULATIVE DUTY (MM BTU/HR)              1.331




PARTIAL MATERIAL AND COST ESTIMATE
(FROM BOTTOM OF FIRST ROW TO TOP OF TOP TUBE ROW)

                              TOTAL       TOTAL
                              AREA        WEIGHT      COST
ITEM                NO        SQ FT       POUNDS      DOLLARS

TUBES               20                    531.2        69.06
FINS                                      258.6        71.28
RETURN BENDS        18                                 72.00
WELDS               40                                227.99
END TUBE SHEETS      2        8.5                     123.89
INTERMEDIATE T.S.    1        2.9          73.3       100.36
WALLS                        23.2                     169.91
HEADER BOXES         2        8.5                      64.88
                                                    ----------
                                          TOTAL      898.59
```

Fig. 10 Convection section optimization printout

Experimental investigation of design and performance relationships for heater convection sections would undoubtedly show substantial economic returns particularly when related to the computer programs. The generally accepted published data on convection heat transfer leave a great deal to be desired in this application to the size and configurations usually found in large fired heaters.

The very considerable maldistribution of heat input around the circumference of a convection tube, and its effect on both tube metal and extended surface tip temperature, has received little consideration to date. The manner in which flue gases are evacuated from the convection section has a substantial influence on the flow of gases in the convection section and the effectiveness of the heat transfer surface. Flue gas baffles, although helpful in providing better flue gas distribution along the length of a convection section, also have a substantial and undefined shielding effect on the surface immediately below them. More definite data are badly needed in all these areas to permit development of correct design criteria that could be used for real optimization of convection section heat transfer surfaces.

An increasing percentage of new heaters use various grades of alloy tubes, either because of the elevated temperature and pressure levels of the service or because of the corrosive nature of the feedstock. In either instance, an accurate determination of the temperature profile within the coil will permit correlating the expected service conditions at any point with a suitable alloy to meet these conditions. With this knowledge at hand, a combination of several alloys (or possibly carbon steel and alloy) is a most effective method for decreasing the cost of heater pressure parts while still providing adequate reliability and service life. It is essential, however, to carefully check the alloy breakpoints for reduced load conditions, since under these circumstances the temperature profile of the pressure parts is often shifted substantially toward the inlet of the heater. Again the computer can be a most useful tool in aiding these determinations with respect to time and accuracy.

16.9 Modular components

An effective and possibly more economical approach to the problem of the heater designer (who requires greater flexibility in his layout) and the field erector (who needs more manageable pieces) is the use of a modular component approach. Instead of heater radiant sections being split into two (or at times three) sections, the heater is broken up into large flat sided components that incorporate both refractory and tubes, and is arranged for ready field bolting.

Actual erection experience with this approach indicates that it may actually result in lower field costs and a more economical heater (on an installed basis) than the use of a smaller number of extremely large sections with their attendant clearance and handling problems.

Whatever route is chosen, there is no question that a major emphasis on shop fabrication is absolutely essential for any heater to be competitive on an installed basis.

PART 2

Operation

Sub Contents
for Operation

1. Direct vs. Indirect Heating

In a direct heating system, heat is transferred from flames or hot flue gas directly to the process fluid, the two being separated by a single tube wall. If the process fluid is temperature sensitive, direct heating has the potential for uneven tube heating, coking, hot spots, and tube failure. With a combustible process fluid, the result of tube failure is uncontrolled fire in the firebox with the possibility of greater damage, hazard to operators, etc.

In an indirect system, an intermediate heat transfer fluid carries heat from the fired coils to the ultimate user(s). The intermediate fluid might be water, steam, a petroleum fraction, a synthetic material such as Dowtherm, or a liquid metal. Indirect heating may avoid process fluid degradation. Remember though, with the indirect method, a flammable heat transfer fluid is still a hazard if a tube ruptures.

The most common indirect heating example is a steam boiler and steam heated process heat exchangers. Condensing steam has a high heat transfer coefficient and gives even heating. It is usually the economic choice if available at the temperature level needed. A steam system at $31.6 kg_f/cm^2$ (238°C) is fairly common; some plants have steam at $56.2 kg_f/cm^2$ (271°C). Carefully selected petroleum oils are suitable up to about 343°C; synthetics go up to 399°C.

In one indirect fire tube heater design, fire tubes and process coil are within the same cylindrical shell, and the heat transfer medium circulates by natural convection. In other systems, the transfer medium is pumped to distant services or moves under its own vapor pressure.

Direct fired heaters are recommended for most refinery and chemical plant applications. A hot oil (indirect) system may be used for several smaller services. Indirect fire tube heaters are recommended for most producing applications; they are safer than direct fired units when running unattended.

2. What Is Combustion?

There are three elements of combustion- fuel, air and a source of ignition. Burning or combustion is a rapid chemical reaction that releases heat and light. The reactants in the reaction are a combustible material and oxygen. The oxygen normally comes from air which is about 21% oxygen and 78% nitrogen by volume. Combustible materials contained in most fuels are carbon, hydrogen and sometimes sulfur. Therefore, the combustion of a mixture of air and fuel combines oxygen (O_2) with carbon (C), hydrogen (H_2) or sulfur (S). Examples of the chemical reaction of the burning of carbon, hydrogen or sulfur in air are as follows:

$$C + O_2 + 3.75N_2 \rightarrow CO_2 + 3.75N_2$$

$$2H_2 + O_2 + 3.7N_2 \rightarrow 2H_2O + 3.75N_2$$

$$S + O_2 + 3.75N_2 \rightarrow SO_2 + 3.75N_2$$

The burning of the air and fuel produces carbon dioxide (CO_2), water (H_2O), sulfur dioxide (SO_2) and heat. The three products of combustion listed above are called chemical compounds. Chemical compounds are made up of molecules in which elements are combined in certain fixed proportions. To form carbon dioxide, each atom of carbon reacts with two atoms of oxygen from the air. To form water vapor, two atoms of hydrogen react with one atom of oxygen from the air.

Most fuels are not composed of simple atoms of carbon, hydrogen, and sulfur. Instead the fuels contain various molecules or compounds that are combinations of these elements. For example, natural gas is approximately 95% methane (CH_4). Methane is a compound composed of one atom of carbon, and four atoms of hydrogen. An equation that describes the combustion of methane in air is shown below.

$$CH_4 + 2O_2 + 7.5N_2 \rightarrow CO_2 + 7.5N_2$$

Note that in this case, one atom of carbon combines with two atoms of oxygen to form carbon dioxide, two atoms of hydrogen combine with one atom of oxygen to form one molecule of water but since four atoms of hydrogen are present in the fuel, and two molecules of water are formed. (1 molecule of methane needs two molecules of oxygen to complete combustion. This, in other words, 1 SCF of methane will need 2 SCF of oxygen or 9.5 (2/0.21= 9.5) SCF of air for complete oxidation.)

In combustion, CH_4 fuel needs more air than H_2 per fuel scf, but reverse per fuel lb. The stoichiometric ratio calculation for fuel gas is easy, compared to fuel oil where composition is somewhat complicated. C/H ratio is used to express element content.

	O_2, scf/fuel scf	Air, scf/fuel scf	Air, lb/fuel lb	Air, scf/fuel lb
CH_4	2.0	9.53	17.2	226
C_2H_6	3.5	16.7	16.1	210
C_3H_8	5.0	23.8	15.7	205
C_4H_{10}	6.5	31.0	15.5	203
H_2	0.5	2.38	34.5	451

Stoichiometric ratio

	Flame speed (m/sec)	Ignition temperature (℃)	Adiabatic flame temperature (℃)	H/C ratio
CH_4	0.46	632	1,949	0.33
C_2H_6	0.70	472	2,066	0.14
C_3H_8	0.85	466	1,988	0.22
C_4H_{10}	0.88	480	1,971	4.80
H_2	2.83	572	2,099	-
CO	0.52	609	2,110	-

Flame speed & flame temperature

The heat that is given off in the combustion process is excess energy which new molecules are forced to liberate because of their internal makeup. The excess energy that is given off during the combustion reaction is normally seen as a temperature difference between the reactants in the reaction.

Example: Air requirements for 1×10^6 Btu/hr

Fuel CH_4 (LHV: 911 Btu/Nft^3)

Fuel required = 1,000,000/911 = 1,098 scfh

$$1,098\ CH_4 + 2,196\ O_2 + 8,257\ N_2 \rightarrow 1,098\ CO_2 + 2,196\ H_2O + 8,257\ N_2$$

Air = 2,196 + 8,257 = 10,453 scfh

Fuel H_2 (LHV: 275 Btu/Nft³)

Fuel required= 1,000,000/275= 3,636 scfh

3,636 H_2 + 1,818 O_2 + 6,836 N_2→ 3,636 H_2O + 6,836 N_2

Air= 1,818 + 6,836= 8,654 scfh

2.1 Fuel gas or fuel oil

Below shows the effect of fuel parameters:

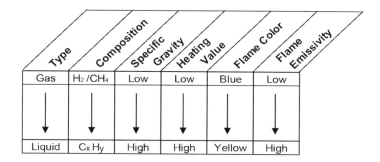

Type	Composition	Specific Gravity	Heating Value	Flame Color	Flame Emissivity
Gas	H_2 /CH_4	Low	Low	Blue	Low
Liquid	$C_x H_y$	High	High	Yellow	High

(1) Refinery fuel gas varies in heating value depending on the status of various plants.

(2) A change in heating value can change the heat transfer in the heater.

(3) A system to blend the fuel gas is installed to maintain a constant heating value. This can stabilize burner operations.

Process heaters may burn gas, oil, or both. The choice depends on several factors, including the source reliability, operator attention, and response time. Process gas (fuel gas) has always been preferred because it is easier to handle, burns cleaner and is less expensive. Many locations use some natural gas. However, most natural gas contracts provide that industrial gas can be cut off on short notice whenever the demand by domestic users exceeds the supply. Consequently, those locations have some provisions for backup oil burning when selected gas is curtailed. Steam boilers are often chosen for supplemental oil firing when the local facilities are subject to possible natural gas curtailment because they are normally attended, generally use larger quantities of fuel, and operate more nearly at constant load. If more units have to be switched to oil, the process heaters usually selected are large users in less severe services with steady loads, like atmospheric crude heaters. Design of oil fired heaters is more complicated than that for gas fired heaters:

(1) The oil requires separate oil guns and steam for atomization, along with separate controls.

(2) The flame for a given heat release is longer and of a less predictable shape than the flame in a gas fired heater, which affects firebox geometry. For example, when heating a temperature sensitive process fluid with gas, we often specify a clearance of 152.4mm between burner centerline and nearest tube surface. With oil the figure is 182.88mm.

(3) Because oil burning produces ash and soot, design of the heater's convection section should include provision for tube cleaning by onstream water washing or soot blowing.

(4) Choice of refractory material may be affected due to chemical attack from components in the oil.

(5) Because of the soot produced with oil, the maximum number of fins per 2.54cm of convection tubes is only three (maximum is five with gas). This affects thermal efficiency and draft.

(6) Based on operating experience, the design excess air with gas fuel is 10%; with oil, 15%. This also affects the draft requirement.

(7) Fuel oil has a higher sulfur content than gas. This may affect the minimum temperature allowed for steel surfaces in direct contact with flue gas. The concern is sulfuric acid condensation on metal surfaces, which may present corrosion and air pollution problems. The table below shows the minimum required steel surface temperature (not flue gas temperature) which is necessary to prevent acid condensation. The temperature varies depending on the sulfur content of fuel.

Sulfur(wt%)	Temperature(℃)
1	163
<1	149
<0.1	135

(8) The vanadium often present in fuel oil may cause severe corrosion of hot metal parts like tube supports.

(9) Oil guns foul quickly, requiring frequent attention.

(10) Oil firing requires much more maintenance time and operator attention.

2.2 Fuel quality

Impurities in fuel oils are expected and we go to great lengths to mitigate their effects. Duplex strainers/ filters and continuously circulating headers are normally provided. Even so, burner fouling and oil drips are just a way of life in oil burning. When the temperature for viscosity control gets too low, it gets worse.

Pipeline-quality natural gas burns clean with very few problems. But those who burn process byproduct refinery fuel gas are sometimes plagued with oil-like troubles in their gas burners. Heavy components in the fuel gas sometimes condense, giving erratic flames and potentially coking and plugging the gas ports. Where this occurs, steam tracing often helps.

Where sour process gas is sweetened in amine units, then used as fuel gas, there is the possibility of contamination of the fuel supply. Amine plant upsets can allow some H_2S breakthrough. Alternatively, the fuel gas may become extremely dirty if sour gas is bypassed around an amine unit. Potential solutions to fuel quality problem are:

(1) Enlarge the burner ports (possibly you get poor flames)

(2) Clean up the gas

(3) Install strainers (they must be kept clean)

(4) Try chemical cleaning, a good solution, if the deposits come from debris accumulated over years.

(5) Try a hardware solution: go to a waste oil burner (huge ports like in a flame thrower) for burning gas. You would get no turndown, excess air would go up, efficiency would go down, and you might have to rebuild the fireboxes so the flames would fit into them. But the burners probably wouldn't plug.

(6) Change from our typical raw gas burners (small ports) to premix burners (larger ports). This may relieve the symptoms, but has plugging problems with premix

burners, too. Also premix burners are intolerant of the large quantities of hydrogen that we often encounter in refinery fuel systems. They backfire and blow out when the hydrogen content is too high.

Unfortunately, no foolproof method of preventing burner plugging has yet been developed. In actual practice operators periodically must clean burners manually to assure even combustion and optimum heater efficiency.

2.3 Flame color

The flame produced by liquid fuel is more luminous than a gas flame. For a given heat release, a liquid fuel flame will be longer and of a less predictable shape than a gas flame. Both blue and yellow flames produce the same products of complete combustion: carbon dioxide and water, so there is no difference in the amount of heat released. Yellow flames result when fuel dissociates (cracks) to its carbon and hydrogen components. In gas burning, this is more likely to happen on the heavier gases like propane and butane, particularly when poorly mixed with air. The higher the hydrogen/ carbon ratio by weight in the fuel, the less prone it is to yellow flame burning. Also the lower the molecular weight in the same series of compounds, the less prone they are to yellow flame burning. The separate burning of hydrogen is very fast and that of carbon slow. The slower burning free carbon is heated to a bright yellow incandescence.

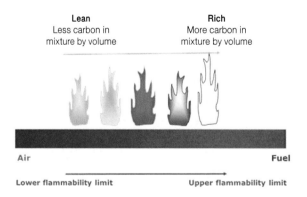

Oil flames are typically characterized by a very bright yellow incandescence because there is so much free carbon around. Somewhat yellow flames are typical of the heavier gaseous fuels, too, but yellow flames in gas burning are more often a tipoff of poor or inadequate air mixing.

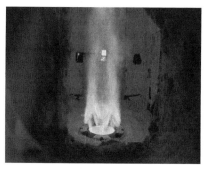

Gas flame

Oil flame

2.4 Fuel effects

Many operators report that the heater problems began when the heater fuel gas source was switched from high hydrogen fuel to refinery fuel. This is a very good observation overlooked by many people.

Hydrogen fuel requires less oxygen to burn. It burns with a hotter flame temperature and a faster flame speed. Hydrogen has a flame speed of approximately 2.1 m/sec versus 0.3 m/sec for most hydrocarbons. The following table is based on an all radiant, single fired, a petrochemical process heater at 27,125 kcal/m^{2}hr flux and a 493°C tube wall temperature.

Fuel	High H$_2$ (@15% excess air)	Natural gas (@15% excess air)	Fuel oil (@25% excess air)
Flue gas rate difference	base	121%	133%
Flame temperature, °C	1,950	1,817	1,749
Heat release	base	105.2%	108.4%

Radiant heat transfer is $T^4_{flame} - T^4_{tube}$. Hotter flame temperatures improve radiant efficiency and reduce the fuel firing rate. Hydrogen takes 18-23% less air to burn than hydrocarbon fuels for an equivalent heat release. The higher flame temperature and less air requirements for hydrogen fuel reduce the flue gas rate dramatically.

The pressure drops across the convection section ratios directly with the square of the flue gas rate. This means the pressure drop is 46% greater for natural fuel than the high hydrogen fuel. It would be 77% higher for oil fuel. This is a major problem when the heater has a stack limitation. The high flame speed of hydrogen results in rapid combustion. This eliminates the haze problems and afterburning problems.

The high hydrogen fuel is Platformer off gas with 92.3% H_2 with a molecular weight of 4.2 and a heating value of 3,373kcal/Nm^3. The natural gas fuel has a molecular weight of 19.0 and a heating value of 8,383kcal/Nm^3. The fuel oil is a 20 API gravity (Sp. Gr. 0.934) oil with a heating value of 9,955 kcal/kg.

It is interesting to note that natural gas has a net heating value of 10,443 kcal/kg while the fuel oil has 9,955 kcal/kg. The heating values are similar because the carbon/hydrogen ratios are similar. The high hydrogen fuel has a heating value of 18,960 kcal/kg. Hydrogen has a higher kcal/kg heat release than carbon.

2.5 Air preheat effects

Increasing the combustion air temperature from 16°C to 204°C for an all radiant heater has the following effects:

(1) The flame temperature increases by 93°C

(2) The Bridge Wall Temperature barely changes by a few degrees

(3) The fuel rate decreases by 10-12%

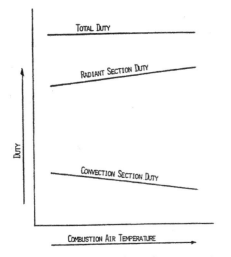

On a radiant convection heater, the duty shifts from the convection section to the radiant section. The radiant flux rate increases by 7-10%. The hotter flame temperature increases the radiant efficiency. Less fuel firing decreases the flue gas rate and the heat pickup in the convection section.

Effect of combustion air temperature on radiation section- convection section duty split in a radiant convection process furnace

2.6 Flame emissivity

Below is flame radiation by Talmor method. The curve shows the flame emissivity for various fuels. Oil flames have flame emissivities that are 3 times higher than gas flames. Oil flames must be located further away from the tubes to prevent overheating from flame radiation due to the high flame emissivity.

(Reference) Radiation heat absorbed (Q_r) = f x ε x σ $(T^4_{flame} - T^4_{tube})$

        Where, f geometry factor

              ε emissivity

              σ Stefan-Boltsmann constant

              T^4_{flame} flame temperature

              T^4_{tube} tube temperature

(Reference) Flame temperature (T_f) = $(C_v + T_{ca})/(m_{cg} \times C_{p,av})$ + 25

        Where, C_v constant volume heat capacity

              T_{ca} combustion air temperature

              m_{cg} (cold) flue gas mass

              $C_{p,av}$ average constant pressure heat capacity

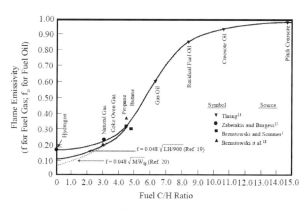

Flame radiation (Talmor method)

2.7 Combustion emission

Even CO_2 is now being considered pollution because it is a greenhouse gas.

NO_x produces ozone in the lower atmosphere.

SO_x produces acid rain.

CO, HC and Soot are produced of incomplete combustion.

RCF (refractory ceramic fiber) is dispersed as insulation degrades.

Burners generate noise. (Noise over 80 db requires hearing protection)

Soot   Plume

2.8 Complete combustion

These heating values assume complete combustion. Carbon may burn completely to carbon dioxide or incompletely to carbon monoxide. In the latter case, less than one-third of the heat of combustion is actually liberated.

(1) $CH_4 + 2O_2 \rightarrow 2H_2O + CO_2 + 11{,}944$ kcal/kg (for complete combustion)

(2) $CH_4 + 1.5O_2 \rightarrow 2H_2O + CO +$ less than 11,944 kcal/kg (for incomplete combustion due to lack of O_2)

(3) $CH_4 + 3O_2 \rightarrow 2H_2O + CO_2 + O_2 + 11,944$ kcal/kg (for complete combustion with excessive O_2)

Combustion is complete if everything combustible burns to carbon dioxide, water vapor and (for sulfur bearing fuels) sulfur dioxide. It is incomplete when the final products contain unburned carbon or combustible gases such as carbon monoxide, hydrogen, methane or other hydrocarbons. Below is a basic combustion reaction for some fuels.

$C + O_2 \rightarrow CO_2 + 7,826$ kcal/kg

$2H_2 + O_2 \rightarrow 2H_2O + 28,641$ kcal/kg

$CH_4 + 2O_2 \rightarrow 2H_2O + CO_2 + 11,944$ kcal/kg

$S + O_2 \rightarrow SO_2 + 2,553$ kcal/kg

Complete fuel burning is the norm in process heaters. Incomplete combustion is very unusual. Generally, there is no such thing as a more efficient burner. Burners typically burn all the fuel.

2.9 O_2 enrichment combustion

Oxygen/fuel combustion is to use oxygen in place of air for combustion of fuel. There are three kinds of method in the integration of oxygen into fired process: enrichment which injects oxygen into combustion air, lancing which strategically introduces oxygen directly to combustion chamber through existing burners, and installing a new burner (a special oxygen-fuel burner tip) which can usually fit between existing air/fuel burners. Currently these technologies have been applied for some refining processes such as thermal reactor in SRU, coke burning in catalyst regenerator of RFCC, and SAR decomposition furnace in Alkylation unit etc. as well as fired heater in process.

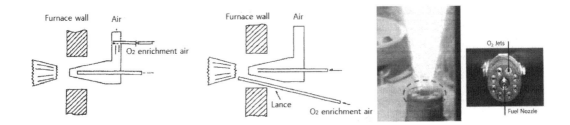

2.10 Heating value

Knowing the heating value of a fuel tells you what fuel rate your fired heater requires. All fuels containing hydrogen produce water vapor during combustion. Such fuels have two heating values depending on the final state of the water. The higher or gross heating value (HHV) is that obtained with the product water present as a liquid.

The lower or net heating value (LHV) assumes the water remains in the vapor state. The difference is the water's latent heat of vaporization. In actual practice, the lower heating value is used in designing and rating process heaters because the product water normally is not condensed on heat transfer surfaces and leaves the stack as a vapor.

For hydrocarbon gases and mixtures of hydrocarbon gases, the lower heating value on a volume basis (kcal/m^3) is linearly related to the specific gravity of the fuel (air = 1). But on a mass basis, the net heating value (LHV) of all hydrocarbon gases is close to 11,101 kcal/kg.

Fuel oil always contains some noncombustible materials like water, ash or metals.

So the heating value of fuel oil is generally a little lower, typically around 9,991 kcal/kg. Energy is often expressed in Btu per equivalent fuel oil barrels (EFO). Note that an EFO is approximately equivalent to 6MM Btu's.

2.11 Estimating flue gas rate

To size the volume of heater, it will be needed to calculate the minimum volume of air required to burn the fuel completely, with no excess oxygen in the gas. It turns out that the minimum air consumption (X_{min}) and the minimum volume of combustion products ($V_{p,min}$) can be represented as linear functions of the lower heating value.

The actual volume of combustion products (V_p) is:

$$V_p = (V_{p,min}) + k (X_{min})$$
Where, k = % of excess air expressed as a fraction
(i.e., for 20% excess air, k=0.20)

When V_p is multiplied by the fuel rate, this gives the total volume rate of combustion products the heater has to handle. Note the fuel rate must be known.

A simple rule-of-thumb is: A typical mixture of hydrocarbon gases and air with the usual 10 to 20% excess air typically has a heating value of about 555 kcal/kg. Since mass in equals mass out, the flue gas production rate must be directly proportional to the heat release, e.g., a fired heater with a heat release of 25.2MMkcal/hr produces about 45,400kg/hr flue gas.

2.12 Rules of thumb in flue gas flow

The following may be used to estimate flue gas flow when firing gas:

(1) It takes 719 pounds of air/ mmBtu (HHV) of gas fired. This is the pounds of theoretical air per "million Btu" fired for natural gas.

(2) Add 9.5% to account for 2% excess oxygen (equal to 9.5% excess air). A good target for gas firing is 2% excess oxygen.

(3) A typical HHV of natural gas is 21,869 Btu/pound.

Multiply the expected boiler or fired heater heat release (in mmBtu) by the ratio of 833 (Pounds flue gas/mmBtu). Based on adding air flow to fuel flow to obtain flue gas flow, assuming gas firing, and 2% excess oxygen, 833 is calculated as follows:

Air Flow	+	Fuel Flow		= Ratio
$\dfrac{719(\text{Lb air}) \times 1.095}{\text{mmBtu}}$	+	$\dfrac{1,000 \text{ Btu}}{21,869 \text{ Btu/Lb}}$	x 1Lb fuel = 833	$\dfrac{\text{Lb}}{\text{mmBtu}}$

Similarly, factors to estimate flue gas flow for oil firing:

(1) It takes 746 pounds of air/mmBtu (HHV) of oil fired. This is the pounds of theoretical air per million Btu fired for liquid oil.

(2) Add 15.8% to account for the 3% excess oxygen. A good target for oil firing is 3% excess oxygen.

(3) 18,240 Btu/Lb is a typical HHV of # 6 fuel oil.

Use the same process to determine how much heat is going up the stack when firing oil. For oil firing, multiply the expected boiler or fired heater heat release in mmBtu by the ratio of 919 (calculated as follows) for liquid oil firing and 3% excess oxygen to get the flue gas flow.

$$\frac{746(Lb\ air) \times 1.158}{mmBtu} + \frac{1,000,000\ Btu \times 1Lb}{mmBtu\ \ 18,240\ Btu} = 918.7 \quad \frac{Lb}{mmBtu}$$

2.13 Rules of thumb in efficiency

When fuels are priced per barrel, the barrel normally means the Equivalent Barrel. In particular, It is very convenient to use Equivalent Fuel Oil (EFO) factor for fuel gas. The EFO factor means that a fuel gas 1 Nm^3 is equivalent to the fuel oil in liter. An Equivalent Fuel Oil (EFO) is defined as follows.

(1) 6,300,000 Btu/Bbl, EFO on a Higher Heating Value (HHV)

(2) 6,000,000 Btu/Bbl, EFO on Lower Heating Value (LLV)

Note that the common industry practice is the use of HHV.

For an existing boiler and a preliminary estimate, use an average, typical steam boiler efficiency of 80% on a HHV basis. For a new boiler, use about 82% (HHV) for gas firing and 85% (HHV) for oil firing.

There are three important rules of thumb worth remembering in waste heat recovery:

(1) For every 4.4°C that a stack temperature is reduced, there is a 1% improvement of

the efficiency for the unit.

(2) For each 12.2℃ that boiler feed water (BFW) is heated, the boiler efficiency is improved 1%.

(3) For each 10℃ that combustion air is preheated, 1% of the fuel can be saved.

2.14 Flame temperature- fuel oil

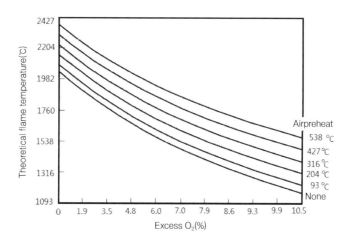

The chart for fuel oil is very similar to fuel gas. Since the fuel oil is difficult to burn fuel, higher excess air levels are required. New gas burners may be capable of operation at 10-20% excess air level while fuel oil burners require 25-45% excess air level.

Fuel oil is composed of 30% cutter stock (kerosene and diesel) and 70% vacuum bottoms. The vacuums bottom has a boiling range of 593℃-816℃. The liquid fuel must be vaporized before it can burn. Combustion is slow and higher air levels are required.

If too low of an air level is run with fuel oil, high levels of unburnt oil fouling and soot fouling occur. The fouling has a larger effect on fuel efficiency than excess air level.

2.15 Flame temperature- natural gas

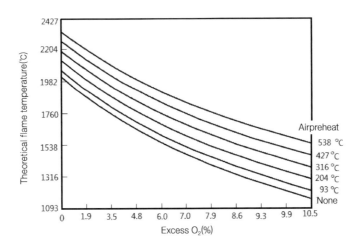

This chart shows the effects of excess air and air preheat temperature. It is very important to control excess air on radiant of only heaters such as catalytic reformers and hydrogen reformers because the heat transfer is flame temperature minus the bridge wall temperature.

For example, take a catalytic reformer heater operating at 15% excess air. The flame temperature is 1,816°C. The heat transfer is 1,816°C - 816°C = 982°C per unit of fuel fired. At 100% excess air, the flame temperature is 1,149°C. The heat transfer is 1,149°C - 816°C = 316°C per unit of fuel fired. The firing rate would be 3 times larger (982°C/316°C) at 100% excess air as compared to 15% excess air.

Now let's look at the effect of excess air on a radiant convection heater like a large crude heater. The high excess air has the same effect on the radiant section. But the increase in excess air produces more flue gas, increasing the heat pickup in the convection section. Increasing the excess air shifts duty from the radiant section to the convection section. The convection section cold end approach would increase, but only by 16°C-38°C. The firing rate would only increase by 2-3% at the higher excess air level. Stack temperature is more important than excess air level for fuel efficiency on a radiant convection heater.

Many delayed coker heater are intentionally run at high excess air levels to shift duty to the convection section, lowering the radiant flux rate and the rate of coking in the tubes. One test run showed a 24°C increase in skin temperature by lowering the firebox O_2 level from 7% to 3%.

2.16 Turndown

Turndown is the ability to operate at lower than design rate. Turndown is not often a problem on the firebox side. Fuel oil burners may be turned down to typically 33% of design capacity, gas burners to 10%.

A heater with a single phase process stream will have less turndown difficulty than one with a two phase stream. This is because low velocity in the two phase case may result in a poor flow regime, pulsating flow, and increased cracking and coking. If the furnace was designed with tube velocity at the upper end of the recommended range, turndown capability will be greater. Steam injection is often added to the radiant coil of a heavy oil heater to raise the bulk flow velocity and increase turndown ability.

If multiple passes are flow controlled, turndown can move the controllers out of their optimum range, leading to pass stalling. Horizontal tubes with flow downward through the coil are most resistant to stalling. Vertical tubes are least resistant.

3. Furnace Operations

The concept that oil firing requires more excess air is simply no longer true. It has been found that gas firing excess air down to less than 10% can also be maintained for oil firing with comparatively few exceptions. This condition exists because theoretical air demand/mm Btu's with methane firing is 27.9 moles, and with #6 oil the air demand is 27.76 moles. If the air supply is adequate for methane, it is also adequate for #6 oil to within less than 1%, and the difference favors oil firing. If identical excess air is not possible with both fuels, a complaint should immediately be made to the particular burner manufacturer for correction.

It is not uncommon to find gas burners operating quite satisfactorily in the range of from 15 to 100% of design heat release in automatic control. In the burning of oil in automatic control a range of operation of from 33 to 100% heat release is more typical. Where automatic control is required, the controls for gas firing are much more routine. However, controls for oil firing are now considered satisfactory if properly designed and operated. But there is a limitation to the heat release per burner as well as to the type of fuel burned. The burner manufacturer should be required to establish these limitations before any automatic control equipment is purchased so that the control equipment can meet the limitations as they are established, to permit proper firing.

The molecular component heat value total is not the sole determinant for the calorific value of a specific compound. Molecular heats of formation can be found in chemical handbooks at f values in kilo-calories per gram-mol. There is 453.59 gram-mol per pound-mol, and a kilo-calorie is equal to 3.968 Btu's. Heats of formation can be either exothermal (carry a minus sign) or endothermal (carry no sign or the plus sign).

Heat-worthy molecular components are carbon (C), hydrogen (H), and sulfur (S).

Each pound of carbon represents 14,093 Btu's, each pound of hydrogen represents 51,602 Btu's, and each pound of sulfur represents 39,083 Btu's. A pound-mol is the number of pounds of a molecular weight; it is also (arbitrarily) 379 scf as a gas. Thus, a pound-mol of methane (CH_4) is 16.043 pounds or 379 scf. Methane contains 12.011 pounds of carbon and 4.032 pounds of hydrogen. The heat of formation (f) for methane is -17.89

The calorific value of a mol of methane is as follows:

$$[(12.011 \times 14{,}093) + (4{,}032 \times 51{,}602)] + (453.59 \times 3.968 - 17.89) = 345{,}131 \text{ Btu's}$$

One pound of methane represents $345{,}151/16.043 = 21{,}513$ Btu's and a cubic foot of methane represents $345{,}131/379 = 910.6$ Btu's.

4. Refinery Fuel Gas System

Over decades of operation, refinery fuel gas systems have become the repository for purge gases from fluidized catalytic cracking, delayed coking, catalytic reforming, hydrotreating, hydrocracking, and other process units. Below Fig. illustrates the various sources of hydrocarbon off gas and hydrogen purge streams in a typical refinery. Flare gas recovery systems also contribute to the build-up of non-methane components in refinery fuel gas. Hydrogen concentrations of 50 vol% (and higher) are not unusual. While most refineries have added conversion and treating units and experience capacity creep over several decades. Refinery fuel gas contains large concentrations of hydrogen, ethylene, propylene, ethane, propane, and heavier paraffinic components.

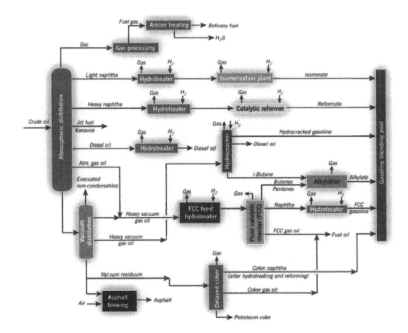

Below Fig. shows a scenario in which multiple off gas streams with high concentrations of hydrogen, olefin, and NGLs (natural gas liquids) are fed to the fuel gas system.

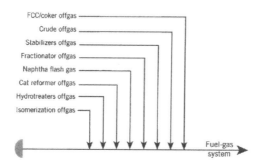

Table below lists several refinery off gas streams and their typical compositions. Large quantities of valuable components such as ethane, propane, ethylene, propylene, and hydrogen are present in these streams.

Off gas source	Crude distillation, mole%	Hydrotreaters, mole%	FCC/Cokers, mole%
Hydrogen	9.00	70.00	10.50
Nitrogen	2.50	-	5.00
Methane	2.500	15.00	40.50
Ethane	10.00	7.500	12.00
Ethylene	-	0.01	7.50
Propane	22.00	3.600	4.500
Propylene	-	0.06	6.50
C_4S	20.50	2.50	0.75
C_5S	12.00	0.75	1.90
C_6+	4.50	0.55	1.65
Oxygen	-	-	0.20
CO	-	-	0.50
CO_2	4.00	0.01	2.00
H_2S	10.00	0.02	6.50

Below Fig. shows a potential process configulaion and its overall material balance. The fluid catalytic cracker and coker off gas stream, containig unsaturated hydrocarbons, is segregated from the saturated hydrocarbons. After retreatment to remove CO, CO_2/ H_2S, mercury, and moisture, the dried stream is sent to a deethanizer to recover C_3^+ component, while the residue gas is sent to the fuel gas system. The pretreatment unit may also contain feed compression. The recovered liquid stream is further fractionated in a depropanizer to achieve 95% recovery of RGP (refinery grade propylene) from the overhead. The remaining C4+ stream is sent to existing fractionation.

Streams with high hydrogen content, such as the catalytic reformer off gas and hydrotreater's off gas, are segregated from other lower hydrogen content streams. The combined high hydrogen stream is routed to a PSA (Pressure Swing Adsorption) unit to recover at least 80% of the hydrogen at 99.9 mole% purity. The hydrogen lean tail gas is compressed and combined with other lower hydrogen content streams.

The combined stream undergoes pretreatment to remove CO_2/H_2S, mercury, and moisture. The pretreatment unit may also contain feed compression. The dried stream is sent to a demethanizer to recover about 80% of the C_2^+ component as NGL product. The residue gas is sent to the fuel gas system.

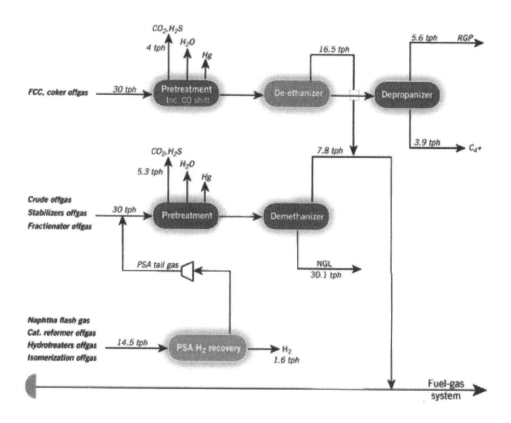

5. Apparatus of Refinery Fuel Gas System

Fuel gas flow: off gases→ pretreatment→ fuel gas→ fuel gas balance drum→ fuel gas knock out drum→ fuel gas strainer→ fuel gas filter (coalescer)→ fuel gas flame arrester→ burner flame→ flame UV detector→ flame observer

① Fuel gas balance drum: collecting & mixing fuel gas from various processes

② Fuel gas knock out drum: settling water in mixing fuel gas

③ Fuel gas strainer: straining a large particles in fuel gas

④ Fuel gas filter (or coalescer): coalescing & eliminating a mist (1-3μm) in fuel gas

⑥ Fuel gas flame arrester: arresting flame from furnace

⑦ Furnace burner: burning fuel gas

⑧ Burner flame detector: detecting & keeping burner flame

⑨ Fired heater camera: observing flame in furnace

① **Fuel gas balance drum**   ② **Fuel gas knock out drum**   ③ **Fuel gas strainer**

④ Fuel gas filter

⑤ Fuel gas filter Coalescer

⑥ Fuel shutoff valve

⑦ Fuel gas flame arrester

⑧ Burner flame UV detector

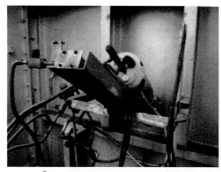

⑨ Fired heater flame camera

⑩ Fuel oil filters

6. Typical Crude & Yield Physical Properties

(1) API/Sulfur/Nitrogen

	°API	Sp.Gr.	Sulfur (wt%)	Nitrogen (wtppm)
Crude	30.9	0.871	1.0	1,342
AR*	16.4	0.957	1.8	2,572
VR**	8.9	1.007	2.3	4,314

* Atmospheric Distillation Residue ** Vacuum Distillation Residue

(2) Merceptan/H_2S/Micro Coke Residue/Asphaltene

	Merceptan (wtppm)	H_2S (wtppm)
Crude	30.9	1.0

	MCRT (wt%)	Asphaltenes (wt%)
AR	8.0	3.9
VR	21.7	8.3

(3) Yield of Crude

Yield	Ratio (wt%)	Details (wt%)
LPG	1	Light End (C_2, C_3, C_4)
Naphtha	13	LSR 4 + HSR 9
Kerosene	12	LKD 7 + HKD 5
Diesel	25	LGO 12 + MGO 8 + HGO 5
AR	48	VGO 25 + VR 23

(4) PONA (Paraffin, Olefin, Naphthene, Aromatics)

PONA (vol%)	LSR	HSR
P	81	49
N	16	39
A	4	11
O	0	0

(5) Metal Contents

Metal (wtppm)	Crude	AR	VR
Cu	0.04	0.2	0.3
Cr	0.01	0.02	0.1
Ni	16.6	28.1	52.3
V	53.0	86.8	160.4
Na	3.7	3.9	7.9
Ca	1.8	5.5	16.6
Fe	2.8	6.7	11.0
Si	1.3	1.8	3.5
Mg	0.2	1.3	2.2

(6) TBP (True Boiling Point) Cut Temperature

Yield		TBP (°C)	Pressure (Torr)
Light End (C_2, C_3, C_4)		Less than 20	
Naphtha	LSR	20 - 82	Atmospheric
	HSR	82 - 143	
Kerosene	LKD	143 - 193	
	HKD	193 - 227	100
LGO		227 - 293	
MGO		293 - 338	10
HGO		338 - 366	2
AR		More than 366	-
LVGO		366 - 454	0.1
HVGO		454 - 566	0.005
VR		More than 566	-

Note) Crude with smaller AR: Champion, Vityaz, AXL, Sepingan, Murban, West Bukha, Troll, Oseberg, Sokol, Shaharan Blend, Palawan Light, Forties

7. Best Practice Operation

The following methods for improving fired heater efficiency: ① Tuning the furnace ② Implementing ACC (automated combustion control) ③ Optimizing burners ④ Minimizing air leaks ⑤ Cleaning heat transfer surfaces ⑥ Adding heat transfer surface ⑦ Repairing insulation ⑧ Others

7.1 Definition

Thermal efficiency of a fired heater is defined as the heat absorbed divided by the heat input. Heat absorbed is heat transferred to all the process streams. These can include hydrocarbon, steam, boiler feed water, and combustion air. Heat input consists of the net (lower) heating value of the fuel burned. In rare cases heat input may also include combustion air preheat supplied by a separate source such as turbine exhaust.

A fired heater must be properly operated and maintained in order to achieve its maximum efficiency. The maximum efficiency that a fired heat can achieve is set by its heat transfer surface area. It is recommended that efficiency be calculated and tracked as a good energy saving practice and also as an early indication of problems in furnace operation or equipment. There is little value in comparing the efficiency of one furnace against another unless the furnaces are identical.

7.2 Calculation of efficiency

Even though the definition of efficiency includes heat absorbed by the process and the energy in the fuel, it is difficult to accurately calculate those values due to errors in the related measurements. A more accurate method for calculating efficiency is to estimate the energy lost. One to two percent of the energy is lost from the furnace's heated surfaces to the environment. It is an industry convention to call this "radiation loss" and assume it is 2 percent. Essentially all the other lost energy is contained in the flue gasses that go up the stack. Fortunately, the stack losses can be accurately estimated using the stack gas temperature and oxygen content.

The below table shows an example for calculating efficiency with stack temperature and stack oxygen. The 2 percent radiation loss is included. The table indicates the results calculated by two efficiency equations, one for use with dry O_2 readings taken with extractive analyzers and one for use with wet O_2 readings taken with in-situ analyzers.

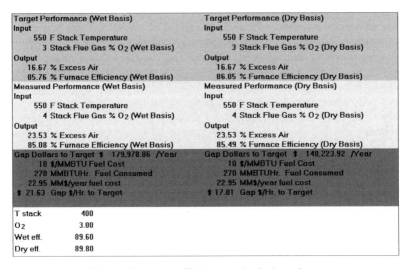

Target Performance (Wet Basis)	Target Performance (Dry Basis)
Input	Input
550 F Stack Temperature	550 F Stack Temperature
3 Stack Flue Gas % O_2 (Wet Basis)	3 Stack Flue Gas % O_2 (Dry Basis)
Output	Output
16.67 % Excess Air	16.67 % Excess Air
85.76 % Furnace Efficiency (Wet Basis)	86.05 % Furnace Efficiency (Dry Basis)
Measured Performance (Wet Basis)	Measured Performance (Dry Basis)
Input	Input
550 F Stack Temperature	550 F Stack Temperature
4 Stack Flue Gas % O_2 (Wet Basis)	4 Stack Flue Gas % O_2 (Dry Basis)
Output	Output
23.53 % Excess Air	23.53 % Excess Air
85.08 % Furnace Efficiency (Wet Basis)	85.49 % Furnace Efficiency (Dry Basis)
Gap Dollars to Target $ 179,970.86 /Year	Gap Dollars to Target $ 148,223.92 /Year
10 $/MMBTU Fuel Cost	10 $/MMBTU Fuel Cost
270 MMBTUHr. Fuel Consumed	270 MMBTUHr. Fuel Consumed
22.95 MM$/year fuel cost	22.95 MM$/year fuel cost
$ 21.63 Gap $/Hr. to Target	$ 17.81 Gap $/Hr. to Target

T stack	400
O_2	3.00
Wet eff.	89.60
Dry eff.	89.80

Table 1 Furnace efficiency calculation sheet

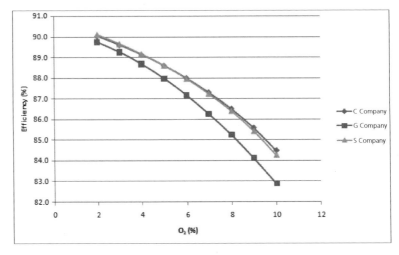

1-1. Furnace efficiency correlation (@ 204°C)

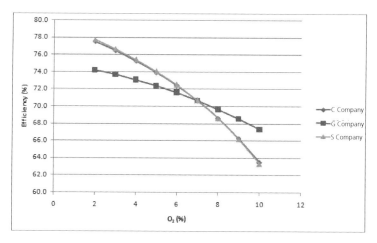

1-2. Furnace efficiency correlation (@ 482℃)

[Reference] Efficiency correlation equation

1. U Company= 101.81971-0.211499*O_2-0.053568*O_2^2-0.056005*FGT (Unit: ℉)

2. C Company= 99.24-((39.3+1.99*(100*O_2/(21-O_2)))*0.0000000000001*FGT^3)-((157.5+1.13*(100*O_2/(21-O_2)))*0.00000001*FGT^2)-(20.6+0.197*(100*O_2/(21-O_2)))*0.001*FGT+(1.17*(100*O_2/(21-O_2))*0.01) (Unit: ℃)

3. S Company= 98.5-(0.001244+0.02117*(1+O_2/(20.9-O_2)))*(FGT-60) (Unit: ℃)

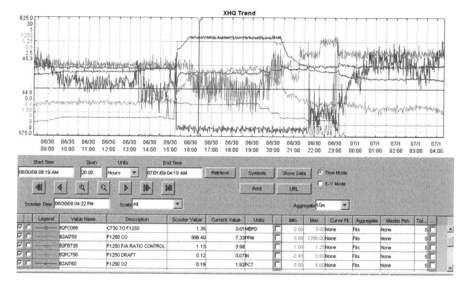

1-3. The trend of operating variables

Be aware that the furnace efficiency equations do not include the energy required to run forced and induced draft fans. When considering adding air preheat to improve heater efficiency, as described later, be sure to include the cost of running the fans in the economic analysis. Large fans such as those required in systems with NO_x reducing SCR's can consume 2 to 4% of the fuel energy.

One error that can creep into efficiency calculations is the mixing of radiant section readings with stack readings. Oxygen analyzers are sometimes mounted in the radiant section to avoid measuring leakage air in the convection section. However, using a radiant section oxygen reading without this leakage air will inappropriately inflate the efficiency. O_2 and temperature readings below the convection section will yield a radiant section efficiency. O_2 and temperature readings in the stack will yield an overall furnace efficiency. Using a mix of radiant section O_2 and stack temperature will yield a meaningless and misleading efficiency value.

7.3 Tuning the furnace

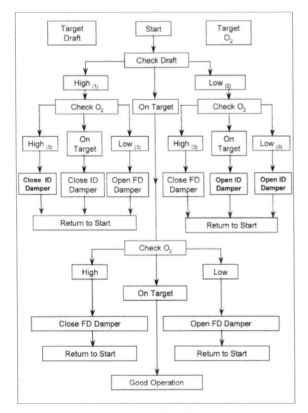

The most cost effective way to improve efficiency is to operate the furnace at optimum O_2 and draft values. Depending on the furnace configuration, O_2 and draft are controlled by stack, duct, fan, or plenum dampers, burner air registers, and in some cases, fan rpm.

It is important to keep the dampers and air registers in operable condition to allow proper tuning of the furnace.

Balanced draft furnace adjustment
(1) High Draft means fire box press. more neg. than target.
(2) Low Draft means fire box press. more pos. than target.
(3) Low or High O_2 means O_2 is above or below target.

Fig. 1 Furnace tuning flow charts

7.3.1 Optimize O$_2$'s

The effect of O$_2$ on furnace efficiency is not linear, however, in general, reducing stack O$_2$ by 1% will increase efficiency by about 0.5% for furnaces with low stack temperature (in the range of 204℃). Reducing stack temperature by 1% will increase efficiency by about 1% for furnaces with high stack temperatures (in the range of 482℃)

In fired heaters, savings from dropping excess O$_2$ by 1% is shown as below chart. And "$/day saved" on X axes is based on difference between 3% excess O$_2$ and 2% excess O$_2$ and $3/mmBtu. Cost is dependent on heater size and stack temperature.

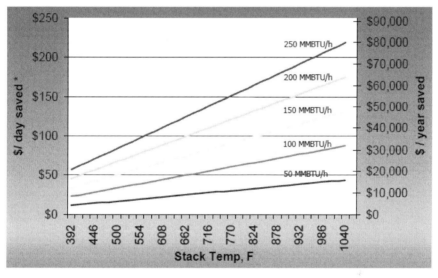

Heater O$_2$ chart

If stack oxygen is reduced too far, the burners will become starved for air and CO will be produced. CO going up the stack represents unburned fuel and can reduce efficiency. If CO builds to high levels, a dangerous fuel rich condition can be created in the firebox. Therefore, O$_2$ can be safely reduced only to a level above the point of CO breakthrough. The low O$_2$ alarm is typically set at 1% above the point of CO breakthrough. The flowchart below is a detailed procedure for using a CO breakthrough test to determine an appropriate O$_2$ operating target.

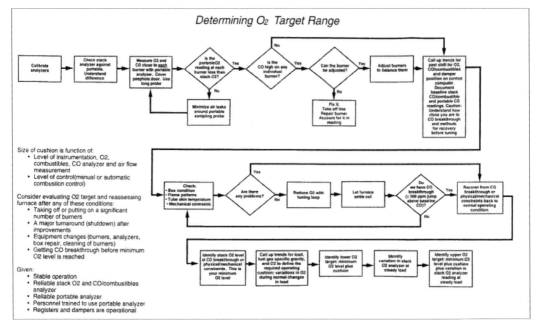

Fig. 2 How to set O₂ target alarm

[Example 1] R Company, Fired Heater, F-1250, June 2009

Background:

① F-1250 is running steadily with good O₂, draft, and CO.

② F-1250 stack damper is stuck at about 40% open.

③ A cold component of the process inlet stream is running at a low rate (MBPD)

Scenario:

① At about 3:00 p.m. the process inlet stream increases from 0.7 to 2.4 MBPD

② The furnace outlet temperature sags.

③ The furnace fires up to compensate.

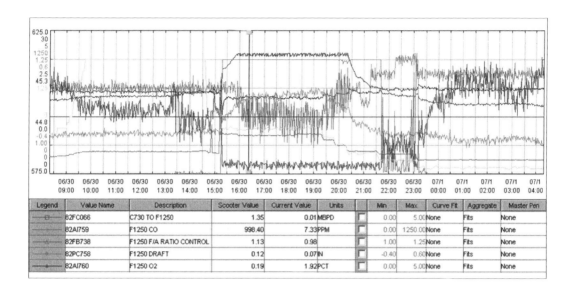

Legend	Value Name	Description	Scooter Value	Current Value	Units		Min	Max	Curve Fit	Aggregate	Master Pen
	82FC066	C730 TO F1250	1.35	0.01	MBPD		0.00	5.00	None	Fits	None
	82AI759	F1250 CO	998.40	7.33	PPM		0.00	1250.00	None	Fits	None
	82FB738	F1250 F/A RATIO CONTROL	1.13	0.98			1.00	1.25	None	Fits	None
	82PC758	F1250 DRAFT	0.12	0.07	IN		-0.40	0.60	None	Fits	None
	82AI760	F1250 O2	0.19	1.92	PCT		0.00	5.00	None	Fits	None

Results:

① Stuck damper didn't respond to need for more air.

② Excess O_2 dropped to 0.2%.

③ Firebox pressure went positive.

④ CO meter pegged at 999ppm.

⑤ Furnace stayed in this condition for 6 hours.

⑥ CO exceeded daily average limit of 400ppm resulting in an environmental violation.

Learning:

① O_2, Draft, and CO should have been in alarm.

② Alarm must be responded to.

③ The results of this event could have been much worse.

Searching for excess O_2 optimized (@ combined oil/gas firing)

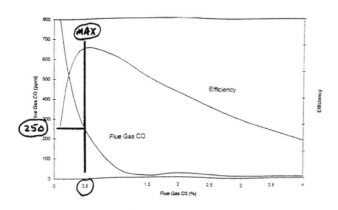

Theoretical furnace efficiency parameters

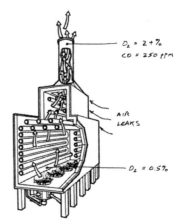

Heater with typical O_2 and CO levels

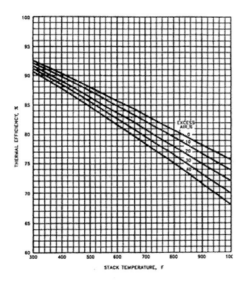

Approximate thermal efficiency

Note: 2% radiation loss is assumed

Gradually decreasing excess O_2 causes CO/efficiency to increase. When the O_2 keeps lowering, CO begins to occur at the O_2 starvation, rises and abruptly increases at any point. The abruptly increasing point is called CO breakthrough. At the CO breakthrough, the efficiency shows the highest point and the excess O_2 is optimum point for the operation. In general, ±0.2% of the O_2 optimum point becomes actually optimized operation range at the given heater. And the optimizing O_2 point is basically different according to the design/operation condition of specific heater such as burner load, draft, fuel kinds and fuel ratio etc. For example, the typical excess O_2 against the draft and fuel is as follows.

	Gas fired	Oil fired
Natural draft	3.5%	4.2%
Forced/Balanced draft	2.7%	3.5%

And excess O_2 is a rough indicator to confirm whether the heater has complete combustion or not in furnace. Because of error by air leak, a refiner is used to utilize CO concentration of bridge wall as additional indicator of optimized heater efficiency. Their heater maximum efficiency is a point of CO 250ppm at the bridge wall. And then a buffered excess O_2 is given to the point with considering the change of operation such as fuel components or burner load etc. Practically a burner maker in their handbook recommends the optimized excess O_2 within maximum CO 100ppm. Therefore, based on the CO value measured at the bridge wall, it is possible to set up the guideline to optimized excess O_2, i.e., ranging from maximum CO 100ppm of 1st target to 250ppm of CO breakthrough.

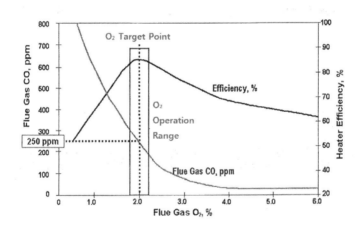

Theoretical furnace efficiency parameters

As shown in Fig. above the highest point of efficiency represents 85% and at that point, excess O_2 is 2.0%, where CO breakthrough shows 250ppm. So operation target of optimized O_2 ranges from 1.8% to 2.2%.

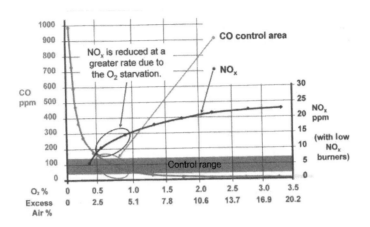

Theoretical furnace NO$_x$ parameters

In NO$_x$ emission, when excess O$_2$ decreases, the NO$_x$ also decreases proportionally. Especially NO$_x$ is reduced at a greater rate at the O$_2$ starvation. The O$_2$ starvation will be the CO control zone where finds out excess O$_2$ optimized. And in oil/gas firing, more oil firing, more NO$_x$ is emitted through stack.

[Test 1] CDU fired heater, 72F-101, oil & gas firing ratio 50% to 50%

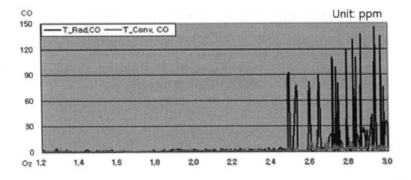

Under enough excess O$_2$ (O$_2$ 2.5~3.0%), radiant CO is high and convection CO is low. So CO combustion is estimated at convection section (CO burning at 608.9℃, flame temperature 2,108℃). In case of CO burning, a damage to tube and refractory can be caused at all times.

[Test 2] CDU fired heater, 62F-101, oil & gas firing ratio 30% to 70%

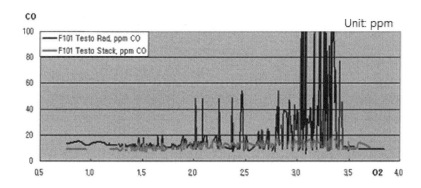

Under enough excess O_2 (O_2 2.0~3.0%), radiant CO is high and convection CO is low. So CO combustion is estimated at convection section too.

[Test 3] CDU fired heater, 72F-101, 14 Dec. 2006, 10:00-20:25

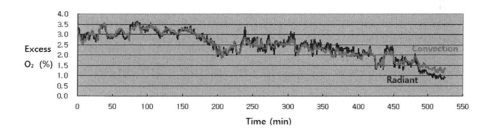

Since the trend of excess O_2 in radiant and convection section is similar each other, it means this heater does not have any air leakage.

[Test 4] CDU fired heater, 62F-101, 1 Sep. 2006, 09:30-16:30

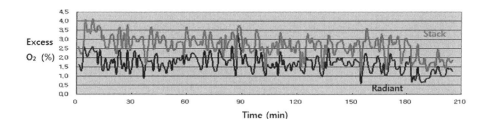

As above trends show the difference of O_2 between radiant and convection section, this means that there is air leakage at somewhere in this heater. For example, such air leak is easy to occur at the points as below.

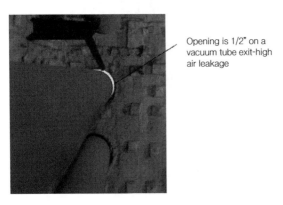

Opening is 1/2" on a vacuum tube exit-high air leakage

Firebox sealing

[Test 5] CDU fired heater, 72F-101, oil & gas firing ratio 50% to 50% (oil normal)

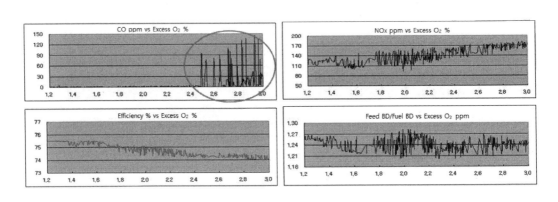

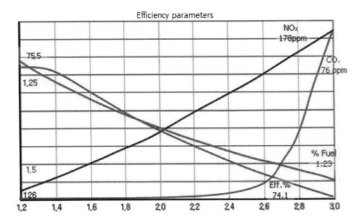

CO occurs a lot at high excess O_2 due to unbalanced loads of fuel gas/oil at burner, which means uneven excess O_2 and draft inside furnace. This can cause the problem of after burning in convection section. And it is difficult to give optimized O_2 guide in this heater.

[Test 6] CDU fired heater, 72F-101, oil and gas firing ratio 75% to 25% (oil max.)

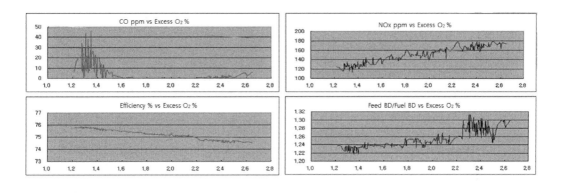

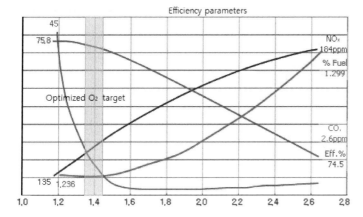

The condition of combustion inside furnace is good and then excess O_2 guide is around 1.4 % at the oil maximum firing ratio.

[Test 7] CDU fired heater, 72F-101, oil and gas firing ratio 25% to 75% (oil min.)

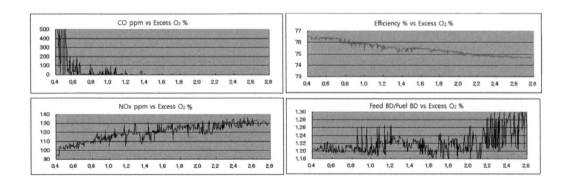

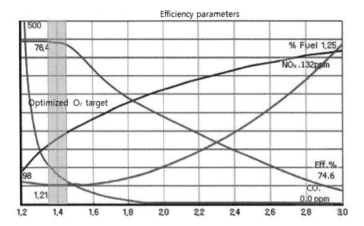

The condition of combustion furnace is good and then appropriate excess O_2 target is also around 1.4 % at the oil minimum firing ratio.

[Reference 1] Status of using analyzer in a refinery

Installation	No of heater	Analyzer type	Location	O_2	CO
Before 2000	50	Extractive Probe	Stack	Y	N
After 2000	18	Extractive Probe	Radiant arch	Y	Y
	2 (VDU heater)	Laser	Stack	Y	Y
		Extractive Probe	Radiant arch	Y	Y

[Reference 2] Type of O_2 analyzer

Item	Laser	Extractive
Method	IN−SITU	Extractive
Principle	Tuneable Diode Laser Absorption Spectroscopy (TDLAS)	Zirconium Oxide Sensor
Response Time	1~2sec	Less than 5sec(0.6 ℓ /min)
Temperature	1,500℃	71℃
Rating	IP66	IP65

[Reference 3] Type of CO analyzer

Item	Laser	Extractive
Measurement	CO	CO equivalent (total gas by incompleted com.)
Method	IN−SITU	Extractive
Principle	Tuneable Diode Laser Absorption Spectroscopy (TDLAS)	Sensor
Min. Range	0~30ppm(ppm CO)	0~500ppm
Accuracy	Detection Limit : 0.3ppm(ppm CO)	+/− 25ppm

[Test 8] CDU fired heater, 62F-201, comparison of O_2, CO and density analyzer

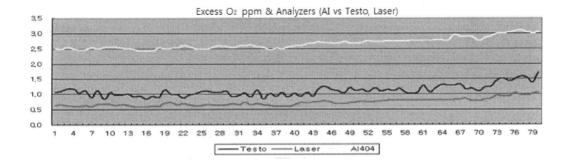

According to the type of analyzer, there are a lot of deviation in measuring excess O_2, i.e., extractive (portable analyzer- Testo, fixed AI404) type and laser type. For reference, the portable analyzer (Testo) is able to measure O_2, CO, CO_2, NO_x, SO_x and draft simultaneously.

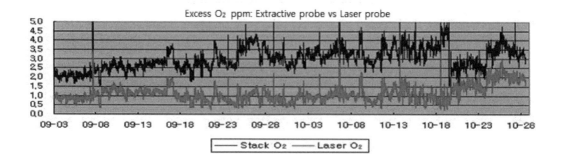

Because extractive type has fouling, the excess O_2 gap between extractive probe (stack O_2) and laser probe (laser O_2) rises according to the running time.

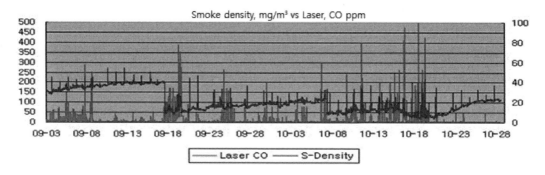

As compared with CO measured by laser, extractive smoke density doesn't have high reliability in confirming incomplete combustion of furnace.

[Test 9] CDU fired heater, 62F-201, fuel oil firing ratio average 55%

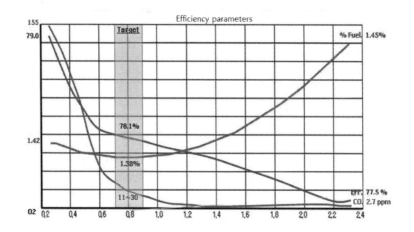

In the O_2 guide test by portable analyzer, optimized O_2 target is around 0.8% at efficiency 78.1%.

7.3.2 Optimize draft

Furnaces always run with draft, or negative pressure, so that any openings will allow ambient air to be pulled into the furnace rather than allow hot flue gas to leak out.

Draft is not uniform throughout a furnace. Draft is high at the floor and decreases with height in the firebox. The point of minimum draft is just below the convection section (called the arch or bridge wall area). It is important to monitor draft at this location. The target range for draft at the arch is −1.27 to −2.54 mm of water. Running with too little draft could allow hot gasses to leak out of any opening in the furnace such as piping penetrations or site ports that don't close tightly. This could hurt personnel and overheat and weaken the furnace structure. Running with too much draft will decrease efficiency by increasing leakage air. Leakage air quenches the flue gas and decreases heat transfer. More fuel is required to compensate for the reduced heat transfer. Therefore, optimum efficiency requires keeping draft within the target range.

7.4 Implementing ACC (Automatic Combustion Control)

Automatic combustion control is the automated control of the flue gas outlet damper (stack or induced draft fan damper) to keep stack O_2 continuously optimized using a computer program in DCS. The efficiency gain due to continuous optimization vs. manual optimization typically pays for the ACC system in less than a year on furnaces over 50 mmBtu/hr.

Below shows schematic to control operation variables for optimization via ACC in natural draft heater.

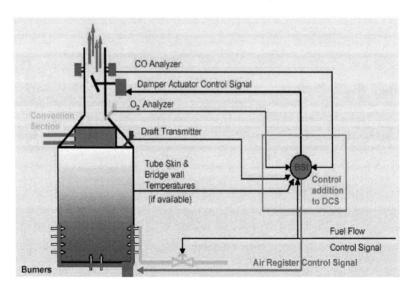

Natural draft heater control strategy

ACC systems monitor several variables and make damper moves which keep the furnace within specified safe limits of O_2, draft and CO. Different refineries use different approaches to ACC. Some systems are simple and some are more complex. The best systems employ a feed-forward feature which increases combustion air before fuel is increased.

Some vendors market ACC systems which control furnace combustion using CO in the flue gas (called CO trim control). Although the theory behind the control scheme is sound, the installed systems do not have sufficient reliability to justify their cost and the effort involved in their upkeep.

[Example] APC control for optimizing operational variables

Following is a schematic of Naphtha Splitter Unit which has a problem in Kero Splitter Reboiler (F103 in Fig.) where has high tube skin temperature and frequent shutdown to remove coke inside tube. In this problem, APC (Advanced Process Control) is applied to the relevant heaters to control optimally operational variables such as excess O_2, draft and tube skin temperature.

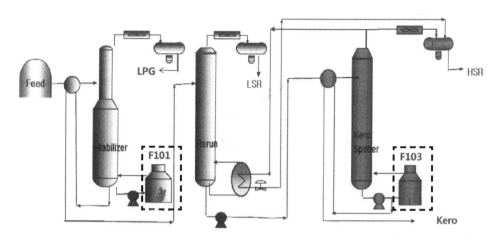

Naphtha splitter unit

In F101/F103 heater, objective to control is the tuning of excess O_2 and draft so that F103 heater's maximum tube skin temperature, 320℃, can keep constantly close to its high limit. Excess O_2 controls lower limit and draft control upper limit respectively for both heaters. Unfortunately, the control of excess O_2 and draft in F103 heater was impossible due to its low firing mode (50% of design) by high tube skin temperature.

Sub: 98HEATER		Master	Last Run Time		Last Load		Why Off	
ON	N	ON ON	20-Jul-2010 09:44:05		7/18/2010 10:44:40 AM			

Independents   Filter None

Name	Description	Combined Status	Oper Srv Switch	Loop Status	Lower Limit	Current Value	SS Target	Upper Limit	Current Move	External Target
98FC062OP	F101 Combustion Air	NORMAL	ON	MV	25.000	30.859	30.475	45.000	-0.014	
98PC064OP	F101 Damper	NORMAL	ON	MV	46.000	50.000	49.330	50.000	-0.024	
98FC082OP	F103 Combustion Air	NORMAL	ON	MV	8.000	9.000	9.606	30.000	0.019	
98PC084OP	F103 Damper	ROC UP	ON	MV	44.			52.000	0.076	
98PC065OP	Common Damper	NORMAL	ON	MV	20.			34.000	-0.030	
98HC150OP	F/D Fan Damper	MIN MOVE	ON	MV	18.			20.000	0.000	

O₂ Control Lower Limit

Draft Control Upper Limit

Dependents   Filter None

Name	Description	Combined Status	Oper Srv Switch	Lower Limit	Current Value	SS Target	Upper Limit		
98AI060APV	F101 O2	LO LIMIT	ON	1.800	1.906	1.800	10.000		
98PC064PV	F101 Draft	HI LIMIT	ON	-25.000	-6.549	-6.000	-6.000		
98TI063PV	F101 Stack Temp	NORMAL	ON	200.000	328.218	327.853	400.000		
98TI062PV	F101 Conv Temp	NORMAL	ON	500.000	631.599	630.532	700.000		
98AI080APV	F103 O2	LO LIMIT	ON	1.500	1.375	1.500	10.000		
98PC084PV	F103 Draft	HI LIMIT	ON	-25.000	-7.885	-8.000	-8.000		
98TI083PV	F103 Stack Temp	NORMAL	ON	200.000	291.185	291.221	400.000		
98TI082PV	F103 Conv Temp	NORMAL	ON	500.000	650.784	650.819	700.000		
98PC065PV	Common Stack Press	NORMAL	ON	-40.000	-20.554	-19.242	-12.000		

(1) Excess O₂ control of F101 heater

The deviation 3% (1.1~4.1) was controlled to less than 0.2% at target 1.8% when APC moves combustion air flow.

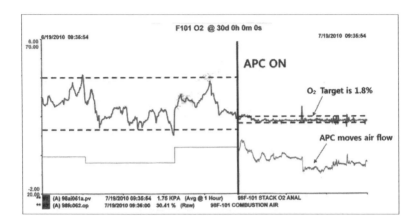

(2) Draft control of F101 heater

As compared to before APC, after APC controlled the draft deviation from 6mmH₂O (-12.5~-6.0) to less than 0.5mmH₂O when APC moves dampers.

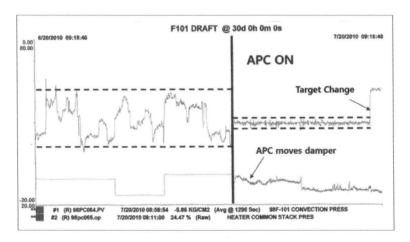

(3) Tube skin temperature control of F103 heater

With application of APC, the deviation of tube skin temperature was reduced dramatically from more than 10℃ to 1℃ at every change of target.

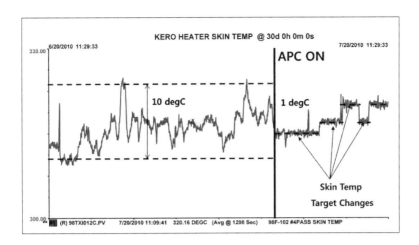

7.5 Optimizing burners

The main goal of optimizing burners is to ensure all the burners are getting the same amount of fuel and air and that the flames have the same shape and appearance.

As discussed above, an important step in increasing efficiency is the reduction of stack O_2. As O_2 is reduced, at some point the burners will run out of combustion air and start to produce CO. It is optimum if all the burners run out of combustion air at the same time. If one burner runs out of combustion air before the others, then CO is produced at that burner and O_2 cannot be reduced further. A burner that runs out of combustion air before the others will prevent the further reduction of O_2 and prevent the related increase in efficiency. To avoid this situation, every burner should be getting the same amount of fuel and air.

To make sure all the burners are getting the same amount of fuel, the fuel tips need to be in good repair with no enlarged holes, and no pluggage. To make sure the flames are shaped properly and the fuel from the various tips is properly interacting, the tips need to be oriented per the manufacturer's specifications. Operators should monitor flame patterns and report any need for burner maintenance. Burner mechanics should make sure the burner tips are clean correctly sized, and correctly oriented.

To make sure all the burners are getting the same amount of air, the burner air registers need to be mechanically intact, workable, and set uniformly. Operators should set the air registers uniformly and request maintenance of any air registers with operability problems. Burner mechanics should repair or replace inoperable air registers.

[Example] 62F-101 horizontal cabin type heater flame test

The heater has 2 problems in operation: one is tube impingement at middle section of radiant and shield zone of convection inlet respectively. The other is flame instability at both end walls. To improve the problems, some tests are carried out as follows.

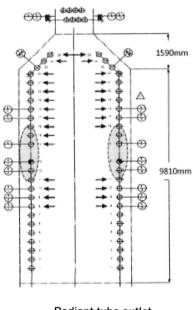

Radiant tube outlet

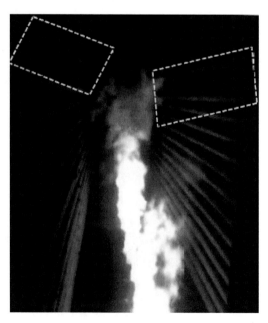

Arch zone

Case 1. Fuel oil vs. Fuel gas ratio (50%: 50%)

(1) Burner condition

BNR No.	1	2	3	4	5	6	7	8	9	10	11	12	13	14	15	16	17	18
Firing Staus	G	O	G	O	G	O	G	O	G	O	G	O	G	O	G	O	G	O
Air Damper OP	5.0	8.0	4.0	7.5	5.0	8.0	5.0	8.0	6.0	7.5	5.5	7.0	5.0	8.0	6.0	8.0	6.0	8.0
FO Ind. V/V OP	0	100	0	100	0	100	0	100	0	100	0	100	0	100	0	100	0	100
FG Ind. V/V OP	100	0	100	0	100	0	100	0	100	0	100	0	100	0	100	0	100	0

(2) Operating condition

F-101 Feed Rate	Kl/h	319.9	Excess O2	%	3.5	A_STM DP	Kg/cm²	2.3	Absorbed Duty	MMkal	31.1
Inlet Temp.	℃	218.7	Arch FGT_S	℃	763.6	FO PCV OP	%	34.7	HTR Load	%	95.9
Outlet Temp.	℃	341.3	Arch FGT_N	℃	708.4	FG PCV OP	%	43.9	AVG BNR Load	%	81.8
Draft.mmH2O_S	mmH2O	-2.9	Conv. FGT_S.	℃	475.3	STM DPCV OP	%	69.4	FO BNR Load	%	81.0
Draft.mmH2O_N	mmH2O	-2.2	Conv. FGT_N	℃	439.4	F-101 H.Release	MMkal	41.5	FO BNR Load	%	82.6
DMPR OP_S	%	51.2	FO Press.	kg/cm²	4.7	FO H.Release	MMkal	20.6	FO BNR per HR	MMkal	2.29
DMPR OP_N	%	43.9	FG Press.	kg/cm²	1.9	FG H.Release	MMkal	21.0	FG BNR per HR	MMkal	2.33

FO Ratio	%	49.5
FG Ratio	%	50.5
FO BNR Firing	BNRS	9
FO BNR Firing	BNRS	9
Rad. Eff.	%	59.2
HTR Eff.	%	74.8

(3) Visual flames

BTM East Side: South to North

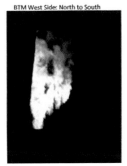
BTM West Side: North to South

MID. Section: North to South

Hip Section: South to North

Overall flames are better than ones before test. The flames of burners close to south and north end wall are unstable. Flame impingement occurs on shield tubes of convection inlet. Around radiant outlet tubes of furnace middle section, flame impingement a little occurs.

Case 2. Fuel oil vs. Fuel gas firing ratio (33.3%: 66.7%)

(1) Burner condition

BNR No.	1	2	3	4	5	6	7	8	9	10	11	12	13	14	15	16	17	18
Firing Staus	G	G	O	G	G	O	G	G	O	G	G	O	G	G	O	G	G	O
Air Door OP	6.0	6.0	8.5	7.0	5.0	8.0	5.0	5.0	7.0	7.0	6.0	8.0	6.0	6.0	8.0	7.0	6.0	8.0
FO Indiv. V/V OP	0	0	100	0	0	100	0	0	100	0	0	100	0	0	100	0	0	100
FG Indiv. V/V OP	100	100	0	100	100	0	100	100	0	100	100	0	100	100	0	100	100	0

(2) Operation condition

F-101 Feed Rate	KL/h	319.9	Excess O2	%	2.4	A_STM DP	Kg/a	2.3	Absorbed Duty	MMkcal	30.2	FO Ratio	%	32.3
Inlet Temp.	°C	218.3	Arch FGT_S	°C	777.1	FO PCV OP	%	24.6	HTR Load	%	92.4	FG Ratio	%	67.7
Outlet Temp.	°C	340.7	Arch FGT_N	°C	713.2	FG PCV OP	%	49.6	AVG BNR Load	%	78.8	FO BNR Firing	BNRS	6
Draft.mmH2O_S	mmH2O	-0.9	Conv. FGT_S	°C	475.6	STM DPCV OP	%	66.2	FO BNR Load	BNRS	76.5	FO BNR Firing	BNRS	12
Draft.mmH2O_N	mmH2O	-1.9	Conv. FGT_N	°C	436.7	F-101 H.Release	MMkcal	40.0	FO BNR Load	%	79.9	Rad. Eff.	%	59.3
DMPR OP_S	%	48.2	FO Press.	kg/a	4.4	FO H.Release	MMkcal	12.9	FO BNR per HR	MMkcal	2.16	HTR Eff.	%	75.4
DMPR OP_N	%	42.6	FG Press.	kg/a	1.8	FG H.Release	MMkcal	27.1	FG BNR per HR	MMkcal	2.25			

(3) Visual flames

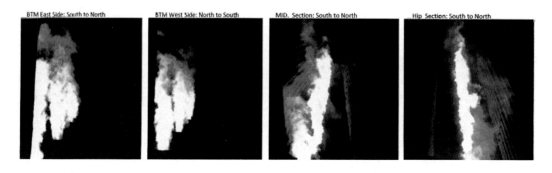

Overall flame impingement seems to be less than Case 1. The oil flame of No.3 burner close to south end wall impinges tube. Less flame impingement than Case 1 occurs on shield tubes of convection inlet. Around the radiant outlet tube of furnace middle section, periodically the flame impinges tube.

Case 3. Fuel oil vs. Fuel gas firing ratio (50%: 50%)

(1) Burner condition

BNR No.	1	2	3	4	5	6	7	8	9	10	11	12	13	14	15	16	17	18
Firing Staus	G.Min	G&O	G&O	G&O	G&O	G&O	G&O	G&O	G&O	G&O	G&O	G&O	G&O	G&O	G&O	G&O	G&O	G.Min
Air Door OP	3.0	7.5	6.0	7.0	6.0	7.0	7.0	6.0	6.5	6.5	6.0	6.0	6.5	6.0	6.0	8.0	6.0	3.0
FO Indiv. V/V OP	0	70	70	70	70	70	70	70	70	70	70	70	70	70	70	70	70	0
FG Indiv. V/V OP	30	50	50	50	50	50	50	50	50	50	50	50	50	50	50	50	50	30

(2) Operating condition

F-101 Feed Rate	KL/h	329.3	Excess O2	%	3.0	A_STM DP	Kg/㎠	2.3	Absorbed Duty	MM㎉	29.4	FO Ratio	%	53.2
Inlet Temp.	℃	223.2	Arch FGT_S	℃	757.8	FO PCV OP	%	11.8	HTR Load	%	89.7	FG Ratio	%	46.8
Outlet Temp.	℃	337.8	Arch FGT_N	℃	703.6	FG PCV OP	%	39.6	AVG BNR Load	%	81.6	FO BNR Firing	BNRS	16
Draft.mmH2O_S	mmH2O	-3.2	Conv. FGT_S.	℃	453.7	STM DPCV OP	%	64.7	FO BNR Load	%	45.9	FO BNR Firing	BNRS	18
Draft.mmH2O_N	mmH2O	-2.5	Conv. FGT_N	℃	439.7	F-101 H.Release	MM㎉	38.8	FO BNR Load	%	35.7	Rad. Eff.	%	59.8
DMPR OP_S	%	39.6	FO Press.	kg/㎠	3.2	FO H.Release	MM㎉	20.7	FO BNR per HR	MM㎉	1.29	HTR Eff.	%	75.7
DMPR OP_N	%	45.7	FG Press.	kg/㎠	0.9	FG H.Release	MM㎉	18.1	FG BNR per HR	MM㎉	1.01			

(3) Visual flames

BTM East Side: South to North

BTM West Side: North to South

MID. Section: North to South

Hip Section: South to North

Overall flames are best of 3 case tests. No.3 burner flame hits tube and the radiant tube outlet has flame impingement, but these show less phenomenon than Case 1 & Case 2.

(4) The constraint of fuel pressure

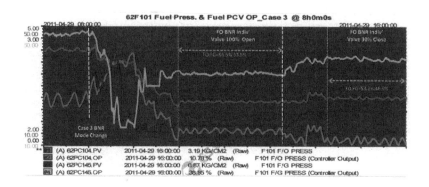

Due to low gas/oil pressure, under their minimum PCV OP, individual valve opening is 50% in gas and 70% in oil respectively (opening 100% impossible). In case of No.1/ No.18 burner of end wall, flame is not good so oil firing is off and only minimum gas firing was kept during the test respectively.

Case 4. Fuel oil vs. Fuel gas firing ratio (66.5%: 33.5%)

(1) Burner condition

BNR No.	1	2	3	4	5	6	7	8	9	10	11	12	13	14	15	16	17	18
Firing Staus	G&O	G&O	G&O	G&O	G&O	G&O	G&O	G&O	G&O	G&O	G&O	G&O	G&O	G&O	G&O	G&O	G&O	G&O
Air Door OP	9.0	7.0	6.5	6.0	6.0	6.5	7.0	6.5	6.0	6.0	6.0	6.0	6.0	6.0	6.5	6.5	7.0	7.0
FO Indiv. V/V OP	100	100	100	100	100	100	100	100	100	100	100	100	100	100	100	100	100	100
FG Indiv. V/V OP	50	50	50	50	50	50	50	50	50	50	50	50	50	50	50	50	50	50

(2) Operating condition

F-101 Feed Rate	Kl/h	329.1	Excess O2	%	3.4	A_STM DP	Kg/cm2	2.3	Absorbed Duty	MMKcal	29.8	FO Ratio	%	66.5
Inlet Temp.	°C	223.3	Arch FGT_S	°C	717.8	FO PCV OP	%	14.9	HTR Load	%	91.6	FG Ratio	%	33.5
Outlet Temp.	°C	338.5	Arch FGT_N	°C	717.5	FG PCV OP	%	34.1	AVG BNR Load	%	78.1	FO BNR Firing	BNRS	18
Draft.mmH2O_S	mmH2O	-3.9	Conv. FGT_S	°C	459.3	STM DPCV OP	%	64.1	FO BNR Load	%	52.0	FG BNR Firing	BNRS	18
Draft.mmH2O_N	mmH2O	-3.5	Conv. FGT_N	°C	447.8	F-101 H.Release	MMKcal	39.6	FO BNR Load	%	25.9	Rad. Eff.	%	60.3
DMPR OP_S	%	47.6	FO Press.	kg/cm2	3.1	FO H.Release	MMKcal	26.4	FO BNR per HR	MMKcal	1.47	HTR Eff.	%	75.1
DMPR OP N	%	48.4	FG Press.	kg/cm2	0.5	FG H.Release	MMKcal	13.3	FG BNR per HR	MMKcal	0.74			

(3) Visual flames

Heat balance inside furnace was improved. In the south and north section, both arch draft and arch flue gas temperature are almost equal respectively. Flame impingement sometimes appears at the middle section of radiant outlet tube. But it was more improved than Case 1/Case 2.

Summary of tests

Flame impingement occurs at the middle of radiant section. The reason will be that as shown in Fig. above (Arch zone), one row tube is omitted there and it has more space between tube and tube. And the instability of flame at the both end-walls may basically come from the configuration of box (or cabin) type heater. So it should be considered in burner design. As the test results, combined gas/oil firing represents less flame impingement than single fuel firing. And Case 4, combined oil/gas firing ratio of 6.5 to 3.5, would be highly recommended in fuel firing mode by showing more balanced draft and flue gas temperature inside furnace.

7.6 Minimizing air leaks

As mentioned above, air which leaks into the furnace quenches the flue gas and requires that additional fuel be burned to compensate for the reduction in process heating.

7.6.1 Finding air leaks

The largest air leaks can usually be found by visually inspecting the outside of the furnace. Common areas of leakage are the following:

① Piping penetrations

② Convection section header box covers and panels

③ Warped explosion doors

④ Poor fitting inspection doors

⑤ Cracks in casing (furnace skin) welds

⑥ Open air registers on out-of-service burners

⑦ Leaky duct expansion joints

⑧ Undetected corrosion in floors, ducts, etc.

Most leakage occurs at process piping penetrations. The penetrations are bigger than the tubes so the tubes can move relative to the furnace on startup and shutdown due to thermal expansion and contraction.

Air leaks in the radiant section cause dark spots or lines on the hot refractory surfaces and can often be found by visual inspection of the firebox during operation.

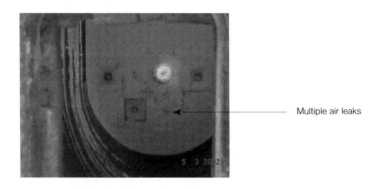

Multiple air leaks

Smoke bomb leak testing fell out of favor for many years, possibly because of environmental concerns, but it has been resumed recently in a few facilities. The local air district must be notified beforehand of plans for smoke bomb testing. The testing is done during a shutdown. The stack damper and burner air registers are closed. The furnace is pressurized with a portable fan and a smoke bomb is set off inside. A team of people mark the locations where smoke leaks out. The leaks are sealed and the procedure is repeated until all the major leaks have been sealed.

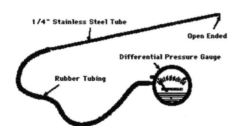

An on-line leak detecting device can be devised using a magnahelic-type differential pressure gauge, as shown left.

The DP gauge is sensitive enough so that the needle will deflect when a small flow of air passes over the open end of the tube. Surveying a furnace involves tracing suspicious seams, holes, etc. with the open end of the stainless steel tube while watching the needle for deflection. A typical 100 mmBtu/hr box type furnace can be surveyed using this method in 2 or 3 hours.

7.6.2 Sealing air leaks

Joints at convection section header box doors and panels can be sealed using silicone RTV high temperature calking. Tube penetrations can be sealed with tube seals. There are a variety of tube seal designs that work adequately. Most facilities use the local insulation contractor to do the tube sealing work on-line.

Explosion doors are no longer installed on new furnaces. If older explosion doors are leaking, they can be taped, bolted, or seal welded shut.

Most burners do not have tight shutoff air registers. If a furnace must run for extended periods of time with many burners out of service, then retrofitting tight shutoff air registers or installing an air-tight seal over the air register opening is advisable.

7.7 Cleaning heat transfer surfaces

There are a variety of deposits that can build up on fired heater tubes and reduce heat transfer. Cleaning heat transfer surfaces typically pays for itself in a few months. In this cleaning there are several methods: Chemical on-line cleaning under operation, on- or off-line sand blasting.

On-line sand blasting is to inject the sand of size 100-150 micron through the furnace peep holes using facilities shown below. The injection pressure to use plant air is 6-8 kg_f/cm^2.

Sand storage tank & blasting port & sand bag

Below photos show the results of on-line chemical cleaning service to remove slag/fouling on the external tube surface while running in convection section and radiant section respectively.

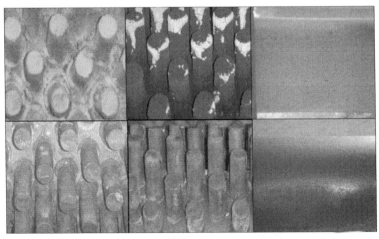

Before (up) and after (down) on-line chemical cleaning service

7.7.1 Deposits

Furnace tube deposits consist mainly of (1) ash and soot from oil burning, (2) dust and debris from insulation, and (3) iron oxide scale on non-stainless steel radiant section tubes.

Oxide scale creates a significant resistance to heat transfer because it has a low thermal conductivity and is often loosely adhered to the base metal. It is typically five times thicker than the base metal that produced it. Oxide scale can be extremely hard and hard to remove.

The deposits on the tube surface typically contains components such as vanadium (24 wt%), nickel (13.81 wt%), iron (7.72 wt%), sulfur (7.33 wt%), sodium (6.29 wt%), and calcium (2.16 wt%) etc. The source of components comes from the fuel oil of crude that each refinery treats. In case of vanadium it is rich into crude from Middle/South America (for example, Venezuela, Mexico) etc. The contents reach more than 500 wtppm in residue fuel oil. Also in calcium, it is rich in acid crude. In any case, these foulants should be removed by either mechanical or chemical treatment.

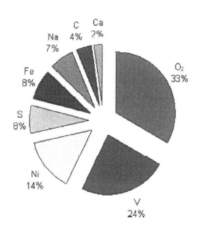

An analysis dada of a scale on external tube surface

Ash, soot, and insulation debris are usually soft deposits that are easy to remove. However, the same material can become hard when mixed with oxide scale particles and packed between the fins of convection section tubes for several years. The interaction of these deposits with trace fuel sulfur and moisture during shutdowns can make the deposits between convections section fins quite tenacious. Sometimes large pieces of refractory from convection section walls clog the spaces between convection section tubes.

7.7.2 Tube cleaning methods

Ash and soot from oil burning are commonly removed from convection section tubes by permanently mounted soot blowers which use steam to blast the convection section once or more per shift. These same deposits can be removed from radiant tubes by steam or air blowing during shutdowns. Theoretically, on-line steam or air lancing of the radiant section tubes would be effective at removing oil burning ash and soot deposits. However, no facility currently practices this.

Harder deposits that have built up over years between convection section fins must be removed by an abrasive method. Sand blasting has been used in the past, however, CO_2 pellet blasting (by Polar Blast) has been used more successfully lately and has the advantage of less debris to remove from the furnace afterwards. After a convection section has been thoroughly cleaned by CO_2 pellet blasting, in some cases it may be kept fairly clean by steam or air blasting during subsequent shutdowns. This technology utilizes CO_2 pellets shot from a lance to clean finned tube convection sections while the furnace is shut down. The advantage over other cleaning techniques is the minimal cleanup required because the abrasive material evaporates and no water is used that will soak the refractory. Below shows the results of before and after Polar Blast service in a refinery.

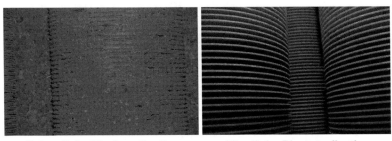

Before Polar Blast application   After Polar Blast application

CO_2 pellet blasting usually penetrates about three tube rows into convection sections because the tubes are staggered (on a triangular pitch). Therefore, to get the best results, the cleaning must be done from above, below, and from the sides where access panels can be removed.

Sand or grit blasting are the typical method used to remove oxide scale from radiant section tubes during shutdowns. Walnut shell blasting using water-cooled lances can be used to clean radiant section tubes on-line. The abrasive walnut shells remove debris and soft oxide scale and then burn up in the furnace.

There are three main chemical methods for cleaning tubes on-line. Three methods use chemicals to react with the deposits to weaken them for easy removal. One method (Crayhurst) uses water-cooled lances to apply chemicals to radiant section tubes and then rinse them off with steam. The second method (CTP) injects chemicals into the flue gas stream and relies on flue gas velocity due to temporarily increased combustion air to sweep the deposits away. This method mainly affects convection sections. The third method (GTC Technology) sprays chemical near to tube. The chemical is sprayed with various nozzles into radiant or convection section through view door or manhole. The sprayed chemical is vaporized/decomposed by high temperature (fire box), dispersed by draft (pressure) and mixed with arising flue gas. The mixed gas circulates inside the furnace. And it contacts scale on the tube surface and instantly reacts each other. Reactant scale dust is blown-off to the stack as the fine dust type (less than 10 micron). For a good mixing of chemical & flue gas, the spraying point targets on near/top flame, near tube and hip section. This is for using flue gas flow pattern of dynamics in temperature, pressure, and velocity.

GS Caltex in Korea, Hawaii in USA, and Reliance in India have experienced damage to castable refractory behind tubes due to chemical and steam sprays. It is unknown if this was due to poor quality castable or poor operation of the spray equipment.

Abrasives, chemicals, and steam cleaning methods all have the potential for damaging insulation. During shutdowns, plywood or tarps can be installed behind the tubes to protect the insulation from damage. On-line methods which spray abrasives, chemicals, or steam onto tubes cannot be used where ceramic fiber insulation is in close proximity to the tubes. Below is the list of heater cleaning vendor.

Tube Cleaning Method	Company	On-line	Off-line	Rad. Tubes	Conv. Tubes	Facility/Experience
Sand or grit blasting	In-house or local contractor		X			Numerous/good
Walnut shell blasting	Cetek (USA)	X		X		
CO$_2$ pellet blasting	Polar Blast (USA)		X	X	X	Pascagoula/Hawaii/both good
Chemical spray onto tubes	Crayhurst (RSA)	X		X		GS Caltex/Hawaii/both mixed, Cape Town/good
Chemical spray onto tubes	ThermaChem (UK)	X		X		Pemmbroke/good
Chemical spray onto tubes	Global A.T.S (France)	X		X		
Chemical spray onto tubes	Technochem (Thailand)	X		X		SPRC/in planning
Chemical spray into flue gas	CTP (France)	X			X	Pemmbroke/good
Chemical spray into flue gas	GTC (USA)	X		X	X	Reliance/HPCL/MRPL/all good

7.8 Enhancing heat transfer surfaces

High emissivity ceramic tube and refractory coatings can enhance the heat transfer in the radiant section. The coatings have paid out in less than a year where tubes were highly scaled.

Below shows catalytic reformer furnace, F-35001, in a refinery after ceramic coatings are applied to tubes and refractory.

Tubes before ceramic coating   Tubes after ceramic coating

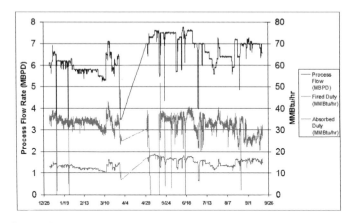

Operating parameters before and after ceramic coatings

Below represents the results of ceramic coating applied to the furnace in another refinery.

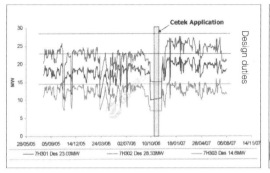

Heater duties

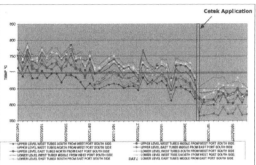

Tube wall temperatures

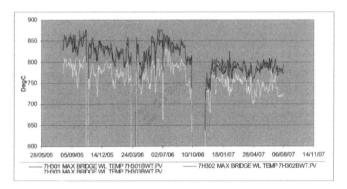

Bridge wall temperatures

Below indicates the completed ceramic coating of refractory and tubes respectively in Platformer process heater.

Insulating brick refractory ceramic coating   Furnace tube ceramic coating

7.9 Adding heat transfer surface

The addition of heat transfer surface can significantly increase furnace efficiency. In most cases, the most economically viable alternatives are (1) additional process heating, followed by (2) steam generation and boiler feed water heating, followed by (3) air preheat. The higher the current stack temperature, the more likely the additional heat transfer area can be economically justified.

Added tubes

Adding tubes to the future rows of convection section

The ability to add radiant or convection section tubes is highly dependent on furnace geometry. As specified by API 560, all convection sections are designed with space for two additional tube rows. A burner configuration change allowed several process tubes to be added adjacent to the lower radiant section walls. The addition of process tubes almost always pays for itself, however, the affect on draft must be taken into consideration when adding tubes to the convection section.

Adding tubes to the convection section has two negative effects on draft. The added tubes increase the flue gas pressure drop across the convection section and the decreased stack temperature decreases the draft generated by the stack. Lengthening the stack to increase draft is often feasible.

7.10 Repairing insulation

Heat loss from furnace casing is usually minor compared to stack losses. Hot areas of casing are typically an indication of a localized refractory problem and a possible reliability risk rather than a significant impact on efficiency. You will recall that casing heat loss does not vary in the efficiency calculation since it is assumed to be 2%. This value is based on the average casing temperature being 93°C or less. Heat loss from the casing is approximately proportional to average temperature so efficiency could be affected if a large section of insulation were in severe disrepair.

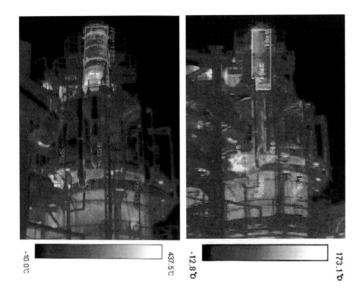

Casing temperature of Rheniformer heater (left before TA, right after TA)

Heater	Section	Direction	Infrared Scan Casing Average Temperature (°C)		Difference (After–Before)	Major Maintenance
			Before TA	After TA		
03F–101	Convection	South West	85	55.8	29.2	Burner replacement, Furnace wall/floor repair
	Duct~Convection~Arch	North East	296.2	91.1	205.1	
04F–101	Stack, Convection	South North	107.5, 115.2	57.8, 57.4	49.7, 57.8	
	Stack~Radiant Top	North	149.5	34.1	115.4	
	Convection, Arch	North West	123, 183	77.9, 70.4	45.1, 112.6	
04F–102	Convection, Radiant	North West	100, 345.5	62.6, 63.8	37.4, 281.7	
	Convection	South East	102	60.4	41.6	
63F–101	Duct Stack, Radiant Upper	North	225, 106(102–133)	62.1, 42.3(40.4–56.8)	162.9, 63.7	Burner replacement, Furnace wall/floor repair, Baffle plate installation
63F–102	Duct Stack, Radiant Upper	North	180, 108(104–132)	52.7, 52.5(52.4–61.9)	127.3, 55.5	
63F–103	Duct Top, Radiant Upper	South East	417, 123(135–421)	65.8, 55.7(65.8–86.3)	351.2, 67.3	
	Stack Upper	East	111(101–230)	87.6(42.9–87.6)	23.4	
63F–104	Radiant Upper	North	83	39.1	43.9	Burner replacement, Furnace wall/floor repair
64F–101	Radiant Upper	North	98	47.5	50.5	
64F–102	Radiant Upper	North	87	47.5	39.5	
64F–103	Convection~Radiant	South West	114.2	76.1	38.1	
	Radiant Upper	North	182	52.6	129.4	

Casing temperatures of Platformer/Rheniformer heater

CCR heater casing temperature

Location	Label	Value
101	SP01	34.1℃
102	SP02	29.6℃
103	SP03	34.8℃
104	SP04	32.4℃

Before on-line chemical cleaning

Location	Label	Value
101	SP01	19.7℃
102	SP02	22.3℃
103	SP03	19.7℃
104	SP04	18.9℃

After on-line chemical cleaning

CCR heater tube temperature

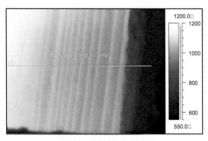

Location	Label	Value
101	SP01	1012.5℃
101	SP02	1005.9℃
101	SP03	1020.3℃
101	SP04	1043.2℃
101	LI01: max	1140.9℃
101	LI01: min	584.2℃

Before on-line chemical cleaning

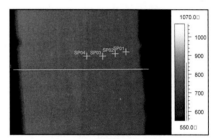

location	Label	Value
101	SP01	726.1℃
101	SP02	734.9℃
101	SP03	753.2℃
101	SP04	763.8℃
101	LI01: max	821.9℃
101	LI01: min	411.4℃

After on-line chemical cleaning

CDU heater casing temperature

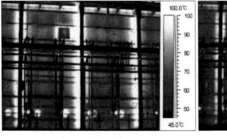

Label	Value	Diff temperature
SP01	70.4℃	14.6℃
SP02	74.6℃	10.4℃
SP03	70.0℃	15.0℃
SP04	75.1℃	9.9℃

Before on-line chemical cleaning

Label	Value	Diff temperature
SP01	57.1℃	27.9℃
SP02	65.4℃	19.6℃
SP03	57.8℃	27.2℃
SP04	64.2℃	20.8℃

After on-line chemical cleaning

CDU heater tube temperature

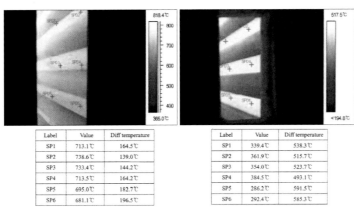

Label	Value	Diff temperature
SP1	713.1℃	164.5℃
SP2	738.6℃	139.0℃
SP3	733.4℃	144.2℃
SP4	713.5℃	164.2℃
SP5	695.0℃	182.7℃
SP6	681.1℃	196.5℃

Label	Value	Diff temperature
SP1	339.4℃	538.3℃
SP2	361.9℃	515.7℃
SP3	354.0℃	523.7℃
SP4	384.5℃	493.1℃
SP5	286.2℃	591.5℃
SP6	292.4℃	585.3℃

Before on-line chemical cleaning   After on-line chemical cleaning

Since most paints oxidize above about 204℃, discolored paint on furnace casing is a sign of insulation problems that should be addressed during the next shutdown. Area of discolored paint should be monitored by IR to assess the actual temperature. Hot areas approaching 538℃ should be cooled by external steam or air lances until the next shutdown. Areas approaching 538℃ are also candidates for on-line refractory repair.

7.11 Other

A few additional efficiency ideas don't fall into the categories above.

Replacing a convection section can sometimes be the economic choice over tube cleaning or refractory repair. An entire convection section replacement can take less shutdown time and accomplish the equivalent of refractory repair, adding heat transfer area, tube cleaning, and renewing tube thickness.

Variable speed fan drives can significantly reduce the utility costs of operating a furnace. The payback is usually attractive.

Furnace efficiency monitoring can be specified by the best practice. Many refineries have developed an on-line page that lists furnace number, firing rate, stack temperature, measured stack oxygen, target stack oxygen, efficiency, and calculated fuel costs or savings due to operating above or below the target. Efficiency gains are best maintained when management reviews efficiency performance and gives positive feedback on improvements.

Operator training on tuning and efficiency is specified by the best practice. It is helpful to remind operators of the relationship between O_2, efficiency, and fuel costs.

A furnace optimization process should be developed for implementation on multiple furnaces within a division or a whole refinery. The process should also be designed for use by a dedicated furnace specialist or 3rd party contractor.

7.12 Economics

The appeal of reducing emissions and energy costs may be the driving force behind the modifications, but a thorough economic analysis needs to be conducted prior to selecting a solution. The primary objective of conducting any combination of modifications is to improve efficiency. To justify the modifications, the immediate savings to the producer will be: (1) Lower fuel consumption (2) Additional service gained while using the same amount of energy (3) Lower CO_2 emission levels (4) Potential for reduction of maintenance outages

To quantify the potential savings related to fuel savings, below is a simplified case study using a medium duty heater, 100 mmBtu/hr, whose efficiency is currently 75%. The cost analysis will be based on improving efficiency to 85%.

$Q_F = Q_A/n$

Where, Q_F fired duty (by the burners)

Q_A absorbed duty (by the process)

n efficiency

Based on above equation, we will proceed with a case where Q_A is 100 mmBtu/hr and the original efficiency n, is 75%.

As such,

$Q_F = 100/0.75 = 133.33$ mmBtu/hr

If the same calculation is conducted for a heater with efficiency of 85%, the results are as follows.

$Q_F = 100/0.85 = 117.65$ mmBtu/hr

The difference between fired duties of 75% and 85% cases is 15.68 mmBtu/hr

(133.33-117.65 mmBtu/hr=15.68 mmBtu/hr). In other words, for every hour the furnace is in operation, efficiency improvement of 10% will reduce the energy consumption by 15.68 mmBtu. In terms of monetary savings, assuming that 1 mmBtu costs $5, the saving in one day is $1881.60 which results in annual savings of $686,784. The payback period will vary depending on the extent of heater modifications to achieve this efficiency improvement. This case study does not include the cost of operating additional equipment that requires energy, such as fans for the Air Pre-heat system.

Economics of scale also help; the chart below demonstrates the savings between low, medium, and high duty fired heaters. The example uses three heaters of various duties (25, 100, and 275 mmBtu/hr) with current efficiency of 75%. Cost savings over one year are plotted for each 1% that efficiency is improved from the existing 75%, again assuming that 1mmBtu costs $5. Although it is more expensive to conduct the modifications on a high duty heater, with higher savings the payback period may be significantly reduced when compared to a low duty furnace.

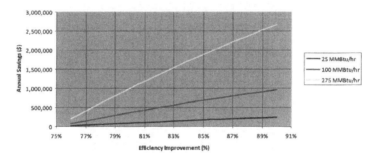

Annual cost savings based on $5/mmBtu

7.13 Solutions

The solutions to efficiency upgrades may be grouped in four categories. The categories are as follows: (1) Operator training (2) Reduction of input energy (3) Reduction of stack flue gas temperature (4) Heat loss reduction

For each possible solution, we also assume that total duty of the heater and process conditions do not change.

7.14 Operator training

Operator training is one of the factors often overlooked when identifying the methods to improve furnace efficiency. Trained operators are able to identify the furnace problems ranging from O_2 and draft levels, to refractory damage. Operator training is also vital to ensure that proper furnace operating logic is in place. As the operators are the eyes and ears in the plant, the training allows them to interpret the heater data, visual or digital, and make the necessary control adjustments and/or mechanical repairs to counter the adverse conditions experienced by the heater.

The operator may believe that the plant's goal is to achieve the maximum throughput per day. While at first it is easy to side with this statement, in reality it should be expanded to include the cost for achieving this goal. In some instances the cost of achieving the maximum throughput may exceed the benefit gained by reducing the efficiency and by the impact left on the environment. Efficient operation of the heater will yield optimum throughput, while reducing the consumption of energy, CO_2 emission levels, and maintenance outages. The training teaches the operators the techniques and controls required to optimize the heater for the most economical and efficient operation.

7.15 Reduction of input energy

The two primary methods of achieving this are: Improvement of burner technology, draft control.

7.15.1 Improvement of burner technology

Over the years, two distinct improvements have been implemented to burners. These are: Reduction of required excess O_2, Improved fuel-air mixing. The ideal fuel to air ratio is established through stoichiometric calculations. Any air that is supplied to the burners above the ideal stoichiometric quantity is heated without purpose. However, achieving the perfect mixing in industrial practice is not possible. In turn, this could lead to unburned fuel being forced into the radiant and convection sections. Aside from not providing adequate heating energy, this is a great source of danger as the presence of unburned fuel combined with a sudden addition of air could lead to explosion. To avoid this problem, excess O_2 is added to achieve complete combustion. Excess O_2, the amount beyond theoretical (stoichiometric), is added to ensure complete combustion of the fuel. The burners in the past employed excess O_2 between 3.5 and 4.2%. New burners, with the help of improved air-fuel mixing (through staged mixing) allow for safe operation of burners with 1.9 to 2.7% of excess O_2.

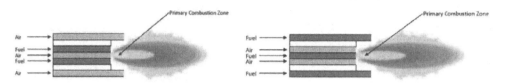

Staged air combustion in oil firing (left) & staged fuel combustion in gas firing (right)

This allows for significantly lower quantity of air being drawn into the system which consequently leads to higher burner efficiency. As such, less excess O_2 decreases exhaust gas quantity exiting the heater so that overall heater efficiency is improved. Sub-stoichiometric combustion is also possible although it would lead to large quantities of CO and fuel being deployed inside the heater. Sudden addition of air could lead to explosion which is the reason fired heater burners should never be designed for this scenario.

The negative aspect of lower excess O_2 is the reduction in turn down ratio. The burners in the past could achieve the turn down ratio of up to 10:1 whereas today's burners are typically capable of achieving 4:1 ratio. The environmental effect of fuel and air staging is the reduction of NO_x. Fuel and air staging reduces NO_x by lowering the flame temperature, and by creating a fuel rich chemistry in the primary flame zone. Flame length, speed, size, and flux distribution also change with the new burners. With numerous conditions to consider, changing of the burners is not a simple task and requires attention of qualified personnel that can take in consideration all characteristics of a given heater prior to selecting a replacement burner which will yield better efficiency and lower NO_x levels.

7.15.2 Draft control

With regards to fired heaters, draft (also referred to as draught) is defined as a negative pressure (vacuum) of the air and/or flue gas in the heater. It materializes because the hot flue gas inside the furnace is lighter and consequently at a lower pressure than the cold air outside. Combustion air is drawn into the burners at which point it undergoes combustion and produces flue gas and energy in the form of heat. Due to buoyancy, the flue gas raises through the radiant section, convection section, and finally the stack. While passing through the convection section, the flue gas encounters friction resistance as it passes between the tubing. This is referred to as draft loss. Stack height must be designed to provide sufficient buoyancy to overcome the draft losses and to ensure the pressure is always negative within the heater. Regardless of whether the furnace utilizes forced or natural draft burners, the stack damper is used to control the draft. As evident on the draft profile Fig., the radiant arch sees the highest possible pressure and is consequently used as a control point for draft regulation.

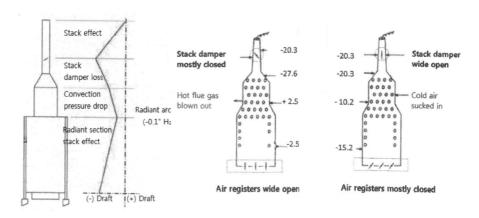

Fired heater draft profile   Adjustment of air register and stack damper

As well as regulating the safe operation of the heater, draft control directly controls the quantity of excess O_2 that is drawn into the heater. Closing the damper minimizes the excess O_2 that is drawn into the heater, but it lowers the draft in the heater. Lower draft jeopardizes the safe operation of the heater as it may not be sufficient to overcome the draft losses, particularly from the convection section. Opening the damper blades increases the excess O_2, but raises the draft within the heater. While the higher draft improves the safe operability of the furnace, the additional O_2 translates to higher emissions and lower efficiency. To maximize efficiency of a given heater, damper blades should be controlled to achieve sufficient draft for safe operation, all while minimizing the amount of excess O_2 that is supplied to the burners. Properly trained operating personnel can help achieve this.

7.16 Reduction of stack flue gas temperature

Aside from reducing the quantity of input energy another method of improving efficiency is to reduce the stack flue gas temperature. Although it is not a linear relationship, as a rule of thumb every 1.7°C reduction in stack flue gas temperature increases the efficiency of the heater by 1%. The reduced flue gas temperature can be achieved in different ways which will be discussed below. These are as follows: Air pre-heat system, convection section upgrades

7.16.1 Air Pre-Heat system

Air preheat system (APH) is employed to improve efficiency of a fired heater. The APH can be looked at as a heat transfer apparatus through which the combustion air is passed and heated by a medium of higher temperature. This medium may be a combustion flue gas or external fluids such as steam, boiler feed water, and process. From here, as the heating medium, the combustion flue gas is considered as it directly relates to the reduction of stack flue gas temperature. This type of APH is commonly referred to as Balanced Draft APH.

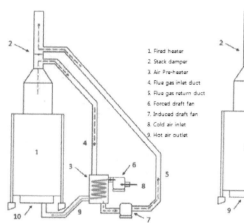

Balanced draft Air Pre-heater system
with internal heat source

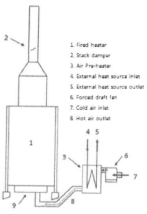

Air Pre-heater system
with external heating source

Direct fired heater with
Air Pre-heater system

The immediate benefits of operating the APH are: (1) Improved efficiency (2) Improved control of combustion air flow (3) Reduced oil burner fouling (4) Better flame pattern control (5) More complete combustion of difficult fuels

The APH recovers the energy (heat) that would be otherwise lost to the atmosphere by heating the combustion air for the burners. In case of Balanced Draft APH, as the combustion gas is heated, the flue gas is returned to the stack. As the heat from flue gas transfers to combustion air, the final temperature of the flue gas in the stack becomes significantly lower when compared to the same furnace without the APH system.

The immediate benefit of implementing the APH is efficiency improvement. From the environmental perspective, Fig. below demonstrates the relationship between APH temperature (combustion air coming out of APH) and the chemical composition of the flue gas. It is evident from the chart that with higher APH temperatures, the quantity of CO_2 also reduces, lowering the environmental footprint left by the operation of the unit.

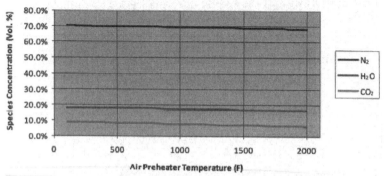

Stoichiometric flue gas composition related to Air Pre-heater temperature (for CH_4)

API 560 (4th Edition, August 2007), section F. 10 established five ways in which the use of APH can impact the environment. Overall, the impact is deemed to be positive. The five ways are identified as follows: (1) Energy conservation- lower energy consumption (2) Stack emissions (3) Reduced noise pollution- due to burners being housed in a plenum (4) Thermal pollution- due to lower flue gas temperature (5) Effluent reduction

Required modifications would vary depending on the heater and APH configuration. If Balanced Draft APH is selected, the new ducting would require the provisions in the stack where the ducts are attached. If the existing burners are of natural draft type, they would also need to be replaced with a forced draft type, regardless of the type of APH selected. However, forced draft burners allow for efficiency calculations to be based on lower quantity of excess O_2 when compared to natural draft burners (API 560, 4th Edition, August 2007). Forced draft burners can also be operated with lower quantity of excess O_2 when compared to natural draft type, thus improving the overall efficiency of the heater.

The majority of the work associated with implementation of APH system can be carried out while the furnace is still operating, thus minimizing the heater's downtime.

7.16.2 Convection section upgrades

Another method of lowering the stack flue gas temperature is to increase the heat transfer surface in the convection section. This can be achieved by extending the tube surface using fins or studs, or by adding tubes in the future rows area of the convection section.

A combination of factors has led to design of convection sections utilizing bare tubing. Low energy prices and lack of emission regulations caused less concern for efficiency. Thus, extended surface was not deemed a priority when designing convection section tubing in the past. Also, the combustion of heavy hydrocarbons produces soot which tends to build up between the fins or studs. The soot, in turn, insulates the tube and reduces the heat transfer effectiveness. Consequently, many older furnace designs contain bare tubing in the convection section. However, with higher energy prices along with technology that allows effective soot maintenance (soot blowers), the tubing in the convection section may, in most cases, contain finned or studded tubes. The extended surface provides greater heat transfer area which in turn reduces the flue gas temperature as it removes additional heat from the energy source. Aside from achieving higher efficiency, the tubes experience lower flux rate which in turn exposes the tube metallurgy to lower thermal stress and, in theory, extended lifetime.

Bare tube vs. Finned tube

Similar to the benefits shown by adding fins or studs to convection section tubes, the same benefits apply to adding new rows in the area designated as future rows. One of differences between the two is the fact that new tubing in the future row may be used for existing process, or for new process such as steam generation. Either way, the flue gas temperature exiting the convection section will be reduced due to additional heat transfer surface area. Consequently the efficiency will improve.

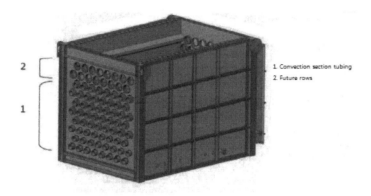

1. Convection section tubing
2. Future rows

7.17 Heat loss reduction

Insulating refractory in the furnace plays a key role in efficiency control. Despite optimizing the entire furnace for the maximum economical efficiency, if the insulation and sealing are not adequate, the producer will not experience the true benefit of the upgrade. It is established that between 1-5% of energy losses are through the furnace casing due to either inadequate design or damages to refractory. Areas where the refractory is a problem are easily found by identifying hot spots. These areas can be identified by using thermographs or through visual inspection. By repairing the areas where the casing temperature exceeds the temperature specified on the fired heater data sheet, the calculated efficiency level can be achieved.

Also, the fired heater is not a pressure tight structure. With numerous observation, access, explosion, and peep doors, along with joints between various heater components, there are many sections where air can get drawn into the heater. This air does not contribute to combustion, and in turn shows up in the flue gas sampling analysis in the stack. This leads to inefficient combustion, waste of energy, and increased NO_x generation. Several precautions that will minimize the leakage of air into the furnace are as follows: (1) Keep all observation, access, explosion, and peep doors closed (2) Ensure proper gaskets are in place on header box doors (3) Ensure there is minimal air leakage through tube guide holes in the radiant section floor.

8. Burner Operation

8.1 Manual (Electric) pilot will not light

Pilot fuels to contain high hydrogen causes the pilot burner burn-back. This happens when flame propagation speed is much greater than the air-fuel mixture discharge speed. Hydrogen makes the propagation speed much higher. For the safety sake of preventing burn-back, during burner operation, intensive care on pilot flame condition should be taken by adjusting pilot mixer air door.

8.2 Fuel pressure does not match capacity curve

(1) The fuel gas composition or temperature does not match that of the design fuel. Check the fuel gas composition against that of the original design fuel. An increase or decrease in hydrogen content or other components will cause the heat content of the fuel to vary and change the fuel pressure required at a specific heat release. A different temperature will result in a different pressure.

(2) The fuel pressure gauge is not located correctly.

(3) The gas tips are not drilled correctly.

Check the ports on the gas tips against the drilling information shown on the Data Sheet. The Data Sheet should show the number and size of the ports. If over time the ports have been enlarged from wear and do not match the information on the Data Sheet, contact the maker to obtain replacements parts. If the drilling information is not shown on the Data Sheet, contact the maker too.

(4) The fuel gas tips and/or risers are partially plugged or damaged.

8.3 Cannot achieve air flow capacity

The method or measuring device used to determine the heat release or fuel flow to the burners is incorrect. Smoky or hazy flames can be the result of insufficient air to the burners. Always check the firebox oxygen level before making any adjustments to the burners. If the firebox is short of air, do not shut off a burner, open the burner damper, or increase the flow of combustion air because an explosion could result. Decrease the fuel rate in steps not to exceed 10%, allowing the oxygen level in the firebox to stabilize between steps until the required excess air level is achieved.

8.4 Oil firing instructions

If the burners are going to be fired on gas only, the oil gun should be removed to prevent damage from heat. In case of a short period gas only firing without removing oil gun, oil gun atomizing steam valve should be open (cooling by steam) to prevent oil tip heat damage.

8.5 Gas tip and orifice cleaning

These ports must be kept free of foreign material that would reduce the effective port size. If the ports become partially or completely plugged, the amount and distribution of fuel entering the combustion zone vary from the design intent, and combustion problems will likely occur. Sources of foreign materials that plug tips and orifices include: (1) Pipe scale and gums from the fuel gas piping (2) Amine compounds from the fuel gas hydrogen sulfide (H_2S) removal process (3) Coking of condensed heavy or unsaturated hydrocarbons in the fuel gas (4) Polymers that form inside the burner heater risers or tips (5) Hydrocarbon mists that vaporize or react in hot risers or fuel tips

The plugging problem of gas tip greatly influences burner heat release, i.e., in case of the plugging, the gas tip port's inside diameter becomes to be narrow, its pressure rises, and the heat release reduces gradually. Accordingly, it is required to clean or replace the gas tip periodically. The reduction of 30% in the tip inside diameter is used to increase the fuel gas pressure by 4 times.

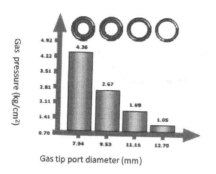

Gas tip port diameter (mm)

Gas tip plugging and pressure

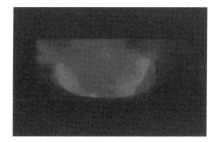

Coke inside tile

8.6 Oil gun positions in regen tile

Proper oil tip location in the regen tile is dependent on the included spray angle, the number of exit ports and the depth of the regen tile. To achieve optimum regen tile function the main body of the spray pattern should clear the exit of the tile but should be close enough to establish full volume recirculation of burning oil mist for stabilization.

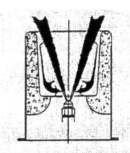

Proper location

The over-insertion of the oil tip in the regen tile will reduce or eliminate the stabilizing recirculation causing combustion initiation to occur further out in the burner and will cause lifting and flameout in turndown conditions. This type of setting will also result in coke buildup on the oil tip and oil dripping due to the loss of the primary air velocity across the tip.

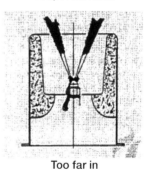

Too far in

An oil tip which has been positioned too far behind the regen tile will often impinge on the exit of that tile. As the distance is increased, the burner flame will progress from smoky edges and sooting of the edges of the outlet, through coke buildup on the tile, to oil splashing and major spills.

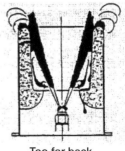

Too far back

8.7 Pilot burners

Pilot burners are premix gas burners. Pilots may or may not be installed on a burner, depending on the requirements of the operator. When pilots are required, they are part of the burner assembly. The pilot may use the same air supply as the main burner, or it may be located outside the main burner air source to ensure stable operations.

The primary function of the pilot is to provide a small source of heat input for the ignition of the main burner fuel. The pilot may be shut off after lighting the main burner, or may continue to operate after the main burner is lit. the pilot must have a stable flame and be located in the correct position near the main burner fuel discharge for ignition of the main fuel.

8.8 Flame patterns

Burner manufacturer can obtain a predictable flame pattern from the burner. The fuel tip drill patterns can be varied based on testing and experience to obtain short, bushy flames or long, narrow flames depending on the requirements of the process heater. The burner tile shape and condition are critical to obtaining the desired flame shape. Round tile shapes provide round flame patterns; rectangular tile shapes provide flat flame patterns. Missing tile or poorly maintained tile with holes and cracks, can cause a poor flame pattern. Substitution of tiles with ones of a different design can restrict air flow, result in poor fuel/air mixing, or cause flame instability because a tile ledge or other feature is missing or incorrectly located.

Oil tile crack   Gas tile crack

The flame patterns and dimensions and the impact of different tip drillings are determined by the burner manufacturer and can be confirmed by performance testing at the manufacturer's test facilities. Today, multi-burner testing is available at some burner manufacturers to determine the effect of the interaction between burners and the flue gas circulation within the firebox on the flame patterns. It should be noted that the flame dimensions observed in a single burner test, particularly a diffusion flame burner, will rarely be observed in a multi-burner firebox. Variation of the fuel and air flow between burners, flame interactions, flue gas circulation currents, and air leakage all act to vary the flame pattern between burners.

The flame pattern from a burner is developed jointly by the heater designer and burner manufacturer. The burner manufacturer selects the gas or oil tip drilling pattern, the type of diffuser (if applied), and the tile shape to achieve the desired flame. Below shown tips all have the same body but the drillings are different each other and therefore the tips are not interchangeable.

Gas tips

There are many different types and shapes of fuel tips. Each fuel tip is drilled with a given pattern to meter the fuel and to inject the fuel into the combustion air. With the tile shape, diffuser, and the fuel tip drilling pattern, the number of burners installed may need to be evaluated. More burners at a lower heat release will provide smaller flames.

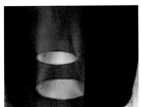 

After modification   Before modification

Any flames that are visually different or unusual, either in dimension or stability, should be investigated and any problem corrected. In the combustion reaction, there are many different types of ions formed, depending on the fuel components

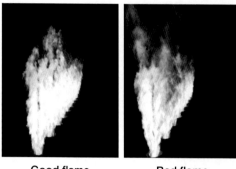

Good flame	**Bad flame**

being burned. The ions are excited by the high flame temperatures and radiate at wavelengths that are visible to the human eye. These visible wavelengths are the definition of the flame patterns on the burners. Sodium ions in the flame radiate in bright yellows and oranges. The color of the flame is affected by the flame temperature, the intermediate ions formed during the combustion reaction, and the amount of carbon particles in the flame. Regardless of the flame color, if combustion is completed, the amounts of heat released and heat available for heat transfer to the process fluid in the tubes are the same. Above left photo shows an example of a good flame within the firebox, while right photo provides an example of a very bad flame pattern.

Good flame patterns alone do not protect against localized tube overheating. Flue gas circulation flows within the firebox and uneven firing of burners will affect the distribution of heat to tubes.

Oil tips

The flame patterns within the heater should be visually inspected as often as necessary, and any change in shape or dimensions of an individual flame should be noted. The flames from all active burners of the same size should be uniform because they all have the same heat release, the same air flow across the burner, and the same fuel flow at each burner. Flame patterns are designed to stay within an envelope that is a safe distance from the process tubes and the refractory. API 560 cites some standards and recommendations as to burner-to-tube and burner-to-refractory. Some users apply greater clearances between burners and between burners and tubes than the API 560 recommendations. The burners installed on existing heaters many years ago may not comply with current API 560 recommendations.

9. Operation Problems

CCR Platformer scale   Crude heater deposit   CCR Platformer oxidation scale

Impingement by afterburning

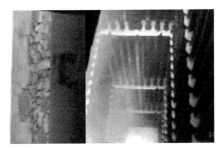

High BWT

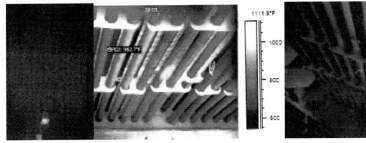

Low BWT   Infrared scan   Bright roof/refractory/support

Color of tube support   Roof tubes of crude heater

Long flame   Flame lift   Flame out

Flame flies   Over firing   Flame leak

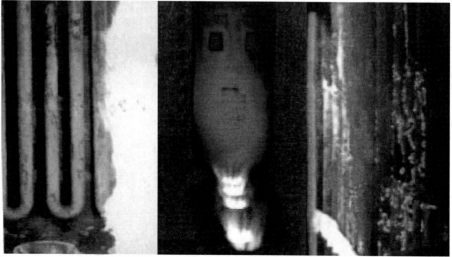

Good oil flame   Good gas flame   CCR Platformer scale

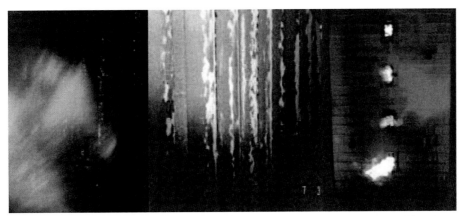

Tube impingement   Oil ash   Burner problem 1

Burner problem 2   Burner problem 3   Burner problem 4

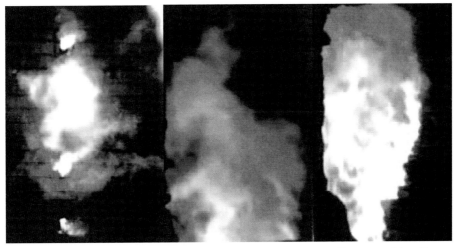

Burner problem 5   Burner problem 6   Burner problem 7

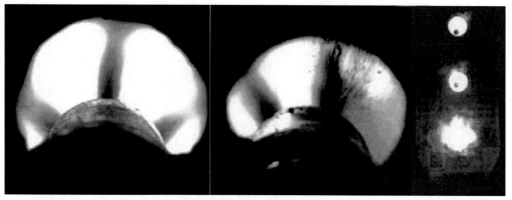

Burner flame (left good, right bad)   Poor combustion

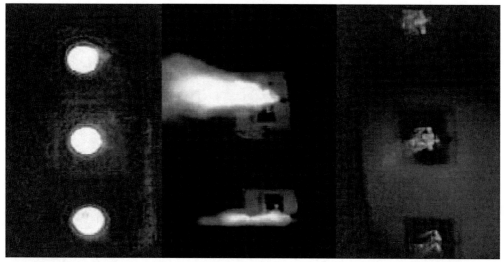

Good combustion   Poor combustion   Good combustion

Good combustion   Good operation   Bad operation

On-line inspection   On-line inspection   On-line inspection

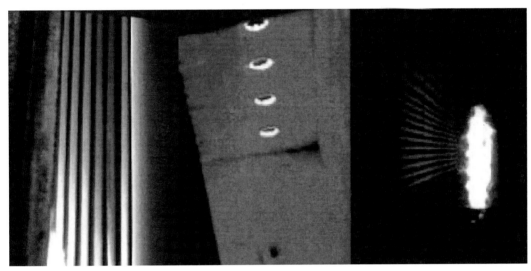

On-line inspection   On-line inspection   On-line inspection

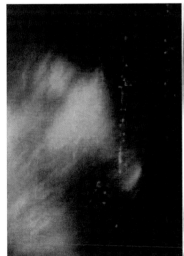

Flame impingement 1   Flame impingement 2   Flame impingement 3

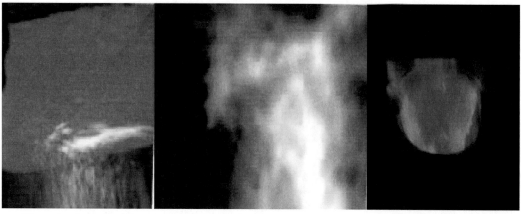

Flame impingement 4   Flame impingement 5   Regentile coke generation

Steam reformer tube hot spot   Heavy tube scale

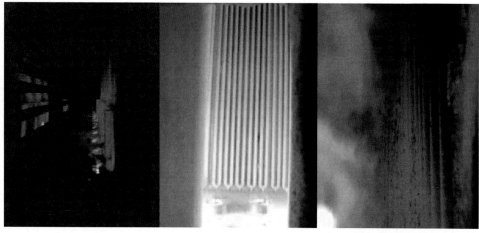

Double fired heater   Over firing   Over firing

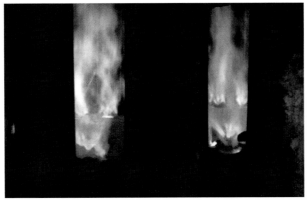

Gas firing   Oil/gas firing

Heavy smoke   Stack plume

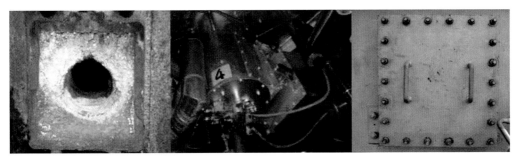

Peep hole door   Burner air register   Inspection door

Fouled air preheater   Cleaned air preheater

10. Heater Operation Manual

Definition of terms

General furnace operations

Preparation for initial start-up

A. External inspection

B. Internal inspection

C. Furnace dry-out

Furnace start-up for natural draft

Furnace shut-down

Some simplified symbols used for instruments

DG-Draft gauge

FG-Flue gas sample connection

FR-Flow recorder

FRC - Flow recorder controller

MCV-Manual control calibrated valve

PDC-Differential pressure controller

PI-Indicating pressure gauge

PIC-Indicating pressure controller

SS-Steam snuffing

SV-Manual shut-off valve

TI-Temperature indicator

TIC-Indicating temperature controller

TRC-Recording temperature controller

AE-Oxygen analyzer

TST-Tube skin thermocouple

10.1 Definition of terms

Brevity makes it advisable to use terms throughout this manual with which the operator may not be familiar. For this reason, some of the more frequently used terms are listed below.

(1) Fluid stream

The term "fluid stream" is used to designate the stream of fluid to be heated in the furnace. The term "fluid" is defined as any substance, liquid or gas or combination thereof, possessing flow characteristics.

(2) Pass

A "pass" is the bank of consecutive tubes through which the fluid travels from the point it enters until it leaves the furnace. Thus, for example, in a two pass furnace, the fluid would divide into two streams at the inlet, flowing separately through their respective tube banks in a parallel manner. A two pass furnace with 32 tubes, for example, would have two flow paths of 16 tubes each, one-half the total flow going through each path. If the same furnace was four passes, there would be four flow paths of 8 tubes each, etc.

(3) Inlet tube

The term "inlet tube" designates the first tube through which the fluid passes after entering the furnace. A two pass furnace would have two inlet tubes, a four pass furnace, four inlet tubes, etc.

(4) Outlet tube

The term "outlet tube" designates the last tube through which the fluid flows before leaving the furnace.

(5) Radiant section

"Radiant Section" is that section of the furnace in which the surface of the tube is exposed to direct radiant heat from the burner flames. The majority of heat transfer in a tubular furnace is affected in this section.

(6) Convection section

The "convection section" is also sometimes referred to as the economizer section. The heat pick-up in this section embodies both non-luminous radiation as well as a convective heat transfer. Either bare tubes or tubes with extended surface, or a combination of both, may be present in this section.

(7) Bridge wall temperature

This temperature is the equilibrium gas temperature or the residual temperature of the gases after they have been cooled by pure radiation. This temperature is measured at the point where the products of combustion enter the convection section. In case of updraft heater design, this point is directly under an arch, if such is present, or at the top of the radiant section.

(8) Draft

The term "draft" refers to the difference of pressure which tends to push or pull the air, fuel and flue gases through the furnace. Draft is usually measured in mm unit of water column.

(9) Excess air

Combustion air supplied in excess of the quantity theoretically required for complete combustion. Excess air is normally denoted as percent of theoretically required air quantity.

(10) Efficiency

The ratio of heat absorbed by the fluid stream to the heat liberated by the fuel

(11) Lower heating value (LHV)

This is also sometimes referred to as Net Heating Value. It is obtained by subtracting the latent heat of vaporization of the water vapor formed by the combustion of the hydrocarbon in the fuel. For a fuel with no hydrogen, net and gross heating values are the same.

(12) Instrumentation

The principal operating variables of process heaters are the rate of feed, the inlet temperature of the charge, the outlet pressure, the pressure drop through the heater, and the quantity and quality of fuel required to maintain the outlet temperature. To properly control these variable parameters, certain instrumentation is necessary to effect proper operation.

10.2 General furnace operation

These instructions and/or precautions are not to be considered as the only procedure which should be followed in the start-up and shut-down of a process heater. All operators should be thoroughly familiar with and should follow exactly the recommendations of refinery safety committees or overall refinery recommendations. The normally accepted operating procedure of most refineries will result in satisfactory performance with horizontal or vertical furnaces. The following will list a series of operating hints which if followed, will allow optimum heater life and minimum maintenance.

(1) Flame condition

Burners must be adjusted so that no flame impingement on tubes is observed at any time. High heat concentration caused by flame impingement results in radically high levels of heat transfer. Even with low temperature hydrocarbons, coke formation inside the tubes may result in subsequent overheating of tubes. In the case of multi-burner installations, the fires should be kept as uniform in length and size as possible.

(2) Excess air

1) Regular check should be made of the flue gas to determine the excess air in the furnace. Normal operation is in the range of 15-20% excess air. If an oxygen analyzer is installed, a level of 3-4% measured in the flue gas is a reasonable rule of thumb.

2) Levels below 15% may be achieved with low excess air burners, using caution due to the possibility of carbon monoxide formation. Generally, it is recommended a

low excess air operation only if the heat load and the fuel calorific values on the unit are very constant.

(3) Overloading

Good operating practice must recognize that while excessive firing is at times necessary and the furnace design allows for a limited percentage of overload on a continuous basis, there is, nevertheless, a definite limit beyond which it is both uneconomical and unsafe to go. Overload capacity varies with the size and type of heater as well as the individual design. If continuous operation above design conditions is contemplated, it is suggested that the proposed operating conditions be forwarded to the heater maker for study so that absolute limitations on heater duty may be established or design changes recommended.

Much maintenance and lost time can be avoided by making adequate design revisions and substituting materials where necessary.

(4) Furnace pressure

1) Standard furnaces are not designed to operate under positive pressure and are generally designed for natural draft, and these instructions are based on this type of operation. In the event that special forced draft units have been provided, separate supplementary instructions are issued.

2) As a general rule, settings are designed so that a negative draft exists at the inlet to the convection section. Burners are specified for drafts predicted at burner level which may vary anywhere from 0.2 inch (5mm) and up. (High levels of draft are commonly found only in vertical cylindrical heaters having long radiant heater tubes).

3) In order to protect the furnace internals, it is important that the draft condition in a furnace be carefully monitored. Any tendency for a pressure to build up at the radiant arch should be cut to a minimum by keeping dampers and/or burner registers adjusted, consistent with good flame conditions.

4) In order to assist the operator in maintaining satisfactory draft conditions in the heater, draft gauges should be installed. The draft at the top of the radiant section should be monitored daily and become part of the permanent operating data of the furnace.

(5) Adequate process flow

1) Flow must be maintained to each heater pass at all times. Commonly, flow indicators and manual control valves are provided at the inlet of each pass on vaporizing heaters. The operator at all times must be watchful of low flow or temporary loss of flow to the furnace.

2) If discrepancies in individual pass temperatures are noted during operation and if it is established that equal flow is maintained to each pass, recommended practice is to adjust burners for pass temperature control. Adjustment of flow to each pass to maintain temperature control is not recommended.

(6) Flame impingement on tubes

1) Flame impingement on tube surfaces will greatly decrease tube life and may result in a serious tube failure.

2) The furnace should be checked for any signs of flame impingement at regular intervals and particularly after any change in load. In gas fired furnaces at high firing rates, the true flame patterns may not be visible to the naked eye. However, local hot spots on the furnace walls, especially if of varying intensity, are indicative of impingement by the non-luminous portion of the flame.

3) Misalignment of the burners, insufficient combustion air or enlargement and corrosion of the burner ports are frequently the cause of incorrect flame patters.

(7) Afterburning

1) Operation of the furnace with insufficient combustion air may result in incomplete combustion. As a result, afterburning may take place where ambient air leaks into the furnace due to the pressure differential. Afterburning in a furnace may not be visible from the outside and the only evidence will be a rapid increase in stack temperature. Permanent damage may result due to overheating of the heater internals and permanent damage to the structural steel.

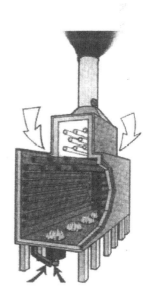

2) The immediate remedy is introduction of more combustion air by opening of the burner dampers in the following sequence: (1) cut firing rate; (2) open burner dampers.

3) If the additional air admitted is insufficient to halt burning, or if the dampers are already in the wide open position when the afterburning occurs, a substantial temporary cutback in firing rate must be made. As soon as the afterburning has ceased and the furnace has cleared itself of carbon monoxide, the firing rate may be increased again. Corrective steps in the case of afterburning must be immediate and drastic to avoid serious damage. Whenever afterburning becomes a problem, it is an indication that a program of more adequate combustion control and flue gas analysis is in order.

(8) Testing tubes

When hydrostatic testing the tube coil, it is necessary to expel all air from tubes. This is best done by circulating water at full main volumes through each pass until a full flow of water is noticed at each outlet. Allow water to run out of outlets to sewer for one-half hour, then shut off drain valves and bring up to test pressure with test pump. With stainless steel tubes, care must be taken to avoid chlorides in the test water.

10.3 Preparation for initial start-up

The first step to be taken is a careful physical inspection of the furnace in order to uncover deficiencies, which can cause problems that may require expensive modification after start-up. The major points to be checked are the following, to the extent that they pertain to a specific job. A piping and instrument diagram is helpful to determine the scope of supply.

[External heater inspection]
(1) Structural

1) Generally inspect the heater structure and verify that all field work has been completed. Critical areas are the main load transfer points, such as stack base, convection wall trusses, convection to radiant splices and radiant bases.

2) Verify that terminals are either free to expand or anchored per design drawings.

3) If heater has an air preheat system, verify that all ducting joints, expansion joints are made up properly and all temporary shipping clips and bracing removed so joints can operate freely.

4) Verify that all blinds and slide gate guillotines (isolation blinds) are operational and open.

(2) Stack & duct dampers

1) On units equipped with dampers, the full open and full closed position should be checked visually and marked on the damper linkage or operating cable. The movement of the damper must be free & unobstructed, and it is imperative that the full open position be in accordance with that shown on the drawings. Partial closing of the damper will materially affect furnace operation and may result in a great deal of difficulty in firing at higher rates. By looking inside the stack or duct, the damper position thus observed should then be checked against the outside indicator. Grooves are provided at each of the damper shafts parallel to the damper blade, to indicate damper position.

2) When dampers are controlled by pneumatic operators, verify that the operator/ positioner move the blades in the correct direction and from full open to full closed. The actuator/indicator should correctly indicate damper position. This should be done from the control room and at the local hand control station. All linkages should allow full stroke or rotation.

(3) Instrumentation

All flow controllers, recorders, thermocouples, draft gauges and safety shutdown devices should be checked prior to start-up. Flow recorders should be properly zeroed. Thermocouple leads transmitting outlet temperatures are sometimes reversed thereby making proper furnace adjustments impossible. Leads should therefore be carefully checked high temperature or low flow shutdown devices should be tested prior to start-up. Flame failure devices should be checked by simulating the hazardous condition.

(4) Blowing-out tubes

Prior to lighting-off, it is important to make certain that tubes are clear and unobstructed. Some water may remain in the tubes after the hydrostatic test, or other foreign matter may have accumulated. Compressed nitrogen/air should be forced through the tube coil at a high enough pressure and in sufficient quantity to assure satisfactory flow. In case of multi-pass units, each pass must be blown independently. The inlet valves to the other passes should be closed and nitrogen/air forced through one pass at a time. A poly pig can also be used to remove water and debris.

Construction records should be checked to determine that an adequate hydrostatic test has been applied to the tubes and that any specified inspection has been carried out. This would include field radiograph techniques or weld hardness checks.

(5) Plug headers
On all furnaces equipped with removable plug headers, it is advisable to remove the header box covers and make certain that all plugs are properly seated and dogs screwed down. The header box covers should be left off during the initial start-up to permit checking the header plugs for possible leaks. Once the furnace has reached its normal operating temperature, the covers may be reinstalled.

(6) Sampling points
All couplings and flange connections not used for draft or thermocouple connections should be sealed-off.

(7) Burners
1) Verify the burner position against the suppliers prints. Make sure the burner tips and/or spiders are placed correctly and that the connecting piping permits proper vertical alignment of the burner core and tip. In many cases, if the tips are incorrectly positioned, improper flame patterns may result with resultant damage to settings or tubes. Furthermore, in the case of oil firing, improper tip position can lead to speedy coke buildup on the burner blocks.

2) Make sure that all air shutters and dampers function freely and handles indicate correct position. It is important that the operators be able to make the necessary adjustments freely after the furnace is in operation.

3) All burner piping should be thoroughly blown-out with compressed nitrogen/air. This is of particular importance with oil burning equipment. Before any oil or steam is circulated, and with the burner tips and mixing orifices disconnected, the burner piping should be blown-out thoroughly. Both steam and oil lines must be treated this way. On multi-burner units fed from a single header, each burner should be blown independently. Control valves for all burners except one should be closed and full pressure and flow applied to each burner in turn. Particles of scale and rust interfere with the proper functioning of the atomizing tips and result in long

flames, dripping and coking. Failure to remove mill scale and rust in this manner prior to start-up can result in severe operating difficulties.

4) If a fuel oil system is installed, fuel oil should be circulated past the heater for a period long enough to bring the fuel oil to the burner manufacturer's recommended temperature.

5) If wet process fuel gas is to be burned, all knockout drums should be blown-down so that no condensate is admitted to the fuel gas burner.

6) Fuel gas lines should be pressure tested for tightness. Gas lines must be purged prior to light-off.

(8) Soot blowers

1) Soot blowers may be either fixed rotary or retractable types. They are usually motor driven and supplied with a grade mounted control panel. Verify that soot blowers are installed per supplier's installation manual.

2) All soot blowers should be dry cycled from local station and panel before start-up.

(9) Air preheaters

There are various types of air preheaters available such as regenerative or recuperative. Where applicable, the supplier's installation manual shall be checked to verify proper installation. All joints should be gas and air tight.

(10) Electrostatic precipitators

1) This equipment is very specialized and the supplier's instructions shall be checked to verify installations.

(11) Fans

1) Heaters may be supplied with either forced draft, induced draft or a combination of fans.

2) Fan dampers shall be verified similar to the requirements. Check blade angle against vendor drawing.

3) Fan bearings shall be checked for compliance with the manufacturer's drawing. When required, cooling oil or water supply shall be operational. Check fan bearings

during run in for overheating.

4) In general, if you close the damper on a fan suction, it unloads the fan and the motor load (amps) goes down. If you close a damper on a fan discharge, it loads the fan, the motor load (amps) goes up and depending on the fan curve, the fan can begin to surge. If surging occurs, check for blockage on the discharge side. If high amps are present when staring fans, check if: ① Voltage is correct at motor ② Inlet damper is closed ③ Damper blades in rotation direction

5) Verify the rotation of all fans by bumping the fan. This is done by pressing the start button and then immediately pressing the stop button. The fan should always rotate toward the discharge side.

6) After rotational check, run fans at speed and check for shaft vibration. This should be done by closing the suction dampers and running the fans up unloaded. Any vibration outside that allowed should be reported for correction.

[Internal heater inspection]
(1) Pressure parts (coils)
1) Verify that all temporary shipping braces, wires and blocking have been removed. Check both radiant and convection ends in header boxes.

2) Verify that all tube supports and guides are properly installed.

3) Verify that sufficient clearance is available to accommodate coil expansions.

(2) Refractories
1) After erection, all refractory areas will not be accessible for inspection. This should have been done prior to erection. Where possible limited visual inspection may be done through stack access doors and openings in the convection walls.

2) Verify integrity of the radiant refractory, make sure all construction joints are packed and that there is no exposed casing or structure.

(3) Burners
1) Inspect each burner to verify all pilot, gas and oil tips are clean, that plastic protection covers and tape protection covers are removed and tips are in the proper position and orientation per burner drawings.

2) Verify that burner tiles are installed properly and are not broken. All burner components must be free of construction debris. Space between the tile and floor refractory must be sealed with ceramic fiber unless burner manufacturer requires that this space be left open. Consult burner drawing.

(4) General

1) All construction materials such as unnecessary scaffolding, ladders, tools, insulating materials, pipe spool drops, structural steel drops, wood forms and boards are to be removed for the operator's safe movement about the furnaces.

2) After final inspection, all manways and access doors may be closed.

[Furnace dry-out procedure]
(1) General

1) Curing the castable begins after it has taken its initial set. The term "curing" as applied to a castable material means either keeping the material wet after installation or keeping the surrounding atmosphere humid. Usually a 24 hour curing period should develop maximum strength. This allows the hydraulic cement bond to react chemically with the mixing water to form a strong structure. The lining may also be cured by the application of a membrane forming compound. When curing by this method, water spraying will not be required.

2) Castable strength will be reduced if wet mixture is exposed to freezing temperatures. After placement, the cast material should be protected against freezing for a minimum of 48 hours.

3) In extremely hot weather, the strength of the finished product will be improved when the fresh castable is kept cooled by water spray or other means during the 24 hours after it is placed. Spray should not be applied until the material has taken its initial set. At least 24 hours should be allowed for the castable to cure before starting to dry out. Allow more time when the temperature is low.

4) Usually air drying is used for about 48 hours before lighting the burner. Air drying of the castable after curing will result in slightly increased strength.

5) When using castables, there is an advantage in curing them at a temperature over 21℃ because they develop a more permeable structure in this temperature range than when the temperature is below 21℃.

6) Following curing and air dry-out, the castable may be heated immediately. However, in most cases, heat drying will not start for some time after curing and air dry-out (consider time from shop installation, shipping, erection, etc.). Consideration should be given to the use of a sealer to protect castable from the element.

(2) Preliminary preparation for dry-out

1) Verify that all external and internal inspections specified have been completed.

2) Verify that proper fire fighting equipment and fire extinguishers are present and operational.

3) All personnel not associated with the dry-out should keep free of the heater area. All hot work should be stopped in the heater area.

4) It is the responsibility of the owners, contractor, sub-contractors and all parties involved, to assure that any and all personnel involved in the operation of fired equipment have read and understand the appropriate manuals and a familiar with the procedures to take place.

5) Either fuel gas or fuel oil may be used for dry-out. However, it is often difficult to operate with heavy fuel oil because of the low firing rates which are required on each burner during the dry-out cycle. With such low rates, fuel oil temperature frequently becomes too low for adequate atomization, thus leading to smoky fires and coke build-up on the burner blocks. However, if adequate fuel preheat can be maintained by a recirculation system, the difficulty with heavy fuel oil during dry-out should be minimized.

6) Blow-down condensate drains on atomizing and steam snuffing lines. Check damper for open position and remove fuel blinds which protected the fuel lines.

7) With respect to instruments only a flue gas thermocouple is required. Use a location at the top of the radiant section of the heater to indicate the firing rate required for the drying cycles.

8) With rare exceptions, it is necessary to circulate some cooling fluid through the tubes so that no overheating will be experienced. Usually steam is used for this purpose, but in some cases air or stable hydrocarbon can also be used. If stream is used for cooling purpose, a small fire should be established in the furnace to warm the tubes and setting before introducing steam into cold tubes. This would prevent water hammer from condensate. Provision to vent steam should be available. The

flow of cooling fluid, especially through the convection section should not be too large, or flue gas will cool and down stream refractory will not dry.

9) Purge the heater of combustible gases before lighting burners by introducing snuffing steam to the combustion chamber. Allow steam to blow at least 15 minutes. If system has fans, purge with fans.

10) After completion of purge, light the first burner in accordance with the burner instruction. If steam is used for tube cooling, delay changing steam to the heater coils until several burners are lit and the flue gas thermocouple registers about 150℃. By this time, sufficient heat is contained in the tubes to prevent steam condensation.

Warning: burner light-off for dry-out may be hazardous.

(3) Start of drying operation

1) In general, three cycles of operation are required to obtain a completely dry insulation. The length of each cycle depends on the type and thickness of the insulating materials which are enclosed by the casing of the furnace. The burners are operated with a large amount of excess air in order to maintain low flame burst temperature and to carry the generated moisture from the combustion chamber to the atmosphere. Stack damper, if any is kept wide open.

2) With all burner registers wide open, burners are lighted and fired at a rate of 20℃/hr to 120℃ in order that a bridge wall temperature of 120℃ at the top of the radiant section is maintained. Hold this temperature for the first 24 hours.

3) During the second day cycle, firing rates may be increased so as to produce a temperature of 300℃. During the third day, a maximum temperature at the bridge wall of 538℃ is recommnded.

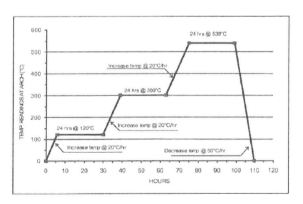

4) These temperatures ae recommended levels, and are subject to tolerance of ± 40℃. Normally, refractories will then be dry enough to permit placing the furnace in operation. However, the most accurate indication of readiness of a furnace for operation is the shell temperature. When the fires are started in a new furnace, the moisture of the outer surface of the refractory evaporates into the combustion gases but much of the moisture held deeper in the refractory is driven back into the insulation next to the shell. During this period, the furnace shell remains cold. Then, as heat density in the combustion chamber is increased, the lining heats through and the moisture from the insulation begins to evaporate. This causes the furnace shell to heat up by the generation of steam on the adjacent insulation. However, when this moisture has been thoroughly evaporated, the shell temperature will again fall. If the operator keeps some progressive account of the shell temperature, he will note this rise and fall during the drying out process and when the shell temperature has decreased to the rated casing temperature, the furnace can be considered ready for operation.

5) Definite requirements for drying out time vary with the type of insulation and the moisture content of the insulation material which, in turn, is affected by the length of time the furnace may have been idle between completion of erection and beginning of operation. The longer a furnace is inoperative, the more moisture is taken up by the refractory and insulation due to atmospheric exposure.

6) As a general rule, a three day dry-out cycle should be followed, during which burners are operated so as to increase gradually the temperature of the insulation. At the end of three days, if proper precautions are taken, furnace refractory should be completely dry. Some recommended drying-out periods are listed in table below.

	Total drying out time, hrs	Time of temperature cycle, hrs
4-1/2″ 1,260℃ IFB, 3″ block ins.	60	20
4-1/2″ 1,093℃ IFB, 2″ block ins.	36	12
4-1/2″ 1,093℃ IFB	30	10
5″ insulating castable	72	24
Ceramic fiber	None	-

7) There are two reasons why castables should not be heated too rapidly the first time. One is that they have a considerably lower permeability than the comparable brick or plastic refractories and it is more difficult for water

to pass through the castable and escape as the refractory is heated. If the moisture is heated too rapidly, high pressure steam will develop within the body of the material which may rupture the new castable lining and ruin it.

8) Another reason for not heating castables too rapidly the first time is that it can cause a pattern of cracks which extend part way through the lining. A rapid heating schedule causes the hot face to dry out and heat up while the rest of the castable is kept cool by the evaporating of the moisture. The hot face expands causing cracks to develop.

9) For dual layer castable lining (dense and less dense materials)- For drying out the double layer castable insulation with various density layers, types and thicknesses, the time required is dependent on the density factors, climatic conditions and type of fuel used. The drying develops a ceramic bond on the inside layer surface and continuous rise of firing rate is required for completion of drying operation. The rate of firing should not exceed 40°C per hour and the third cycle in the 400°C range is extended until the shell temperature registers the drop well below the 100°C water boiling temperature. It is recommended that the operator record the shell temperatures in order to properly complete the operation.

10) For dual laver lining (castable and block insulation), the block insulation has a variety of mineral wool products, all of which result in greater insulating characteristics. The care necessary in drying out this type of construction is similar to that of dual layer castable materials although the overall time required can be somewhat reduced.

11) After the initial dry-out, it is good practice to shut down, cool down and inspect the lining for damage. After making repairs and properly curing, the unit may be started.

12) After the unit has been dried out, it is important to prevent entrance of excessive moisture into the furnace. Close tightly the stack damper, access and peep doors and all burner air registers. If the heater is to be shut-down for an extended period before initial start-up and if the stack has no rain hood, provide a canvas stack cover to eliminate rain water infiltration.

13) Frequently, furnaces will contain two types of linings. For example, side walls lined with insulating ceramic fiber, but the furnace arch or floor constructed of castable. In these cases, select the longer dry-out period from the table above.

14) Warning- when heater includes an air preheat system care must be taken not to exceed the design temperature of downstream equipment (air preheater, precipitators, ID fans, etc). Flue gas entering the glass section of an air preheater should not exceed 205℃.

10.4 Furnace start-up for natural draft

With the dry-out completed above, the actual furnace start-up can proceed. The following outline will give the proper sequence for this operation.

(1) Preliminary preparation

1) Verify that all fuel valves are closed.

2) Verify that the burner primary, secondary and tertiary air registers, as applicable, are 100% open.

3) Verify that the stack damper is open.

4) As gas is brought into the area on gas fired units, leak test of valves and blinds for combustibles.

5) Warning- purge heater firebox adequately prior to pilot light-off. If possible, check radiant flue gas outlet for combustibles with an explosion meter. A minimum of 4 volume changes is recommended.

(2) Start flow to furnace

1) The circulation rate should be as close to design as possible. In addition, on multi-pass coils, the operator must be sure that circulation exists in all passes by checking their flow indicators or controllers. If there is a by-pass around the furnace, it must be sure that the by-pass valves are closed. Flow distribution between the passes should be adjusted as closely as possible before the burners are ignited.

2) Establish adequate charge circulation by opening fully any valves in the charge lines leading to and from the heater.

3) The above procedure is particularly important in the case of all liquid heaters, or heaters where partial vaporization is accomplished in the heater coil. With low flow in one or more passes, excessive vaporization may then take place in that pass. This promotes high pressure drop in that pass which further tends to limit flow.

(3) Placing firing control on manual operation

1) All firing during start-up must be done on manual control. Upsets in various areas of a plant frequently occur during these periods and with the fuel feed system on automatic control, any upset which causes the temperature of the fluid passing through the heater to drop suddenly will actuate the automatic fireman. The fuel valve will then open wide and permit severe overfiring. A similar condition will result if the charging rate is suddenly increased. Sometimes, a furnace is lined out satisfactorily at flow rates considerably below design. If the feed rate is stepped up rapidly on automatic fire control, overfiring will again be the result.

(4) Light-off burners and hold on low fire

1) With flow established, light burners following the burner manufacturer's instruction. It is important here to point out that as many burners as practical should be fired, thus promoting even heat transfer to the tubes even during times of low heat release. The tendency of lighting one or two burners and firing these units to the limit of capacity before cutting in additional burners, should be discouraged. This practice can impose a severe thermal shock on those sections of the heater affected by the burners. Therefore, maintain ignition on as many burners as practical during start-up.

2) Keep a fireman stationed at or near the heater during start-up. Frequently, during low heat load operation, burners may be extinguished without warning.

3) Partially close the stack damper to reduce draft at the burner. Air registers and dampers may be partially closed for the same reason.

4) Commence lighting of burner pilots

① As soon as pilot fuel flow is started, the pilot gas pressure regulators should be re-adjusted to the specified pressures.

② As each pilot gas burner is ignited, the operator should adjust the air door (wheel) in the pilot burner mixer to attain good flame color. Operators observing from the side observation ports will be able to assist in confirming pilot flame stability, size and color.

③ All pilots should be ignited prior to main burner ignition.

5) Commence lighting of main burners

① Increase burner firing slowly and maintain proper fuel pressure at the burner.

② Adjust burner air registers to obtain good flame color and pattern.

(5) Check outlet temperature on each pass

① Immediately after lighting-off, check the outlet temperature of each pass by noting the thermocouple readings. In some cases where inlet temperatures are low, a more rapid and purely qualitative check can be made by placing the hand very briefly near an outlet flange to determine any tendency toward a temperature increase. If any outlet tubes remain cold, the fire must be immediately extinguished until the reasons for the blocked pass are determined.

② Experience has shown a wide variety of causes for flow obstructions. On gas heaters or steam superheaters with a low pressure drop, accumulations of condensate or water remaining from a hydrostatic test frequently block one of the passes. On new furnaces, blinds are sometime left on flanges. Frequently, waste material, rags, etc., are carried through the system and deposited at the inlet valve or in a return bend.

(6) Bring fire up gradually once flow has been established

The circulation of the feed stock through the furnace should be brought up to design as rapidly as possible. However, to prevent overfiring and overheating of the fluid streams the firing rate should be brought up gradually. During this period, heat must be supplied both to the charge stock and the cold setting. An attempt to reach design outlet temperatures too quickly can result in overfiring. In furnaces where the fluid stream is partly vaporized while passing through the tube coils, some surging between passes frequently occurs as the temperature of the stream is brought into the vaporizing range. The outlet temperatures of the passes should be watched carefully and the various fluid streams trimmed accordingly. If one pass overheats or vapor locks, the flow to this pass should be temporarily increased until normal flow is restored. As charging rate and temperature approach design conditions, the pressure drop through the furnace will increase and the unit tend to the line-out. Bringing up the firing rate too quickly does not permit making the necessary intermediate adjustments and may make it impossible to line-out the furnace at full rate. Depending on the type of process and the furnace design involved, the time required to reach design outlet conditions will vary. The generally recommended increase in outlet temperature of the feed stock during start-up of a furnace of standard design is 38℃/hr maximum.

(7) When outlet temperatures reach normal operating level, switch to automatic control

When the furnace has been brought up to expected operating level, the swing to automatic control may be attempted. In some cases, it is advisable to line-out the furnace on hand control before switching, particularly on new units where the automatic firing control may not have been completely adjusted. On a routine start-up, however, where the instrumentation is reliable, considerable time may be saved by making the swing from manual to automatic control as the outlet temperature approaches the desired level. Under such circumstances, the automatic controller may stabilize furnace outlets more quickly than would be the case on hand control.

10.5 Furnace shut-down

Furnace shut-down procedure varies with the type of furnace and the reason for shutting down. High temperature furnaces or units operating under heavy overloads must be brought down more slowly than standard units operating at design. The following rules will serve as a general guide. In case of doubt, detailed instructions may be obtained from the engineering department of heater maker.

(1) Normal shut-down

On normal shut-down, the reverse procedure of start-up operation is followed. All control is shifted to the manual position and the firing rate per burner gradually decreased. Retard fires and lower the outlet temperature to a point 10℃ below normal operating temperature. When outlet temperature has dropped to 38℃ below normal operating temperature, all burner fires may be extinguished with charge circulation continued until the heater cools to 93℃ below normal outlet temperature. At this point, shut-off circulation through tubes and introduce steam into tubes via steaming-out line. Continue steaming-out for at least one hour, then check outlet drain for oil.

1) With high temperature furnaces (above 427℃ outlet temperature), temperature must not be above 371℃ when steam is introduced to the tubes.

2) With some heavy oils, or in case of deposits on the inside of tubes, the furnace can be fired during the steaming-out period until the steam reaches a temperature of 427℃ and then cooled down. This will insure that the tubes are completely dry under all conditions.

(2) Emergency shut-down

If a shut-down is necessitated due to emergency other than furnace trouble, shut-off all burners and shut-off fluid circulation, introducing steam to tubes simultaneously. Continue steaming-out until outlet temperature drops to steam temperature.

(3) Tube failure shut-down

In case of rupture, shut-off fires, and stop circulation, introducing steam to tubes at the same time. Open snuffing steam to firebox to extinguish fire and cool setting.

The above shut-down is for natural draft operation. Shut-downs that occur during force draft and air preheat operation should follow a similar procedures also.

11. Commercial Fired Heater Efficiency Test Run

In order to compare efficiency between 54% and 10% gas firing, a test run was accomplished for Naphtha/Kero Hydro Treating Unit (Hydrobon) heater having CO analyzer of laser type. The heater is a horizontal cabin type of natural draft. And Burner is total 8 burners and they are oil/gas duel firing at end walls. Before test, O_2 analyzer is calibrated and oil gun had steam purging. The test was implemented at the same operation condition as possible as can, for getting more exact evaluation.

11.1 Heater design data

Feed rate	Heat absorption	Coil Inlet temp	Coil outlet temp	Excess O_2	Draft	Radiant flue gas temp	Convection flue gas temp	Efficiency
kl/hr	MMkcal/hr	℃	℃	%	mmH$_2$O	℃	℃	%
125.9	3.3	260	309	5.0	-2.5	740	400	76.7

Note: Excess O_2 is adjusted with air register/stack damper and Gas/oil firing ratio is tuned by cock valve and control valve.

11.2 Heater operation condition during test period

Gas firing ratio	HSR feed	Kero feed	Total feed	Kero	Coil Inlet temp	Coil outlet temp	Fuel gas component	Crude switch	Others
	kl/hr	kl/hr	kl/hr	%	℃	℃	wt%	vol%	–
Case 1 54%	52	23	75	31	264	316	(Note 1)	(Note 2)	(Note 3)
Case 2 10%	50	25	75	33	264	316			

(Note 1) Case1: 44.0%H$_2$/14.9%C1/23.3%C2/9.3%C3/2.0%C4/0.5%C5/0.2%C6$^+$
Case2: 40.8%H$_2$/14.7%C1/22.3%C2/9.4%C3/2.6%C4/0.3%C5/0.3%C6$^+$
(Note 2) Case1: Champion 40%/ Ras gas C. 36%/ Doba 15%/ Duri 7%/ Dolphin 1%
Case2: L.Zakum 64%/ Murban 11%/ Forties Blend 10%/ Ras gas C. 8%/ AXL 4%
(Note 3) Feed ratio change of Kero to HSR: 23 vs 52 to 29 vs 46kl/hr

11.3 The result of test run

At the same operation condition such as CO, CIT, COT, and feed rate, the 10% gas firing (90% oil) case showed the higher efficiency by 1% than that of 54% gas firing case for given heater. The results of test are shown as below.

Gas Firing ratio	CO	Excess O_2	Radiant flue gas temp	Convection flue gas temp	Radiant efficiency	Convection efficiency	Total efficiency
	ppm	%	℃	℃	%	%	%
Case 1 54%	3.7	2.7	795	508	56.3	16.1	72.4
Case 2 10%	3.7	3.0	785	487	56.7	16.7	73.4
Difference	0	-0.3	10	21	-0.4	-0.6	-1.0

Note: Analyzing data using 1minute average value is grouped to the same feed rate, is classified to the same gas firing ratio again, and lastly only data with the same CO value was selected for detailed evaluation.

11.4 Heater TI's points & analyzer position

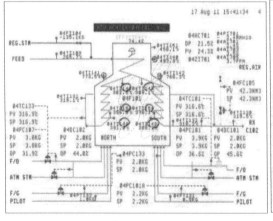

TI's points

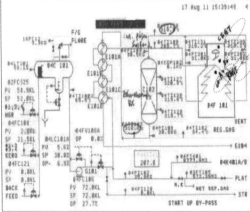

Analyzer position

11.5 The trend of main operation variables

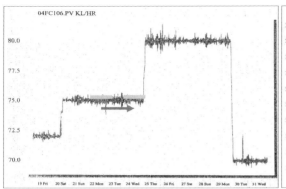

HSR feed rate (kl/hr)

Coil inlet temperature (℃)

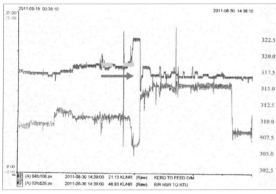

Kero feed rate (kl/hr)

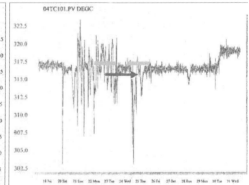

Coil outlet temperature (℃)

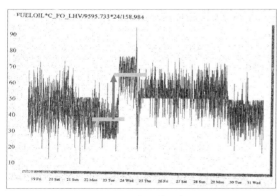

Fuel oil (Bbl/day)

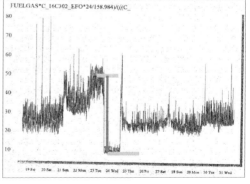

Fuel gas ratio (%)

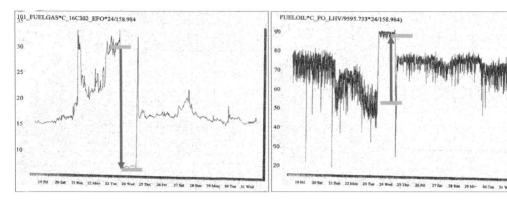

Fuel gas (Bbl/day)   Fuel oil ratio (%)

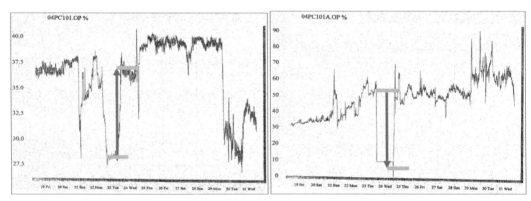

Fuel oil control valve opening (%)   Fuel gas control valve opening (%)

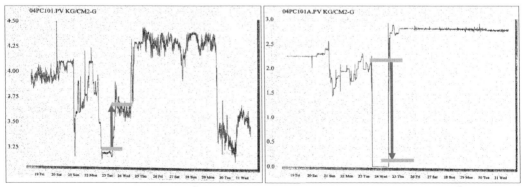

Fuel oil pressure (kg/cm^2)   Fuel gas pressure (kg/cm^2)

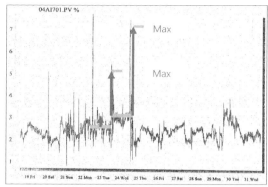

Excess O$_2$ (%)   Draft (mmH$_2$O)

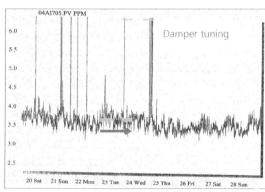 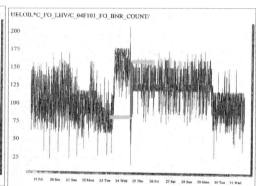

CO emission (ppm)   Oil burner load (%)

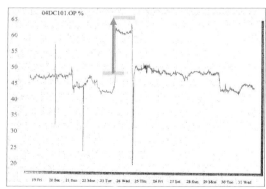 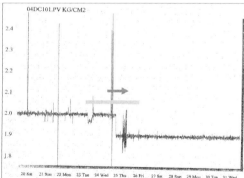

Atomizing steam DC op. (%)   Atomizing steam DP (kg/cm^2)

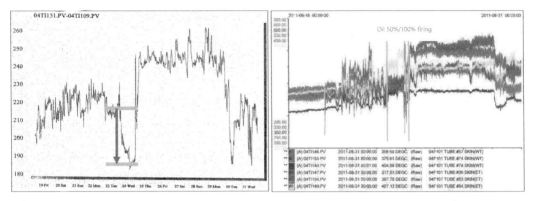

Cold end approach temperature (℃)   Tube skin temperature (℃)

11.6 The monitoring of operation data

Fuel Color — 2011-08-23 11:16

Section	Description	04F 101 EAST	04F 101 WEST	04F 102 STRIP	04F 102 SPLIT
In/Out	Feed Rate. KL/Hr		75.0	110.1	110.5
COT(Delta) Temp.	Feed Rate. B/D		11,324.7	16,615.2	16,676.2
3.2	H2 Recycle Gas Flow. NM3/Hr		6,830.4	-	-
Rad. Out #1/2 Coil Pt	Inlet Temperature. ℃	262.5	262.5	201.0	160.3
30.9	Outlet Temperature. ℃	318.2	315.0	216.4	166.6
30.9	In/Outlet Delta Temp.℃	55.6	52.4	15.3	6.4
Pass Flow	Reboiler #1 Pass	-	-	55.0	55.0
	Reboiler #2 Pass	-	-	55.0	55.5
Furnace	Draft. mmH2O	-3.2	-3.2	-1.7	-1.7
	Actual Excess O2. %/CO	4.9	3.8	2.5	
	측정 Excess O2 가이드	1.8			
	Design Excess O2 Guide.%	2.6	2.6	2.2	2.2
	Damper Opening.%	32.3	32.3	25.9	25.9
	Rad. Out FGT(Arch).℃	788.0	779.5	661.4	636.3
	Rad. Out AVRG FGT(Arch).℃	783.7	783.7	648.9	648.9
	Conv. Out FGT ℃	510.0	510.0	447.3	447.3
Efficiency	Cold-End Approach.℃	223.6	223.6	266.7	266.7
	Radiant Efficiency. %	55.6	55.6	63.3	63.3
	Convection Efficiency. %	72.2	72.2	76.0	76.0
	Conv.- Rad.= Delta Eff. %	16.7	16.7	12.2	10.8
Fuel Supply	Fuel Oil Press. KG/㎠g	3.6	3.6	5.0	4.8
	Fuel Gas Press. KG/㎠g	2.1	2.1	2.6	2.0
	Atom. STM DP. KG/㎠g	2.0	2.0	2.4	1.5
	Fuel Oil CV Open %	33.1	29.9	16.5	38.5
	Fuel Gas CV Open %	54.9	75.8	36.7	12.4
	Atom. STM DPC OP	44.6	44.6	78.3	47.6
Steam	Steam/Oil Ratio. Kg/h	0.00	0.00	9.20	9.20
Consumption	Steam Consump. Kg/h	147.5	147.6	188.2	62.6

Section	Description	04F 101	04F 102
Fuel	Fuel Oil . LTR/Hr	337.7	52.6
Consumption	Fuel Oil. B/D	51.0	7.9
	Fuel Gas.NM3/Hr	224.6	170.4
	Fuel Gas.B/D	32.6	24.7
	HTR Total Fuel. B/D	83.5	32.6
	HTU Fuel Consump.	116.2	B/D
H-Release	Fuel Oil	3.2	0.5
MM Kcal/hr	Fuel Gas	2.1	1.6
Fuel Ratio.%	Fuel Oil	57.3	17.8
	Fuel Gas	42.7	82.2
HTR Load	Design Firing Duty.MMKcal/hr	5.5	
	HTR H-Release	5.3	2.1
	Design H.R vs HTR Load %	96.2	21.3
	Design A. Duty.MMKcal/h	3.65	
	Absorbed Duty.MMKcal/h	3.8	1.6
	Design A.Duty vs HTR Load %	105.1	38.1
Burner Status	Burner Numbers		
	Fuel Oil Firing	1	9
	Fuel Gas Firing	4	4
	Fuel Oil Firing Ratio%	50.0	56.3
	Fuel Gas Firing Ratio%	50.0	25.0
Burner	H-R/BNR LHV MMKcal/hr(Oil)	0.81	0.06
Performance	Oil Burner Load %	117.4	7.4
	H-R/BNR LHV MMKcal/hr(Gas)	0.52	0.39
	Gas Burner Load %	75.0	51.6
	오일vs. 가스 버너의 Load 분자.%	68.1	44.3
	버너 평균 LOAD.%	96.2	21.0

Tube Skin Temperature.℃

04F101 Radiant Section	Tube No.	Temp. ℃	04F102 Radiant Section	Tube No.	Temp. ℃
West	#57	304.8	Stripper	#20	251.7
	#73	360.6		#20	246.1
	#74	409.2	Splitter	#20	188.4
East	#36	313.7		#12	182.2
	#53	381.4		#20	202.8
	#54	415.0			

11.7 Analysis of test result

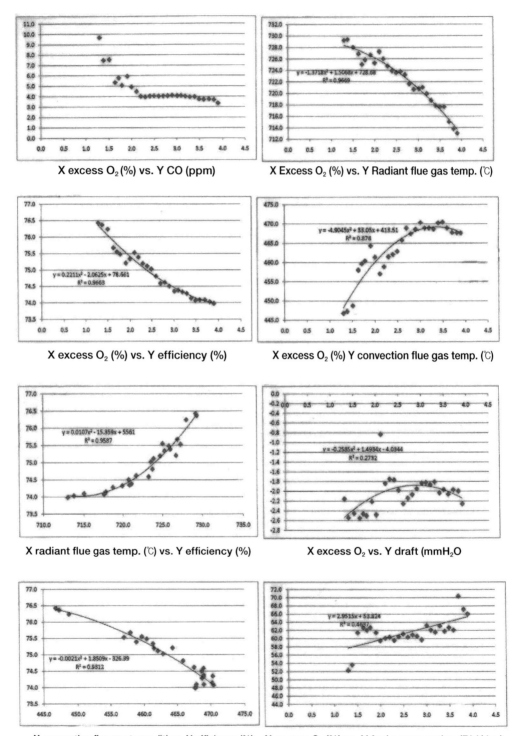

X excess O_2 (%) vs. Y CO (ppm)

X Excess O_2 (%) vs. Y Radiant flue gas temp. (℃)

X excess O_2 (%) vs. Y efficiency (%)

X excess O_2 (%) Y convection flue gas temp. (℃)

X radiant flue gas temp. (℃) vs. Y efficiency (%)

X excess O_2 vs. Y draft (mmH$_2$O

X convection flue gas temp. (℃) vs. Y efficiency (%)

X excess O_2 (%) vs. Y fuel consumption (Bbl/day)

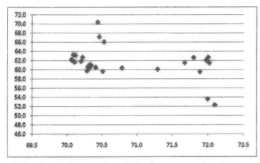

X feed rate (Bbl/day) vs. Y fuel consumption (Bbl/day)

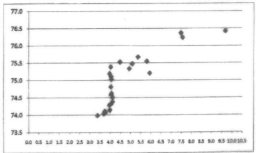

X CO (ppm) vs. Y efficiency (%)

12. Heater Operation Control

12.1 Process flow

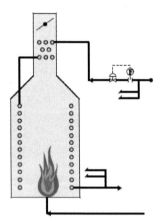

① Flow control to each pass in vaporizing service
② Low flow stops heater firing
③ Minimum stop on valve prevents no flow
④ Total flow rate used for heat balance
⑤ Flow control not required for all vapor applications

Flow control is reuired on all two phase flow systems. The flow enters the heater as a liquid and exits the heater as two phase liquid and vapor. The flow is set at design rate for all operations. A company puts minimum stops on the flow valve to prevent no flow hazop situations. The low flow is detected by the high temperature alarm on the outlet of each heater pass.

12.2 Process pressure

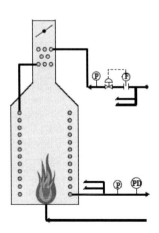

① Measured on individual passes to check flow distribution
② Measured on individual passes to check for coking and/or fouling

The pressure gauge downstream of the control valve is very important on black oil services such as crude, vacuum, coking, and visbreaking to measure pressure drop on each pass.

12.3 Process temperature

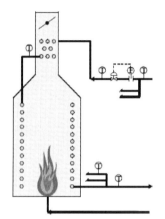

① Inlet temperature used for heat balance
② Outlet temperatures on each pass for flow distribution
③ Total flow outlet temperature used for fuel control

Temperature measurements at the crossover between the radiant section and convection section are only used when the process fluid is liquid (vacuum heaters, coker heaters). These measurements are useful in determining the amount of process heat absorbed in the convection section.

12.4 Combustion air flow

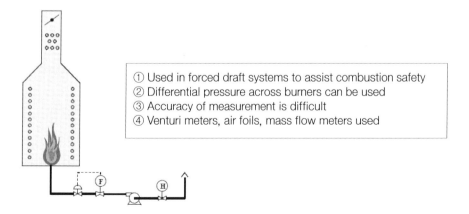

① Used in forced draft systems to assist combustion safety
② Differential pressure across burners can be used
③ Accuracy of measurement is difficult
④ Venturi meters, air foils, mass flow meters used

Venturi meters are more accurate than pilot tube measurements.

12.5 Pilot gas

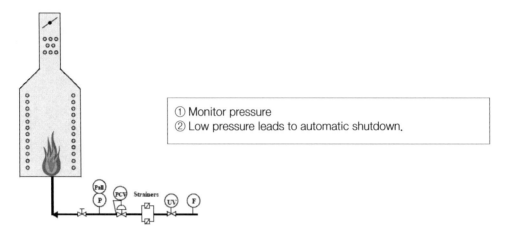

① Monitor pressure
② Low pressure leads to automatic shutdown.

The ST-1S pilot is a premix burner that is prone to flameout at fuel gas pressure below $0.42 kg_f/cm^2$.

12.6 Fuel gas

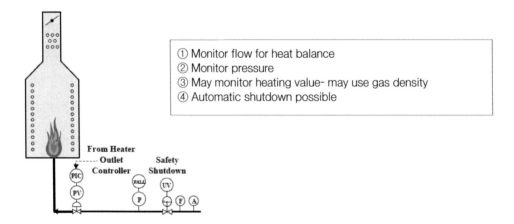

① Monitor flow for heat balance
② Monitor pressure
③ May monitor heating value- may use gas density
④ Automatic shutdown possible

A refinery has added density compensation to all fuel gas systems. The cost is only $1,500 for the density compensation. Any refinery with a coking unit of FCC can experience variations in heating value. The density compensation is an inexpensive way of handling these variations. Density compensation is a refinery practice on all fuel gas systems.

12.7 Fuel oil

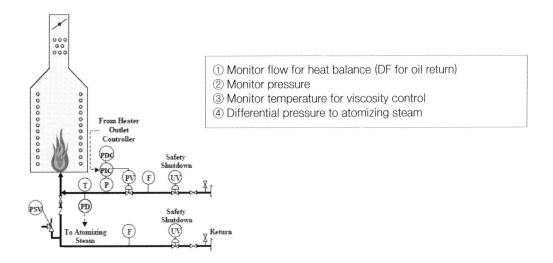

① Monitor flow for heat balance (DF for oil return)
② Monitor pressure
③ Monitor temperature for viscosity control
④ Differential pressure to atomizing steam

Coriolis mass flow meters have proven reliable in fuel oil systems. Each heater should have a local temperature gauge on the fuel oil line.

12.8 Atomizing steam

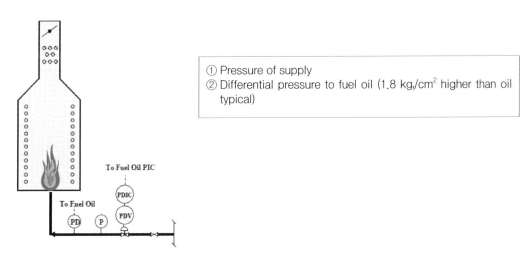

① Pressure of supply
② Differential pressure to fuel oil (1.8 kg_f/cm^2 higher than oil typical)

Higher steam differentials can often improve flame quality.

12.9 Fuel input control (feed forward system)

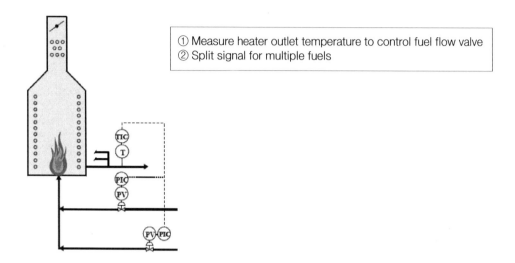

① Measure heater outlet temperature to control fuel flow valve
② Split signal for multiple fuels

A refinery uses a feed forward system to calculate the heat input into the heater. Each fuel is metered. The flow rate is converted into a heat input. The air rate is set by the heat input.

12.10 Fuel input control (lead lag system)

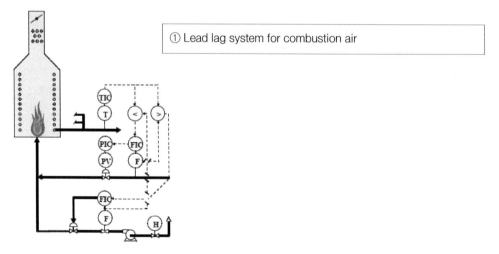

① Lead lag system for combustion air

A refinery uses a feed-forward system to calculate the heat input into the heater. Each fuel is metered. The flow rate is converted into a heat input. The air rate is set by the heat input. This system cuts fuel first before lowering the air rate. It raises air first, before fuel is increased.

12.11 Draft

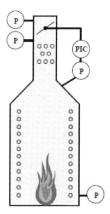

① Measure draft at burners, bridge wall, top convection, below damper and stack
② Use stack damper to control draft at bridge wall.

A best practice is to measure all the heater drafts automatically and record them on the DCS. The stack damper is pressure controlled when there is a forced draft fan. A refinery specifies HIC actuators on large natural draft heaters.

12.12 Flue gas monitoring

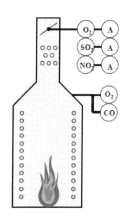

① Depend on local requirements
② Manual vs. automatic monitoring
③ Oxygen monitoring at bridge wall for control, stack for emission monitoring
④ Combustibles monitoring at bridge wall
⑤ Emissions monitoring in stack
⑥ Oxygen/CO reset of fuel control can be done but analyzers are not reliable enough

Environmental authorities require that pollution monitoring analyzers be located in the heater stack.

12.13 Flue gas temperature

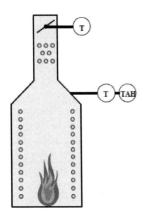

① Bridge wall temperature- velocity thermocouple better accuracy but costly and not normally used
② Stack temperature

A refinery uses a portable TESTO flue gas analyzer to check O_2 levels. This instrument has a velocity thermocouple which gives accurate readings on bridge wall temperatures. The stack temperature is required for efficiency calculations.

12.14 Fan controls

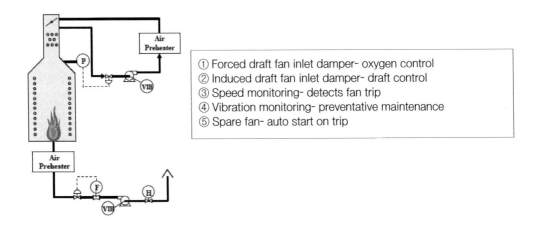

① Forced draft fan inlet damper- oxygen control
② Induced draft fan inlet damper- draft control
③ Speed monitoring- detects fan trip
④ Vibration monitoring- preventative maintenance
⑤ Spare fan- auto start on trip

Refer to API 556 for additional details.

12.15 Combustible control

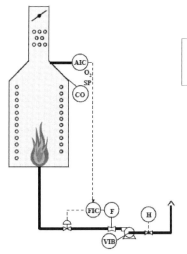

① Carbon monoxide trims fuel by minimizing oxygen set point. Typically used only on large boilers but few heaters

One large oil refinery is experimenting with the Bambeck system. This system uses a fast responding measurement of CO to adjust the air rate. The system has worked well at times, but there are many problems to work out.

12.16 Air Preheater instruments

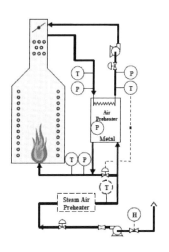

① Cold end temperature measurement
② Cold air bypass controls

Skin thermocouples are very effective on the air preheater to monitor acid dew point.

12.17 Steam system instruments

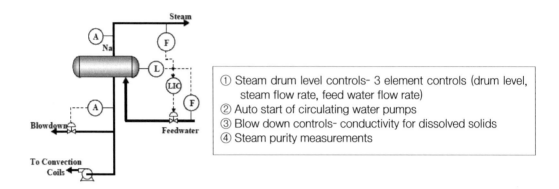

① Steam drum level controls- 3 element controls (drum level, steam flow rate, feed water flow rate)
② Auto start of circulating water pumps
③ Blow down controls- conductivity for dissolved solids
④ Steam purity measurements

Conductivity analyzers are used on high pressure steam systems.

12.18 Miscellaneous instruments/controls

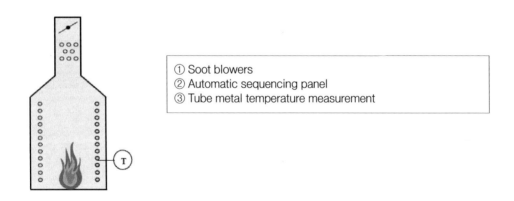

① Soot blowers
② Automatic sequencing panel
③ Tube metal temperature measurement

Dry steam is critical for good soot blower performance.

12.19 Burner management controls

① Safety of operations
② Flame scanners
③ Automatic start up sequencing
④ Prefer manual control during startups
⑤ Automatic trip systems possible

Many companies are installing burner management systems on new heaters to insure that the firebox has been purged properly before lighting. Statistics show that 90% of heater explosions occur during light-off. A refinery has committed to installing these systems on every heater in their system.

13. Control System

13.1 Heater fuel control system

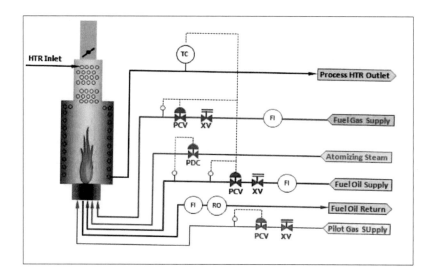

The coil outlet temperatures of all fired heaters are controlled with each pressure (PCV) of fuel oil or fuel gas. In the past, there is an experience that controlled the coil outlet temperature by fuel flow (FCV) at the Rheniformer stabilizer reboiler, but due to instability by flow meter error it was revised to the current pressure control in most fired heaters. In addition, the fuel oil return flow is controlled by fuel oil PCV opening without particular regulation.

In case of steam boiler, because of a bigger capacity as compared to fired heater burner, the steam load adjustment is accomplished with the flow (FCV) of fuel oil and fuel gas respectively.

13.2 Balanced draft control system

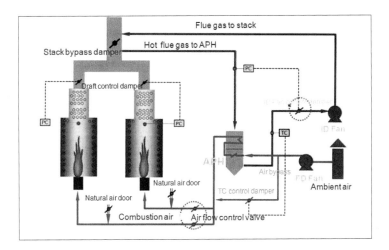

In case of FD Fan failure, individual damper should hold because furnace draft may fluctuate if the damper is open instantly. And if only ID Fan runs it will also have overheating. Therefore ID Fan stops, the stack bypass damper opens and the natural air door must open as well, which can cause the shortage of excess O_2. Meanwhile, when ID Fan fails, it is already made up with the logic which changes to natural draft mode automatically and then opens individual damper. This case also requires the holding time of the individual damper without abrupt opening to avoid the severe fluctuation of furnace draft.

13.3 Xylene column reboiler control system

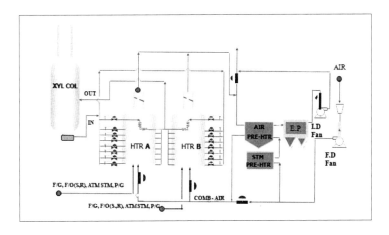

(1) Control logic and setting value

No.	Tag	Description	Alarm	Trip	Current value
1	N5HS-114	Emergency S/D Switch			
2	N5HS-109	F.D/I.D Fan Switch			
3	N5PC-148	Draft H,H – A	5,0mmH2O	30,0mmH2O	–2mmH2O
4	N5PC-159	Draft H,H – B	5,0mmH2O	30,0mmH2O	–2mmH2O
5	N5PI-140	Atomizing Steam L,L – A	4,5kg/cm2	2,0kg/cm2	7,2kg/cm2
6	N5PI-151	Atomizing Steam L,L – B	4,5kg/cm2	2,0kg/cm2	7,2kg/cm2
7	N5TI-151	Flue Gas Inlet H,H	370℃	410℃	355℃
8	N5TC-153	Flue Gas Outlet H,H	220℃	250℃	138℃
9	N5PI-161	Combustion Air L,L	65Nm3	15Nm3	155Nm3
10	N5FC-111A	Pass Flow L,L – A	48KI	36KI	80KI
11	N5FC-111B	Pass Flow L,L – A	48KI	36KI	80KI
12	N5FC-111C	Pass Flow L,L – A	48KI	36KI	80KI
13	N5FC-112A	Pass Flow L,L – B	48KI	36KI	80KI
14	N5FC-112B	Pass Flow L,L – B	48KI	36KI	80KI
15	N5FC-112C	Pass Flow L,L – B	48KI	36KI	80KI
16	N5PI-218	Xylene Column Vapor Pressur H,H	7,8kg/cm2	8,7kg/cm2	6,38kg/cm2
17	N5PI-220	Pilot Gas Pressure L,L		0,7kg/cm2	3,4kg/cm2

13.4 Depentanizer reboiler fuel control system

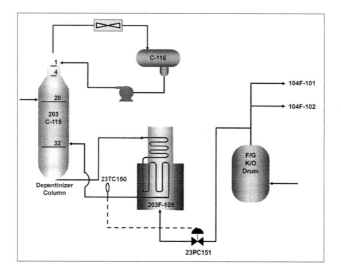

Cascade mode: this is a method that in feed-back control system, the output signal of one controller controls by varying the target of different controller. (Example) In Fig. above, by the setting of 23TC150 (auto), 23PC151 (cascade) is controlled consequently.

Auto mode: this is control to be executed automatically by controller. (Example) Operator adjusts the setting value of 23PC151 (auto).

Manual mode: directly or indirectly to be operated by operator (Example) operator directly to adjust the value of 23PC151 output

13.5 API standards and practices in gas fired heaters

13.5.1 Instrumentation: Temperature, pressure, flow and analytical

(1) Temperature

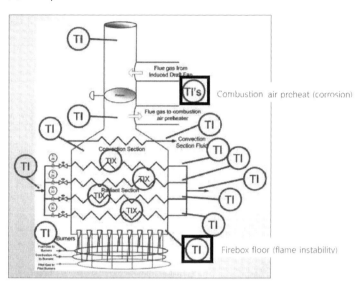

(2) Pressure

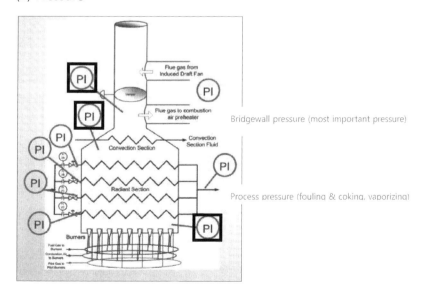

(3) Flow

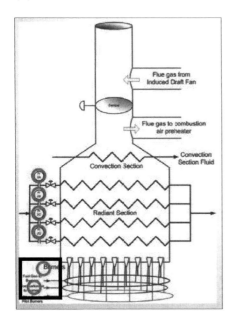

Fuel gas mass flow (mitigate effects of composition changes, coriolis for mass flow recommended)

(4) Analytical

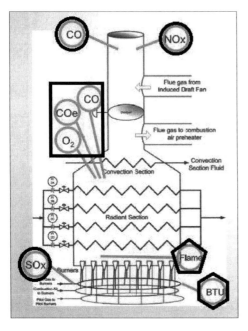

Emissions (local codes & standards, SO_x can be on fuel gas drum- sulfur content, CO for CEMS new infrared)

Oxygen & combustibles (one O_2 & one Coe per 10m of firebox length, tunable diode laser technology for CO added)

Flame sensors (installation consideration)

Heating value (varying amounts of inert compounds in fuel gas)

13.5.2 Controls: Heater firing, charge flow, air fuel ratio, firebox draft

(1) Heater firing

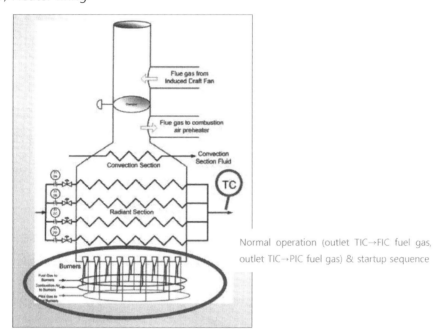

Normal operation (outlet TIC→FIC fuel gas, outlet TIC→PIC fuel gas) & startup sequence

(2) Charge flow

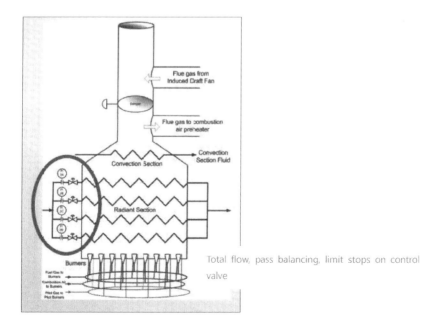

Total flow, pass balancing, limit stops on control valve

(3) Air fuel ratio

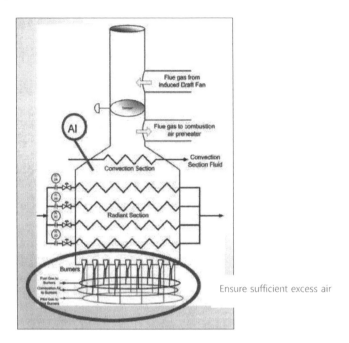

Ensure sufficient excess air

(4) Firebox draft

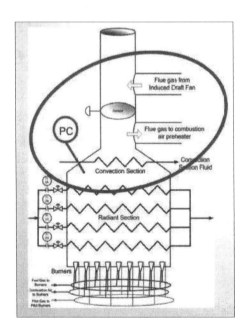

13.5.3 Protective systems

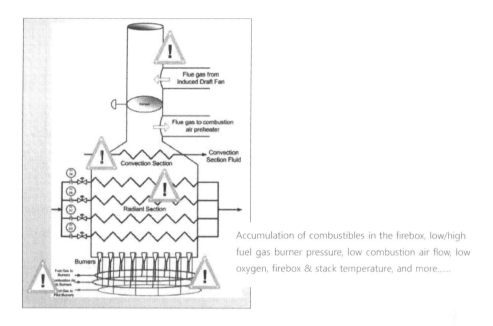

Accumulation of combustibles in the firebox, low/high fuel gas burner pressure, low combustion air flow, low oxygen, firebox & stack temperature, and more......

Typical alarms and shutdown initiators (fired heaters)

Description (see Note 1)	Typical Time Delay To Trip (seconds)	Close Pilot Gas Safety Shutoff Valves	Close Fuel Gas Safety Shutoff Valves	Close Fuel Oil Safety Shutoff Valves	Pre-Shut-Down Alarm	Section Reference
Manual trip	0	X	X	X		3.9.2
Low fuel gas burner pressure	0–4		X		X	3.9.3
Low fuel oil burner pressure	0–4			X	X	3.9.3
High fuel gas burner pressure	0–4		X		X	3.9.4
High fuel oil burner pressure	0–4			X	X	3.9.4
Low pilot gas pressure	0–4	X			X	3.9.5
Low fuel oil temperature					X	3.9.6
High stack temperature					X	3.9.7
Complete loss of flame (see Note 2)	0–4	X	X	X		3.9.8
Partial loss of flame	0–4	(see Note 3)	(see Note 3)	(see Note 3)	X	3.9.8
High heated-fluid temperature					X	3.9.9
High tube-skin temperature					X	3.9.10
Low/high draft pressures					X	3.9.11
High firebox pressure (see Note 2)	0–30	X	X	X	X	3.9.12
Low firebox pressure					X	3.9.13
High combustibles in stack					X	3.9.14
Low charge flow	0–15		X	X	X	3.9.15
Loss of motive power					X	3.9.16
Low atomizing steam pressure	0–30			X	X	3.9.17
Loss of induced-draft fan and/or preheater drive (see Note 2)		C	C	C	X	3.9.18
Loss of forced-draft fan		C	C	C	X	3.9.19
Low combustion air flow (see Note 2)	0–30	C	C	C	X	3.9.20

Notes:
C = conditional on design of fired heater (see Table 2).
Note 1: All shutdown functions shall be annunciated.
Note 2: May not apply to natural-draft heaters.
Note 3: See 3.9.8.

Typical action on fan or preheater failure

Heater Equipped with	Failure Type	Designed for Alternate-Draft Operations	Designed for Forced-Draft Operations Only
Forced-draft fan only	Loss of forced-draft fan	Open air doors (see Note 1) Reduce firing (see Note 2) Open stack damper	Shutdown
Induced-draft fan only	Loss of induced-draft fan	Shutdown (see Note 3) Open stack damper Reduce firing (see Note 2)	Not Applicable
Forced-draft and induced-draft fan	Loss of forced-draft fan only	Open air doors (see Note 1) Reduce firing (see Note 2) Open stack damper	Shutdown
	Loss of induced-draft fan only	Shutdown (see Note 3) Open stack damper Reduce firing (see Note 2)	Shutdown (see Note 4)
	Loss of both fans	Shutdown (see Note 3) Open air doors (see Note 1) Reduce firing (see Note 2) Open stack damper	Shutdown
Forced-draft fan, induced-draft fan, and preheater	Loss of forced-draft fan only	Open air doors (see Note 1) and bypass induced-draft fan and preheater Reduce firing (see Note 2) Shutdown (see Note 3)	Shutdown
	Loss of induced-draft fan and/or preheater only	Shutdown (see Note 3) Bypass induced-draft fan and preheater Reduce firing (see Note 2)	Shutdown (see Note 4) Bypass induced-draft fan and preheater.

Note 1: For natural-draft operation (when heater is so equipped); otherwise, shutdown is required. If the natural draft air doors fail to open, automatic shutdown after a short time delay is required.
Note 2: Reduced firing (automatic or manual) is recommended. Reduced firing is not required on switch to natural-draft if sized for maximum duty.
Note 3: Shutdown is not required if heater has enough natural draft to continue operation.
Note 4: Shutdown is not required if heater is designed to operate with the induced-draft fan bypassed.

14. Air Preheaters (APH)

14.1 APH type and characteristics

APH recovers heat from existing hot flue gas and transfers to combustion air. So, it increases heater efficiency and reduces heater fuel input (or, gets more duty). APH adds to combustion air and flue gas circuit pressure drop- thus, most often needs forced draft (FD) and induced draft (ID) fans and related ducting arrangement. Considerable capital expenses needs economic justification by energy costs saved.

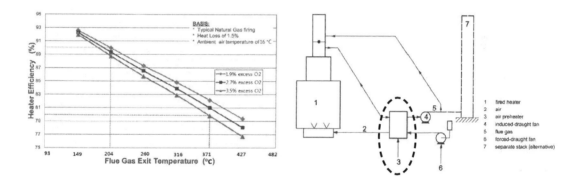

APH has 3 types of regenerative, recuperative, and heat pipe. Regenerative type uses moving elements, picks up heat from flue gas, and releases heat to combustion air. Recuperative type is basically a heat exchanger and this can be plate type or tubular (cast) type. Heat pipe type uses a medium to transfer heat from flue gas to combustion air.

Recuperative type APH's plate type is made by alternating plates that are clamped together in a casing. The plate type is fairly common in fuel gas fired services and air leakage into flue gas is negligible. However small spacing between plates makes it prone to foul and difficult to clean. Also it is prone to flow mal-distribution issues.

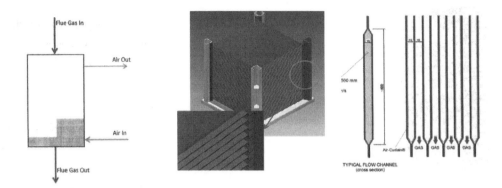

Recuperative type APH's tubular type is made by alternating tubes that are clamped together in a casing. The tube type is fairly common in fuel oil and fuel gas fired services and air leakage into flue gas is negligible. A larger internal spacing makes it less prone to foul and easier to clean. Also the tube sealing is critical. Tubular type can be water-washed on-line and have glass tube section for better efficiency.

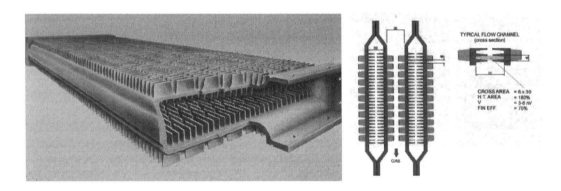

14.2 Acid dew point corrosion issues

Fuel gas and fuel oil often have sulfur which burns to SO_2 and SO_3, and becomes acidic. When flue gas is cold enough, moisture in flue gas combines and make acidic mixture. The acid causes corrosion (sometimes severe).

$SO_2 + H_2O = H_2SO_3$ (Sulfurous acid)

$SO_3 + H_2O = H_2SO_4$ (Sulfuric acid)

Carbonic acid corrosion can occur at temperature below water dew point. Corrosion is possible in APH as well as any other heater component.

$CO_2 + H_2O = H_2CO_3$ (Carbonic acid)

Acid dew point (ADP) temperature is when this condensation occurs. ADP temperature is different from moisture dew point and mostly depends on sulfuric content in fuel, excess air amount, and moisture content in flue gas. Correlation to estimate ADP temperature exists as well. Acid corrosion does not occur where temperature is above ADP temperature. The corrosion happens at colder end. To avoid ADP corrosion, it is important to maintain metal temperature higher than ADP temperature.

Various clients have their required minimum margin to maintain. To be recommended is minimum metal temperature to be kept 25°F above ADP. In case of flue gas temperature, it shall be about 50 to 75°F above ADP. The same corrosion issue can also be that of downstream such as ID fan, stack, and damper.

H_2O vol%	O_2 vol%	Excess O_2 (%)	Fuel	Sulfur (ppmwt)	Sulfur (wt%)	SOx(SO_2) (ppmv)	New (clean) SO_3(ppmv)	ADP(°C)	Old (dirty) SO_3(ppmv)	ADP(°C)	Corrosion rate(mm/yr)
				0.1	0.00001	0.006	0.0003	61	0.0005	66	-
				1	0.0001	0.06	0.003	77	0.005	82	-
			Pipeline gas quality	10	0.001	0.6	0.026	94	0.053	99	-
				20	0.002	1.1	0.048	98	0.096	103	0.03
				50	0.005	2.2	0.095	103	0.19	108	0.05
17	1.7	1.9		100	0.01	4.5	0.19	108	0.39	114	0.10
				200	0.02	9.0	0.39	113	0.77	119	0.20
			Sour gases (low H_2)	500	0.05	23	0.96	121	2.0	126	0.61
				1000	0.1	45	1.9	126	3.9	131	0.79
				2000	0.2	90	3.7	130	7.7	136	0.89
11.2	2.6	2.7	Oil #1	5000	0.5	250	12	134	26	139	1.40
			Oil #2	10000	1.0	520	27	138	59	144	2.11
9.8				20000	2	1000	48	143	110	149	2.90
9	3.3	3.5	Oil #4,5,6	50000	5	2600	100	147	260	154	4.50
8.5	4.0	4.2		100000	10	5400	170	148	520	160	5.99

API 560 recommends minimum metal temperature to maintain as the following.

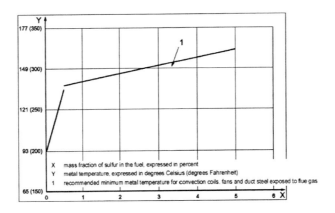

X	mass fraction of sulfur in the fuel, expressed in percent
Y	metal temperature, expressed in degrees Celsius (degrees Fahrenheit)
1	recommended minimum metal temperature for convection coils, fans and duct steel exposed to flue gas

The solution to the issue is to increase minimum metal temperature at design stage.

Plate type air preheaters use "air curtain" to reduce heat transfer at cold ends. Tubular type APH has different type of block design for cold ends. Basically heat transfer coefficient reduced at cold end on air side.

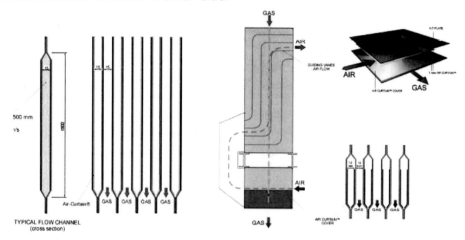

In addition, tubular type APH has a block of glass tube for protecting the cold zone from corrosion.

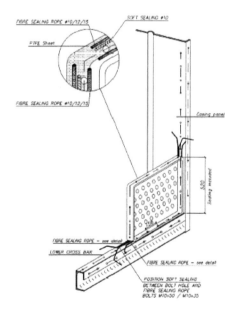

APH is provided with air bypass duct with control dampers. At lower heater capacity operation, APH becomes "over surfaced". In this case, air needs to be bypassed to keep minimum metal temperature above ADP. APH cold elements are provided with TSTC's for monitoring purpose. Also, it is required to maintain flue gas temperature above minimum stack temperature.

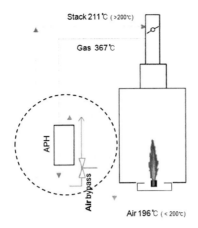

APH air bypass operation (Example)

In operation, APH can be a high maintenance item if not operated properly. Most issues are related to corrosion, fouling or air leakage. Careless washing or start-up/shut down can cause damage. In design, corrosion control starts with proper selection and design of APH. Correct fuel oil/gas sulfur specification is the key. APH design needs to be suited to fuel being used. APH designer must provide proper washing facilities. Hopper or similar construction should be provided for drainage.

Cold end gets attacked first, therefore, TSTC's are to be provided at appropriate places. Air bypass damper must be provided. It is necessary to control flue gas outlet temperature and to maintain TMT above minimum metal temperature. Note, air by-pass damper needs to be tight shut off (TSO) type and avoid heater efficiency loss. And proper air and flue gas distribution at APH inlet needs to be ensured during design.

14.3 APH startup & washing & troubleshooting

In monitoring, minimum metal temperature should be kept at cold end by controlling air bypass damper. And a limit alarm has to be, based on minimum metal temperature. Pressure drop across APH on air and flue gas side. The improper combustion in firebox (say too low excess air) can cause fouling/afterburning in APH.

During start up and shut down it is important not to raise/reduce APH temperature too fast. Sudden expansion or contraction can harm sealing arrangements, besides APH elements. Start up/shut down should follow design and operating manual guideline. Afterburning in APH is uncommon, but good to be aware of, particularly in fuel oil firing. DCS can be provided with simple software to warn of potential afterburning.

APH washing frequency depends on cost of washing and benefit of higher efficiency. Most of APH are not suitable for on-line water washing. Glass tube selection is usually suitable for on-line washing- but only with warm water. Before proceeding water washing, flue gas temperature at section entrance and temperature of water supply should be checked. Also off-line washing with soda water is commonly recommended. Washing requirement is an important parameter to consider at design stage.

In troubleshooting APH, it can be tricky since usually not well instrumented and needs to monitor inlet/outlet temperatures for air and flue gas. Partially open stack or air bypass damper can make analysis complicated. It is necessary to measure O_2 profile on flue gas side with portable analyzer to get air leakage estimate. Performance ratio trend (ratio of temperature drop in flue gas and temperature increase in air) helps to locate if there has been any sudden change.

14.4 Air preheater design case

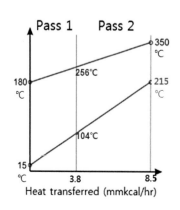

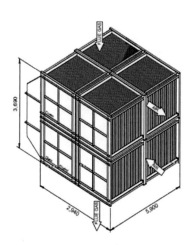

(1) Input data
Fired heater
Fuel oil 4wt% sulfur
Flue gas 187,000 kg/hr/ Air 174,000 kg/hr
Flue gas ΔP 100 mmH$_2$O/Air ΔP 80 mmH$_2$O
Air in 15℃/Flue gas inlet 350℃
Excess O$_2$ 3% / H$_2$O 13%/ SO$_x$ 1,000ppm
Acid dew point 147℃

(2) Output data
2 passes cross flow 8 blocks
Absorbed heat 8.5MMkcal/hr
Heat transfer area 2,616 m^2
Minimum metal temp 178 ℃
Flue gas ΔP 96 mmH$_2$O/Air ΔP 68 mmH$_2$O
Air out 215℃/Gas out 180℃

14.5 Gas & air flow configuration

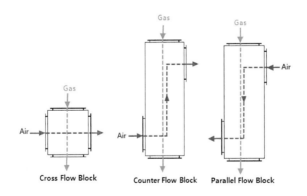

Cross Flow Block   Counter Flow Block   Parallel Flow Block

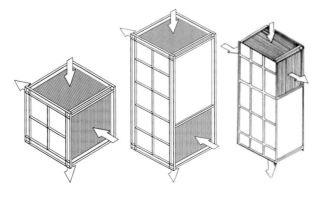

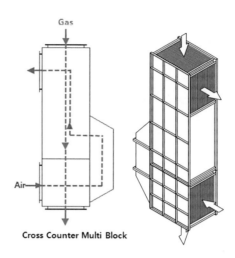

Cross Counter Multi Block

14.6 Design example of plate type air preheater

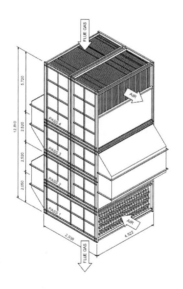

The unit consists of glass coated tube block for cold pass and other 3 passes all made of carbon steel plates. And pass 1, pass 2 and pass 3 are cross flow type and pass 4 is counter flow type. Cold block (pass 1) is made of glass coated tube to protect the corrosion and cold ends in pass 2 is not covered by air curtain due to having enough minimum metal temperature, 132℃, above acid dew point, 110℃. In the dimension the unit is designed to increase the height 12,810mm (about 3 times of original) and the length 2,938mm (a little due to structure) to meet required gas outlet temperature, 150℃. But the width is fixed to 4,522mm of original.

14.7 Problems of air preheater

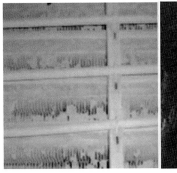

Top fouling   Cold end corrosion

Air inlet corrosion   Cold out cracks

Cold in deformation   Gas in corrosion

15. NO_x Control

Three basic mechanisms for NO_x formation are prompt NO_x, fuel NO_x, and thermal NO_x. Prompt NO_x is quick reaction via hydrocarbon radical mechanism is as below.

$$C_xH_y \rightarrow CH + N_2 \rightarrow N \underset{O_2}{+} HCN$$
$$NO \longleftarrow N + CH$$

This happens at earliest stage of combustion, mechanism not well understood still, and not much control on this source. Meanwhile, fuel NO_x is formed from chemically bound nitrogen in fuel (oil, gas, coal); mainly formed from coal and oil, and also produced from fuel gas. Depending on nitrogen content of fuel, fuel NO_x contributes to roughly 50% of total NO_x in fuel oil firing and roughly 80% in coal firing. Lastly, thermal NO_x is produced by high temperature oxidation/fixation of nitrogen.

$$[NO] = k_1 \exp(-k_2/T) [N_2][O_2]^{1/2} t$$

Where, k_1/k_2 constants, T temperature, and t residence time

Thermal NO_x is predominant in gas firing. Roughly 90% is NO and rest NO_2. NO_x is always NO_2. Thermal NO_x is highly dependent on flame/box temperature and also highly dependent on oxygen concentration. Several potential ways to reduce emission are to: (1) reduce excess air/staged air burner (O_2) (2) reduce flame temperature/staged gas burner (3) reduce flame temperature/add flue gas recirculation (4) reduce flame temperature/add inert (steam). Low NO_x burner technology can reduce emission to 20ppm or even lower.

In NO_x control, there are 2 kinds of combustion control technology and post combustion control technology. The former is to reduce emission by using latest burner technologies, whereas the later has SCR (selective catalytic reduction) and SNCR (selective non catalytic reduction). From here SCR shall be more emphasized than SNCR.

15.1 Combustion control

There are several technologies in Low NO_x burner selection as shown in below.

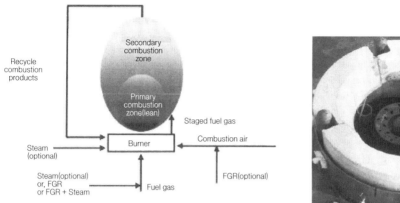

Staged fuel gas burner

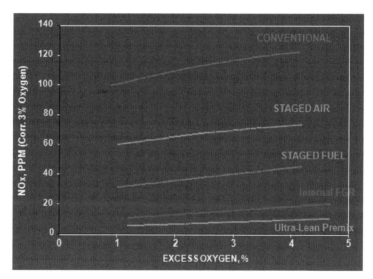

Burner History

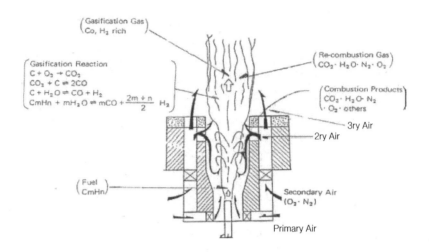

Conventional Low NO$_x$ principle

For the NO$_x$ control, staged air and staged fuel have 30-40% air/fuel gas in primary combustion zone respectively. External FGR (flue gas recirculation) has 25-30% flue gas recirculation and steam injection used to use 0.5-1.0 lbs of steam/lb of fuel. These methods have different potential in NO$_x$ reduction: staged air is 25-30%, staged fuel 30-40%, internal FGR 40-50%, external FGR 50-60%, air/fuel stage + FIR (fuel induced recirculation) 55-75%, and air/fuel stage + external FGR 60-80%. From here, FIR has been mostly in boilers and is not common in process burners.

15.1.1 Burner retrofit
New generation low NO$_x$ burners have been popular. Sub 10ppm NO$_x$ burners would be available as next generation's ULNB. Though they are commercially available, but actual potential is yet to be established. Several factors to impact ULNB's performance are: (1) adding inert on top of staged air/gas (more improved) (2) shifts duty from radiant to convection section

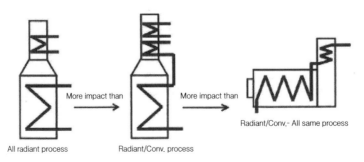

ULNB may have larger burner tile diameter compared to conventional burners. And New Generation ULNB have tiles that are single piece and heavier than convention burners. These all are the structural consideration to be reviewed for burner retrofit. In general, burner manufacturer claims, with ULNB, achieving 15ppm NO_x emission based on fire box temperature of 871℃ at ambient air and natural gas firing.

15.1.2 Fire box geometry

Diluting the fuel/air mixture with furnace flue gas results in lower burning rate and increases the flame dimensions considerably. Heater geometry and direction of firing may impose restriction on design and selection of ULNB. Burner to burner interaction, in order to ensure a proper furnace flue gas recirculation into the combustion zone of a burner certain minimum distance, need to be maintained between the centerline of two adjacent burners. Linear heat intensity and floor area or hearth heat intensity beyond certain design limit is possibility of burner to burner flame interaction. If the burners are too close to each other, there is possibility of low pressure zone towards the center of the heater. This may cause collapsing or merge of the flames. A merged flame two or more burners have considerably taller flame than a single burner.

① **Linear heat intensity (LHI):** fired duty/unit length of furnace floor

Example: [10(burners) x 10 mmBtu/hr]/ 50ft = 2.0 mmBtu/hr/ft (good)

Note: LHI > 3.0-3.5 mmBtu/hr/ft (burner to burner interaction possible)

② **Hearth heat intensity (HHI):** fired duty/furnace floor area

Example: [10(burners) x 10 mmBtu/hr]/ 400 ft^2 = 250,000 Btu/hr/ft^2 (good)

Note: HHI > 350,000-450,000 Btu/hr/ ft^2 (burner to burner interaction possible)

In a well designed furnace as the heat is transferred from the hot combustion gas to the process fluid it cools down and there is flue gas flow recirculation between the tubes and the outer refractory wall towards the middle of the furnace where burners are located.

 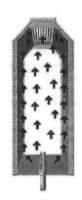

Normal heater recirculation pattern Heater with plugged flow pattern

15.1.3 The other performance impacts of ULNB's

① Furnace geometry/condition dictates reduction possible.

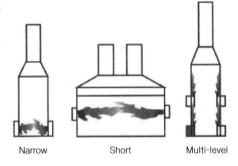

Narrow Short Multi-level

② Dependent on firebox temperature

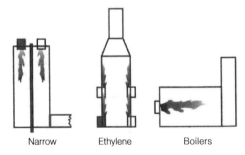

Narrow Ethylene Boilers

③ Presence of tramp air has adverse impact. (Seal leakages are major issue in old equipments at times)

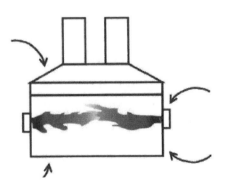

15.1.4 Process consideration

Longer flame length may change heat transfer or flux in the firebox and can have impact on the process. This aspect is particularly important in case of reaction furnaces like Coker, Visbreakers, Steam Methane Reformers. This can cause high fire box or bridge wall temperature causing change in the split of duty between radiant and convection section, higher tube metal temperature, and reduction in process absorbed duty. Turn down ratio of ULNBs may have limitation as compared to conventional burners. In terms of higher CO emission at fire box temperature below 704°C, special consideration shall be given for Hydrotreater and Hydrocracker unit furnaces. If operated at low turn down ULNBs may cause flame instability and flame out.

15.1.5 Excess air (O₂) control

Higher than design excess air means more NO_x and lower than design excess air means possibility of incomplete combustion and CO formation. And flame instability and tramp air ingress should be prevented. Accordingly it needs good control of excess air as the following: (1) automatic control of combustion air by individual damper control or by jack shaft (2) automatic draft control (3) oxygen and CO automatic control/monitoring (4) additional tube skin thermocouples (5) additional temperature indicators for fire box and furnace floor (6) proper flue gas pressure and temperature control

15.1.6 Fuel treatment

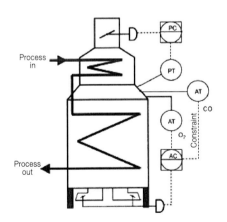

Conventional burners have larger tip size (1/8" or larger) as compared to ULNBs (1/16"). Small tips are prone to plugging. Plugged tip can cause flame instability and unsafe combustion. And pipe scales from carbon steel piping and liquid condensate from fuel gas can plug the tips frequently. So it is necessary to consider the use of flue gas filter/coalescer to remove liquid aerosol, the use of stainless steel, and the proper heat tracing and insulation or even fuel gas heater. Burner tip plugging causes forming bad flame pattern. The bad flame negatively affects performance by getting flame longer, and then causing impinging.

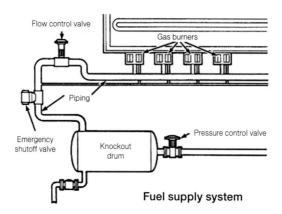

Fuel supply system

15.1.7 CFD (Computational Fluid Dynamics) modeling

Due to higher sensitivity in terms of flame shape and size with ULNBs as compared to conventional burner's proper investigation on flue gas recirculation pattern inside the fire box, heat flux profile, tube metal temperature profile etc. is very important particularly for retrofit application. The CFD modeling of the fire box, combustion air distribution duct or air plenum shall be done for ULNB retrofit application.

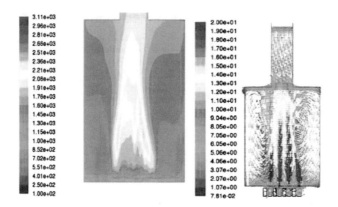

Temperature distribution profile (°F)   Velocity vector (m/sec)

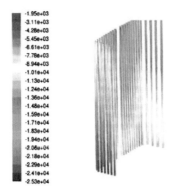

Heat flux profile (Btu/ft^2 hr)

15.2 Post combustion control

15.2.1 SNCR (Selective Non Catalytic Reduction)

This uses ammonia (NH_3) or Urea [$CO(NH_2)_2$] to react with NO_x and their basic reactions are as below.

$2NO + 4NH_3 + 2O_2 \rightarrow 3N_2 + 6H_2O$ (for ammonia based)

$4NH_3 + 5O_2 \rightarrow 4NO + 6H_2O$

$2NO + CO(NH_2)_2 + 1/2O_2 \rightarrow 2N_2 + CO_2 + 2H_2O$ (for urea based)

$4NH_3 + 5O_2 \rightarrow 4NO + 6H_2O$

From here, NH_3 reacts with O_2 at high temperature and makes NO to be counterproductive.

SNCR needs high temperature (871-1,204℃) for reaction to occur and NO_x reduction efficiency is about 40-75%. Also, SNCR needs reaction zone to have uniform temperature and reactant injection and allowable NH_3 slip is about 5 to 20ppm. SNCR is sensitive to load changes and has potential corrosion and fouling at downstream equipments.

15.2.2 SCR (Selective Catalytic Reduction)

Some of the SNCR drawbacks are addressed by using catalyst. This application is possible at wide range of temperature (149-593℃) and has very high NO_x reduction efficiency (90% and higher). SCR uses ammonia (NH_3) to react with NO_x as the following basic reactions.

$4NO + 4NH_3 + O_2 \rightarrow 4N_2 + 6H_2O$

$6NO + 4NH_3 \rightarrow 5N_2 + 6H_2O$

$2NO_2 + 4NH_3 + O_2 \rightarrow 3N_2 + 6H_2O$

$6NO_2 + 8NH_3 \rightarrow 7N_2 + 12H_2O$

$NO + NO_2 + 2NH_3 \rightarrow 2N_2 + 3H_2O$

From above, first reaction accounts for most of the conversion (90% of NO_x is NO). Two basic ammonia introduction systems are aqueous and anhydrous type.

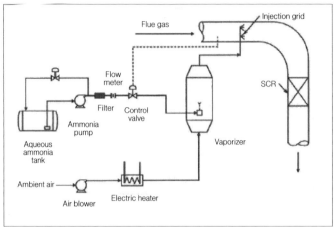

Aqueous ammonia injection system

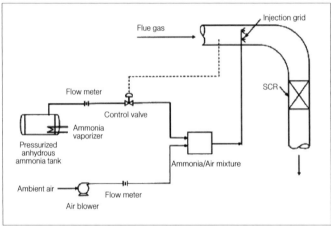

Anhydrous ammonia injection system

Basically three types of catalysts are available: platinum based, vanadium-titania based and zeolite catalyst. Among them, vanadium–titania based catalyst is more common and available forms are pelletized, monolithic, plate catalyst and corrugated. Below is typical temperature ranges for SCR catalysts.

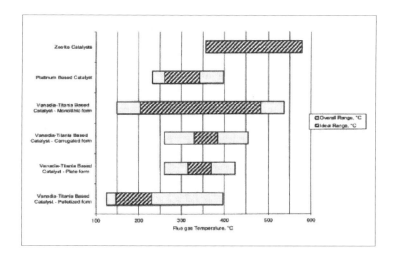

Vanadium–titania catalyst has three temperature ranges: low temperature catalyst (pelletized), medium temperature catalyst (monolithic, plate, corrugated), high temperature catalyst (monolithic, plate, corrugated)

① Low temperature catalyst

	Unit	Typical	Possible range
Operating temperature	℃	178-232	149-360
Pressure drop (WC)	Inch (")	2-3	as low as 1
Performance			
NO_x conversion	%	90	can be 95+
NH_3 slip	ppm	10	as low as 2

In low temperature catalyst, effect of fuel gas sulfur content is:

H_2S, ppm	Minimum allowable flue gas temperature, ℃
0	163
6	177
12	204
130	232

Ammonium sulfate [$(NH_4)_2SO_4$]/Ammonium BiSulfate [$(NH_4)HSO_4$] form and they deposit on downstream equipment.

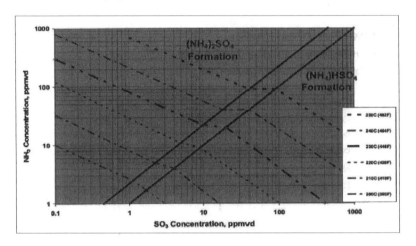

② Medium temperature catalyst

Vanadium-titania based catalyst is coated on ceramic honeycomb or metallic substrate (5-10wt% WO_3, 0-4wt% V_2O_5, 80-90wt% TiO_2).

	Unit	Typical	Possible range
Operating temperature	℃	288-399	246-427
Pressure drop (WC)	Inch (")	2-3	
Performance			
NO_x conversion	%	90	can be 92+
NH_3 slip	ppm	10	as low as 2

③ High temperature catalyst (zeolite catalyst)

No active metal is present. This catalyst gets damaged above 649℃.

	Unit	Typical	Possible range
Operating temperature	℃	454-510	427-679
Pressure drop (WC)	Inch (")	3-4	
Performance			
NO_x conversion	%	90	
NH_3 slip	ppm	10	

Depending on the level of heat recovery from flue gas, the low/medium/high temperature SCR applications are shown as below.

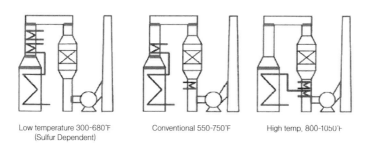

Low temperature 300-680°F   Conventional 550-750°F   High temp. 800-1050°F
(Sulfur Dependent)

Comparison of NO$_x$ removal options are below.

	Typical reductions (%)	Lowest possible emission (ppm)
Use of low NO$_x$ burners	60-75	20-75
Use of latest generation ULNB	70-85	10-20
SNCR on fired heaters	40-75	50
SCR on fired heaters or gas turbines	90-95	2-9

16. Heater Operating Mode

If the fired heater is properly operated, all the internal parts are under negative pressure. This negative pressure is particularly important for safety reasons since the possibility of the flame or hot flue gas to be forced outside of the heater must be definitely avoided.

On fired heaters operated under natural draft, the required draft to maintain the system under vacuum conditions and to overcome the pressure drop of the whole system is provided by a properly sized stack. On forced fired draft heaters, the draft is provided by an induced draft fan, but it follows a similar pressure profile.

Since the entire fired heater is under vacuum conditions, a negative pressure of $2.5mmH_2O$ ($-2.5mmH_2O$) is normally set at the arch level in order to minimize air infiltration. This is the point of the fired heater where minimum vacuum occurs. This pressure is controlled by the stack damper which is placed at fired heater outlet or by the induced draft fan control, whichever is applicable.

Buoyancy provided by the hot flue gases in the radiant section increases the draft available at the heater floor level and it is utilized to induce the combustion air to flow through the burners. This buoyancy provides the proper mix of fuel and air required for good combustion.

The pressure drop through the convection section is balanced by the draft provided by the stack or by the induced draft fan, as applicable. It is normally sized for extra capacity in flue gas flow. For good fired heater operation, it is necessary to adjust the damper located either on the stack or induced draft fan to provide the required draft at arch level, and to properly trim the air register at the burners to maintain the excess air at the design value.

The control of combustion air using only the stack damper is unsafe. This is because a partial closure of the same to reduce the pressure drop available at burner level may result in positive pressure at arch level with possible leaks of hot flue gases through the fired heater casing endangering personnel.

For a balanced draft system (i.e., whenever an induced draft fan is required to discharge the flue gas exhaust to the atmosphere), the fired heater is operated similarly. A forced draft fan provides the required air to overcome the pressure drop through the burners, ducting and air preheat systems as applicable. The draft at arch level is maintained by properly setting the induced draft fan control damper.

17. Heater Efficiency

An important parameter in the fired heater operation is the efficiency(η), which is defined as the ratio:

$$\eta = Q_a/Q_f$$

Where, Q_a heat absorbed by the process

Q_f heat fired

The difference between Q_a and Q_f represents the heat losses which are obviously higher if the fired heater efficiency is lower. These losses (not considering unburned components which, for proper combustion, are normally very close to zero) can be split as follows:

(1) Radiant heat losses through fired heater walls

(2) Heat losses due to the discharge of hot flue gases to atmosphere

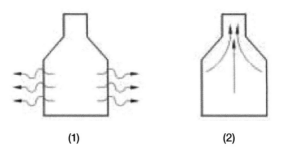

(1) (2)

The normal value of heat losses through fired heater walls are 1.5 to 2% of heat fired while flue gas discharge losses are much higher dependent upon the fired heater efficiency.

It is clear therefore that increasing fired heater efficiency by reducing the flue gas quantity and/or flue gas temperature at the fired heater outlet, is much more effective than minimizing heat losses to atmosphere through fired heater walls.

In order to better understand the importance of proper fired heater operation, there is an example of a properly designed fired heater firing fuel oil having 9,500 kcal/kg LHV operated with 3.5% excess O_2 and designed for a flue gas outlet temperature of 400℃. The corresponding efficiency is 81.5%. If the fired heater is not carefully operated and the excess O_2 is not properly controlled and kept at 6%, which is often considered as "normal" value particularly where old design burners are installed, the fired heater efficiency decreases from 81.5% to 78.7%.

For an atmospheric process unit of 70,000 BPSD the relevant fired heater has a duty of around 52 MMkcal/hr. Considering fuel oil cost of 0.2 Euros/kg the annual saving can reach as much as 250,000 Euro as shown in Table below.

Items	Unit	Excess O_2 3.5%	Excess O_2, 6.0%
Unit capacity	BPSD	70,000	
Heater duty	MMkcal/hr	52	52
Flue gas outlet temperature	℃	400	400*
Radiant losses	%	1.5	1.5
Gross efficiency	%	81.5	78.7*
Net efficiency	%	79.0	77.2*
Heat fired	MMkcal/hr	65.8	67.4
Δ firing	MMkcal/hr	1.6	
Fuel oil saving	Kg/hr	162	
Annual saving (8,000hrs)	Ton/hr	1,296	
Fuel oil cost	Euros/kg	0.2	
Yearly saving	Euros	259,200	

Note: *Not considering higher flue gas temperature at fired heater outlet due to the higher flue gas quantity. This varies depending on convection section design but typically a 1% reduction in efficiency results in a 20°C increase in stack temperature.

18. Parameters Influencing the Fired Heater Operation

The parameters governing fired heaters operation and the effects of variation on certain conditions are as follows.

18.1 Heat load

An increase of fired heater load increases the heat flux and the average flue gas temperature with a consequent decrease of thermal efficiency. A decrease of fired heater load has the opposite effect.

18.2 Excess O_2

An increase of excess O_2 reduces thermal efficiency and switches a portion of heat absorption from the radiant to convection section. A reduction of excess O_2 increases the efficiency. However, going below the minimum required quantity might causes incomplete combustion with the possibility of post firing (after burning), mechanical damage of convection tubes and tube supports, and the fouling of heat transfer surfaces.

18.3 Heater fouling

The fouling of external heat transfer surfaces results in insulation of surfaces and an increase of tube skin temperature, with the consequent reduction in heat transfer and fired heater efficiency. To maintain constant absorbed heat, a firing rate increase is required with relevant heat flux increase. The increase of each of the parameters above results in a significant performance change of the fired heater operating conditions.

Fouling & Deposit

Slag & Scale

18.4 Maximization of process unit profitability

The above is an introduction in order to better understand what can be done to maximize the profitability of an existing fired heater. The increase of profitability of a process unit can be achieved by several different routes: Increase process unit capacity, reduce utilities consumption, maximize on stream time, thus minimizing scheduled and unscheduled shutdowns, minimize operational cost, optimize fired heater performances, and reduce emissions to atmosphere

18.4.1 Increase unit capacity

In many cases after several years of operation, the fired heater becomes a bottle neck for the capacity of the process unit. The limitation of fired heater operation can be due to: Limited heat transfer surfaces, not enough draft (positive draft), flame impingement, high metal tube temperature, difficulties in firing control, coke formation inside the process tubes, and Limited firing capacity

18.4.2 Limited heat transfer surfaces

Taking into consideration low fuel costs, fired heaters in the past were designed to have low efficiency in order to reduce the capital investment cost. This normally meant large radiant and small convection sections. Considering the installed original surfaces, an increase of capacity means an increase of heat flux. In many cases, this is not compatible with the installed tube metallurgy (high tube temperature) and high film temperature of process fluid (possibility of coke formation). To overcome these problems, different options can be considered.

Install extra surfaces in the radiant section. The generous sizing of the radiant section in the old fired heater designs allows, in many cases, the installation of additional heat transfer surface.

Adding or modifying convection section surface. The low amount of heat recovered in the convection section can be substantially increased by adding or modifying the installed surfaces where the radiant and convection sections are linked. The additional heat recovered in the convection section decreases the radiant duty and heat flux, which can be maintained at the original value in spite of the increased fired heater load.

18.4.3 Not enough draft (positive pressure)

The importance of the draft for a reliable fired heater operation has been already described previously. A high draft requirement can be due to the following: ① Excessive firing due to fired heater over loading ② High excess O_2 due to poor burner operation ③ High flue gas quantity due to air infiltration through fired heater casing ④ Low fired heater efficiency due to heat transfer surface fouling ⑤ High pressure drop through convection section due to fouling and plugging of free area around heat transfer surfaces

The problem can be solved with the reduction of flue gas quantity by: ① Excess O_2 reduction (by modifying or replacing the burners) ② Infiltration (by minimizing openings in the fire box) ③ Efficiency (increased efficiency means less fuel firing and less flue gases, i.e., at the same firing heat release, increased efficiency means higher heat adsorption to coil side) ④ Proper maintenance (by cleaning the convection section and radiant section surfaces)

18.4.4 Flame impingement

Flame impingement on heat transfer surface is very dangerous. The local overheating of the tubes results in higher process fluid film temperatures with consequent coke formation. This increases the tube metal temperature and can necessitate unscheduled fired heater shutdowns for decoking or limit the unit throughput. The flame impingement can be due to: ① Excessive firing through fired heater overload ② Too low fired heater efficiency ③ Improper burner operation ④ Fouling of heat transfer surface ⑤ Draft shortage ⑥ Incorrect distribution of heat from firing

The flame impingement can be eliminated by: ① Reduction of heat firing, increasing fired heater efficiency ② Cleaning heat transfer surfaces ③ Improving firing quality by replacing the burners ④ Improving firing quality by modifying existing burners, repairing tips, adjusting air registers, providing the required draft ⑤ Providing an even distribution of the firing of all the burners ⑥ Providing the correct excess O_2 to all the burners ⑦ Improving fuel oil characteristics, such as pressure or viscosity, and by providing atomizing steam at appropriate conditions

18.4.5 High metal tube temperature

High metal temperature of heat transfer surfaces is normally due to: ① Low heat transfer coefficient of the process fluid ② Low velocity inside tubes, combined with high heat fluxes, high coke layer inside tubes, high heat flux, and flame impingement

The low heat transfer can be improved as follows: ① Changing the coil size arrangement or number of passes ② Process conditions- increase flow, modifying coil geometry in critical areas, injecting steam if possible to increase the fluid velocity ③ Temperature increase due to coke inside tubes can be minimized by removing fouling inside tubes (by steam air decoking or by pigging) ④ Removal of outside fouling i.e., scaling or surface fouling

The high heat flux, particularly in the case of uneven firing distribution can be eliminated with proper trimming of the burner dampers and improving the fired heater efficiency. The same actions are effective to minimize flame impingement.

18.4.6 Difficulties in firing control

The difficulties experienced in firing control can be due to several reasons such as: ① Excessive firing ② Low draft available for a correct burner operation ③ Difficulties in excess O_2 control ④ Poor characteristics of fuel oil and atomizing steam ⑤ Excessive pressure or too low pressure of fuel gas ⑥ Fuel tips in poor conditions ⑦ Fuel control valves not operating properly

The excessive firing might be due to fired heater operation actually above design condition or fouling of heat transfer surfaces. Both these reasons result in a higher firing rate. The extra firing associated with poor combustion might result in difficulties in the firing control. Also, if the burner air registers are not easy to operate, this could result in difficulties in controlling combustion O_2 requirement and a proper flame shape.

The same problems could occur: ① If the viscosity of the fuel oil is too high ② If the atomizing steam pressure is too low ③ If the molecular weight of the fuel gas is very different from the original design condition ④ If the fuel control valves are not properly sized

Many options, depending on the type of firing difficulties, can be considered to improve the fired heater performance: ① Reduce the firing rate and increase the fired heater efficiency by installing extra surface in convection section and/or installing an air preheating system ② Change burner tips if not adequate for the fuel characteristics ③ Replace the burners if in bad condition or not adequate for the actual heater operation. The burners and the relevant flame characteristics, similar to the one indicated in the pictures below, either with (a) fuel gas or (b) fuel oil operation, play a very important part in reliable fired heater operation.

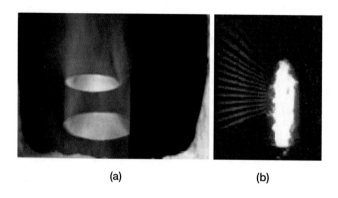

(a)   (b)

④ Improve the fuel characteristic and the firing control system ⑤ Install a forced draft combustion air system to allow automatic control of combustion air and excess O_2.

18.4.7 Coke formation inside the process tubes

Coke formation inside the process tube is due to an excessive localized process fluid film temperature in the hottest zone (generally the radiant section) of the fired heater. The high film temperature can be due to low velocity of fluid inside tubes, high heat flux, flame impingement or maldistribution in the heat transfer surfaces.

The design of fired heaters with low fluid velocities in the process coil to minimize heater pressure drop was quite normal in the older design of fired heater. This incorrect approach was the main cause of inefficient heater operation especially in case of reduced capacity. The increase in throughput in many cases has reduced the problem even at increased fired heater load. The change of coil size in the most critical zone as well as the installation of extra surfaces to reduce heat flux can be considered.

As previously mentioned a good flame shape with no impingement on the heat transfer surface is mandatory to minimize the coke formation. In addition, good heat flux distribution to the fired heater heat transfer surfaces is a good method to reduce the peak heat fluxes in critical areas and consequently limit the possibility of coke formation. Below shows the coke that came from tube pigging in Naphtha Splitting Unit reboiler heater.

18.4.8 Limited firing capacity

Sometimes- in spite of margins available on fired heater components- it is not possible to increase the fired heater load due to limitations in the firing capacity of the burners. The limitations in firing capacity can be given by either the incorrect size of the burner tips, limited draft available or the flame shape at high firing rates.

Limitations in firing capacity can be solved by increasing the heater efficiency to reduce the firing load, or changing the type and the size of the burners. Often the replacement of natural draft burners with forced draft burners is a good way to solve the firing limit and at the same time, to improve fired heater performance and the operability of the process unit.

18.5 Utility consumption reduction

Historically, as the cost of fuel was very low and investment costs were a controlling factor when compared with: (1) Utilities operational costs (2) The fact that the design of the oldest fired heaters were based on low efficiency concept.

In addition to the low design efficiency (i.e., very low fuel price made it possible), the fouling of heater surfaces over time and poor burner operation were additional reasons for further fired heater efficiency reduction. The increase of fired heater efficiency can be achieved through the following different routes: (1) Heat transfer surfaces addition/increase (2) Air preheating system installation (3) Heat transfer surfaces cleaning.

In most cases, low efficiency of the old refinery fired heaters was due to the use of small convection sections designed with limited bare surfaces and consequently with high outlet flue gas temperatures. Upgrading the convection sections has the effect of increasing the heat recovery, with resultant reduction in the radiant heat absorption benefits for heat fluxes and firing rate, are demonstrated in table below.

Items	Unit	Original design	Additional convection surface	Additional radiant and convection surface
Unit capacity	BPSD	46,000	46,000	60,000
Heater duty	MMkcal/hr	33	33	43
Net efficiency	%	70.0	83.0	81.0
Heat fired	MMkcal/hr	47.1	39.8	53.1
Peak film temperature	℃	393	387	387

When the replacement of the convection section is not feasible, the installation of an air preheating system based on suitable return of investment could be the most interesting solution as shown on Table below. Also, the replacement of the original burners with others which are more efficient, and designed for lower excess O_2, might be a good way to improve fired heater efficiency. The installation of devices to maintain clean convection section heat transfer surfaces such as soot blowers is an additional effective method to maintain fired heater performance.

Items	Unit	No APH	With APH
Unit capacity	BPSD	70,000	70,000
Heater duty	MMkcal/hr	52	52
Net efficiency	%	72	90
Heat fired	MMkcal/hr	72.2	57.8

18.6 Maximize on stream time

The profitability of a process unit is heavily affected by the shutdown periods required to perform routine maintenance to the unit. If there are unforeseen reasons to shut down a unit outside a scheduled maintenance turnaround, the consequent production losses heavily affect the profitability of the unit itself. The shutdowns must be minimized acting on: (1) Proper operation of the unit (2) Proper scheduled maintenance activity (3) Careful check of the fired heater components during turnaround (4) Scheduled and programmed replacement of items creating operating problems

It is well known that careful operation of a process unit without exceeding the design parameters allows extended equipment life, thereby minimizing failure during normal operation. In addition to the above, if a comprehensive check and maintenance of fired heater components is regularly performed during the scheduled turnaround, this can minimize the possibility of failure during normal operation.

18.7 Minimize operational cost

Modern refinery operation is based on the concept that a few individuals can take care of several process units. The high degree of automization allows the plant operation from control room with very limited activity at site. To achieve this target it is required to have a proper instrumentation improvement program to allow most changes to the operation to be carried out from the control room. In addition, automatic devices and/ or computer programs can be installed to optimize the firing rate at any capacity.

Furthermore, it is important: (1) To fire the cheapest available fuel (2) To regularly carry out cleaning internal heat transfer surfaces by pigging, decoking, descaling or removal of fouling (3) To regularly carry out cleaning external heat transfer surfaces by on-line chemical cleaning service

It is also important to clean the external surface by steam blowing during a turnaround or in the case of fuel oil firing regularly operate soot blowing systems to reduce to minimum convection section fouling and increase or maintain heater efficiency.

18.8 Optimize heater performance

Fired heater performances are strictly related to the process unit needs which are variable with respect to the feed stocks utilized and the quality of final products.

It is important to change the operation of the fired heater or the feed stocks flow rates and duties charge to optimize the heater performance.

18.9 Emissions reduction

Many refinery and petrochemical complexes are limited in capacity by the overall amount of pollutants legislation allows to be discharged to the atmosphere. Pollutants reduction can be achieved by increasing fired heater efficiency, thus reducing the firing rate and by replacement of the installed burners with state-of-the-art low NO_x type.

The emissions normally considered are NO_x and SO_2. While the SO_2 is a direct function of the fuel fired, NO_x changes can be made by the type of burner installed. Latest state-of-the-art burner technology can produce a reduction of more than 50% of NO_x emission if compared with an older burner design. The efficiency increase brings a further reduction in NO_x at stack outlet. An example of the reduction of NO_x emission for a fired heater of 52 MMkcal/hr properly upgraded is shown in Table as follows.

Items	Unit	Normal burners	Low NO_x burners	Low NO_x burners with APH
Unit capacity	BPSD	70,000	70,000	70,000
Heater duty	MMkcal/hr	52	52	52
Net efficiency	%	77.2	79	90
Excess O_2	%	6	3.5	3.5
Heat fired	MMkcal/hr	67.4	65.4	57.7
Flue gas quantities	kg/hr	142,790	119,340	104,750
Flue gas quantities	Nm^3/hr**	92,142	90,045	79,040
Unit rate emissions*	mg/Nm^{3}**	600/250	350/75	400/90
NO_x quantity (fuel oil)	kg/hr	55	32	32
NO_x quantity (fuel gas)	kg/hr	23	7	8

Note: * 100% fuel oil/100% fuel gas, ** dry basis @ 3% O_2

19. Improper Heater Operation

All the parameters influencing the fired heater operation regarding maximization of the unit profitability have been described previously. In this paragraph fired heater deficiencies as well as incorrect operation are highlighted together with appropriate remedies to overcome or reduce various problems.

Fired heater deficiencies can be due to weaknesses of the original design features or changes between design and current conditions.

19.1 Original design

Many fired heaters included in older units were designed with a low velocity inside the tubes and therefore were more sensitive to the possibility of coke formation in the case of process side or firing maldistribution and convection section fouling. The low velocity inside the tubes of each pass, if not controlled in flow, can result in a maldistribution of the process fluid with consequent increase of possibility of coke formation in the coil at low throughput.

The old natural draft burners can be difficult to trim properly and to operate with the correct excess O_2. In the case of dual fuel firing with old burners, an even distribution of heat can also be very difficult to achieve. The main stack damper is normally a simple blade type and control of the draft inside the firebox is difficult to obtain, especially in case of a very low flue gas velocity through the damper section.

Sometimes the draft provided by the stack, even if generously sized, is not enough to provide the required draft inside the firebox if the burners are operated with too high excess O_2 or, in case of significant infiltration of air through the heater casing.

19.2 Current fired heater conditions

Even if properly designed and after many years in operation, a fired heater normally operates in conditions different from the original design. This is due to many reasons, such as: (a) Heater transfer surfaces fouling (b) Air infiltration through fired heater casing, header boxes and inspection doors (c) High excess O_2 through burner registers not easily operable (d) High heat losses through the heater casing, due to poor conditions of the internal insulation (e) Maldistribution among various passes if not separately controlled due to uneven distribution of coke inside the process coils (f) Uneven distribution of flue gas flow through the convection surface due to distortion of tubes and/or failure of refractory at the lateral wall

PART 3

Maintenance

Sub Contents
for Maintenance

1. Fuel Ash Corrosion

When the fuel burns in heater, the impurities accumulate on the tube surface. The deposit hinders the heat transfer into tube and the deposit deteriorates the efficiency of heater. The problems encountered on the hot radiation are slag and corrosion, which are both due primarily to the same causes: deposition of oxides and sulfates of vanadium and sodium. Temperatures in radiation are typically higher than the melting point, the oxides and sulfates are molten and sticky on the tube surfaces. As the layers of slag build up on the tubes, the heat transfer into the tubes inside is reduced. The heater operator must increase fuel consumption to compensate for this loss of heat transfer or reduce process throughput. According to the using fuel, foulant's component and amount is different, for example, fuel gas contains sulfur, nitrogen compound and carbon ashes, while fuel oil includes many metal components in the contaminates on the tube surface. Even from fuel oil, the metal component and level is different each other according to processing crude and treating process method (residue; AR, VR, cracked-NCC, FCC & hydrocracker).

Source	Crude oil	Vanadium, wtppm	Nickel, wtppm	Sodium, wtppm
Africa	1	5.5	5	22
	2	1	5	-
Middle East	1	7	-	1
	2	173	51	-
	3	47	10	8
United States	1	13	-	350
	2	6	2.5	120
	3	11	-	84
Venezuela	1	-	6	480
	2	57	13	72
	3	380	60	70
	4	113	21	49
	5	93	-	38

Vanadium, Nickel and Sodium content of residual fuel oils

Fraction	Distillation range, °C	Total sulfur, wt%
Crude oil	-	2.55
Gasoline	51-123	0.05
Light naphtha	125-149	0.05
Heavy naphtha	153-197	0.11
Kerosene	207-238	0.45
Light gas oil	247-269	0.85
Heavy gas oil	281-306	1.15
Residual oil	309-498	3.70

Sulfur content in fractions of Kuwait crude oil

Impurities such as vanadium, sulfur, and sodium in furnace fuels can produce fuel ash that causes corrosion of furnace components. Fuel oil, which has higher impurity levels than natural gas, is most likely to lead to corrosion. During combustion, compounds containing vanadium, sulfur, and sodium are melted or vaporized. They then deposit on contact with lower temperature surfaces. These deposits react with the protective oxide scale on the metal and expose new material to corrosion by oxidation and sulfidation. The process is self-perpetuating and can result in severe metal loss. Among other variables, the amount of vanadium and sodium present and the temperature of the metal determine the severity of corrosion. When the sodium plus vanadium content of the fuel oil exceeds 10-15wtppm, corrosion can be severe. Significant corrosion can occur at temperatures above 649°C when impurity levels are high. At low impurity levels, corrosion may not be significant unless temperatures exceed 843°C.

Various components exist in the given ash sample such as vanadium (24 wt%), nickel (13.81 wt%), iron (7.72 wt%), sulfur (7.33 wt%), sodium (6.29 wt%), and calcium (2.16 wt%) etc. They exist as oxide type, and among them especially since vanadium oxide (V_2O_5) is very corrosive, it can cause corrosion of tube in radiation zone or of refractory in burner. So all components analyzed should be removed from on the tube surface because they prevent heat transfer as insulation as well as corrosion. The source of components comes from the fuel oil of crude that each refinery treats. In case of vanadium it is rich into crude from Middle/South America (for example, Venezuela, Mexico) etc. The contents reach more than 500 wtppm in residue fuel oil.

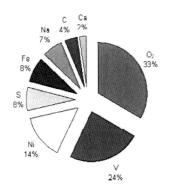

Also in calcium, it is rich in acid crude from Africa (for example, Doba crude in Chard). In any case this foulants should be removed by either mechanical or chemical treatment.

Analysis data of a scale on external tube surface in a refinery

Following is general elements that cause a slagging in the furnace of high temperature.

Periodic table	Elements
Transition metal	Fe, V, Ni, Cu, Co
Alkali metal	Na, K
Alkali earth metal	Ca, Mg, Ba
Non-metal	S, Si, O, C, N, O

Metal V, Ni, Fe, Na
Sulfur compounds
Cracker residues
$(Al_2O_3)_x (SiO_2)y$

$+$

Oxygen (O_2)
Airborne soil (example SiO_2)

$=$ Slag

The kind of slagant in fired side is as follows.

		A type	B type	C type	D type
Oxide (wt%)	V_2O_5	10.2	15.9	39.5	29.5
	NiO	3.5	19.5	14.8	10.7
	Fe_2O_3	10.7	14.7	12.5	39.4
	SiO_2	41.6	17.8	2.9	4.1
	Al_2O_3	34.0	32.0	1.7	
	SO_3			10.1	16.3
	MgO			5.2	
	CaO			7.7	
	Na_2O			5.7	
	Nature of slag	hard	friable	hard	fused

Note: Being hard and brittle when slag is cold

1.1 Mechanism of high temperature corrosion

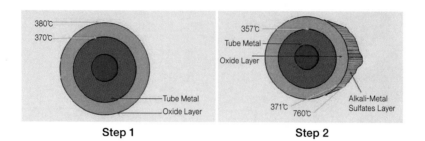

Step 1   Step 2

Step 1. Thin oxide layer (for example, Fe_3O_4) is formed on the metal at 415-426°C.

Step 2. Alkali-metal sulfate is formed on the oxide layer.

This is regarded as that alkali-metal reacts with SO_3 surrounded together with ash melting. And then it is precipitated into oxide layer.

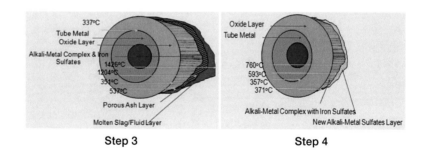

Step 3   Step 4

Step 3. Alkali-metal sulfate layer grows and it becomes to be sticky. Its surface temperature increases until the ash starts to adhere on itself. Since ash layer is porous and low heat transfer, its thicker layer brings higher temperature and it makes outer surface to be fluid. In this process, SO_3 emits. The SO_3 penetrates ash/alkali-metal sulfate, reacts with oxide layer, forms iron sulfate [$(Fe_2(SO_4)_3$]. Again oxide layer reforms and tube thickness becomes thinner.

Step 4. If slag layer accumulates continuously, it becomes to be dropped later. Again a new alkali-metal sulfate starts to be generated on the surface. Process like this is repeated to cycle.

Slag compounds and melting points of complex sulfates

Complex Sulfates	Melting Point (℃)
$K_3Fe(SO_4)_3$	618
$K_3Al(SO_4)_3$	654
*$KFe(SO_4)_2$	694
$Na_3Fe(SO_4)_3$	624
$Na_3Al(SO_4)_3$	646
*$NaFe(SO_4)_2$	690

Note: * In high SO_2 atmosphere

Slag compounds and melting points of vanadium compounds

Common Name	Vanadium Compound	Melting Point (℃)
Eutectic of $5Na_2O.V_2O_4.11V_2O_5$ And $Na_2O.V_2O_5$	–	532
Gamma sodium vanadyl vanadate	$5Na_2O.V_2O_4.11V_2O_5$	577
Sodium pyrovanadate	$2Na_2O.V_2O_5$	620
Sodium metavanadate	$Na_2O.V_2O_5$	630
Beta sodium vanadyl vanadate	$Na_2O.V_2O_4.5V_2O_5$	659
	$Na_2O.3V_2O_5$	668
Vanadium pentoxide	V_2O_5	673
	$Na_2O.6V_2O_5$	702
Sodium orthovanadate	$3Na_2O.V_2O_5$	850
Vanadium trioxide	V_2O_3	1,971
Vanadium tetraoxide	V_2O_4	1,971

Note: $3MgO$ (M.P. 675℃) + V_2O_5 = $3MgO.V_2O_5$ (Magnesium orthovanadate, M.P. 1,243℃)

In furnaces burning fuel oil, corrosion occurs first on tube supports, which are hotter than the tubes. Most common alloys suffer fuel ash corrosion. A 50% Cr/ 50% Ni alloy, IN 657, provides the best resistance to corrosion. IN 657 has resisted corrosion in furnaces burning fuel oil with up to 250ppm sodium + vanadium. Note that this alloy has lower creep strength than common support alloys, and support redesign may be necessary.

1.2 Flue gas condensation and corrosion

SO_2 and SO_3 are present in flue gas as by-products of burning fuels containing sulfur. At temperatures below the dew point, these sulfur containing gases can combine with water to form sulfuric and/or sulfurous acid. Furnace components corrode rapidly under these conditions. Corrosion can be avoided by always staying above the acid gas dew point during operation.

Flue gas condensation and corrosion typically occurs in furnaces that burn fuel oil, which is likely to contain high-sulfur levels. To prevent such corrosion, water washing is required to the furnace fireboxes during shutdowns to remove sulfur deposits.

2. Furnace Problems

2.1 Hot side problems

The two biggest problems encountered on the hot side, slag and corrosion, are both due primarily to the same causes: deposition of oxides and sulfates of vanadium and sodium. Vanadium and sodium are two chemical contaminants found in crude oil. They come from the organic materials that were laid down long long ago that ultimately formed petroleum. Crude oils from various regions of the world contain varying amounts of vanadium. Venezuelan crude oils contain some of the world's highest levels of vanadium. Sodium can also be added during the various transportation activities moving crude to the refinery and also during the refining process itself. As the crude oil goes through the refining process, these two elements and others that are present, are concentrated in the residual oils that typically become the fuels burned in heater. When these two elements are burned (as well as all other metallic elements) they combine with oxygen present to support combustion and with sulfur which is also present in the fuel to form oxides and sulfates. Vanadium oxide (normally present as the pentoxide) has a melting point of 675°C. When combined with sodium sulfate a very low temperature melting point eutectic can be formed (as low as 525°C). The graph below shows the temperatures where the various eutectics are "sticky." These temperatures can be at or below the operating temperatures of a furnace.

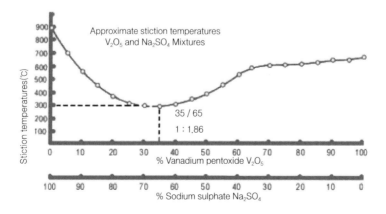

2.1.1 High temperature slag

Slag is the term applied to the metallic components that deposit on furnace tubes and surfaces. Temperatures inside the furnace are often higher than the melting point of the vanadium/sodium eutectics mentioned earlier. When the temperature is above the melting point, the oxides and sulfates are molten. When these materials are in the molten state, they are sticky. Putting this into terms that are more familiar, think of a glass of water. When the contents of a glass of water are thrown against a wall, the water coats the surface- it sticks to the wall. Now perform the same experiment with an ice cube. Very little of the ice will stick to the wall. This is the situation inside of the furnace. When sodium and vanadium are present they will form low melting point compounds. When the temperature is high enough, these compounds will be molten and sticky. They deposit (stick) on the furnace surfaces.

As the layers of metallic materials deposit on the tube surfaces, the outer layers insulate the inner layers from the heat of the flame. The inner layers cool and harden. But as the inner layers cool, they also insulate the outer layers from the cooling effects of the cooling fluid tubes. The outer layers get hotter still and more material will become molten and stick to these layers. This process repeats as long as the untreated furnace is in service. The layers of slag build up. As the slag builds on the tubes, the heat transfer into the fluid tubes is reduced. Often the furnace operator must increase fuel consumption to compensate for this loss of heat transfer. From theoretical calculations the following graph demonstrates the magnitude of the losses possible as slag thickness increases.

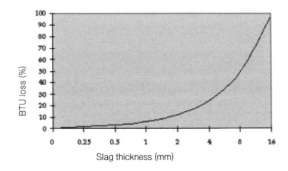

Btu loss relative to slag thickness

As slag thickness increases and heat transfer into the desired activity of fluid is reduced, the heat from combusting oil is lost into the back end of the furnace. Very often this heat is lost out the stack. It has been found that for each 4.4°C increase in stack temperature, a 1% of the efficiency for the furnace could be lost. These temperature changes in a furnace operation are useful indicators of a slag problem.

2.1.2 High temperature corrosion

High temperature corrosion results from the deposition of the same metallic components- vanadium and sodium. For this problem, these compounds must remain molten. Most metals form a protective oxide layer to protect against corrosion. This layer is formed from the outermost layer of metal atoms. The metals used in furnace tubes do this too. While vanadium and sodium are in a molten state they cause destruction of this protective oxide layer. Vanadium oxide in particular dissolves this protective oxide coating. As this protective coating dissolves and is removed, a layer of the tube metal is removed. A fresh protective coating forms from the next layer of tube metal atoms, is subsequently dissolved, and a new layer forms. This process continues, much like the peeling of an onion until the tube thins to a critical thickness. When the tube is thinned sufficiently and if still in service, a blow-out can occur. The overall effect is called corrosion.

2.2 Cold end problems

Just as the high temperature problems are caused by chemical elements, so are most cold end problems caused by chemical elements. In the case of the cold end the offending element is sulfur. Sulfur occurs naturally in the crude oils that are refined. The level of sulfur is concentrated into the residual fraction and finds its way into nearly all furnace fuels at varying levels. The level encountered is normally related to the specification level of the fuel purchase contracts.

When sulfur is burned in the presence of oxygen (required to support combustion) it forms sulfur dioxide. Normally about 1-2% of the sulfur dioxide is further reacted with additional oxygen to form sulfur trioxide. More or less may be formed based upon the conditions found in the furnace. For example levels of excess air/oxygen (higher, more formed); vanadium, nickel or iron (higher, more formed); sulfur in the fuel (more present, more formed); size of furnace (larger, more formed); temperature of firebox (higher, less formed); and the residence time in the furnace (longer, more formed). All of these factors are competing at the same time making prediction of the actual amount of sulfur trioxide that will be formed difficult. The following Table shows the expected amount of sulfur trioxide based upon excess oxygen and sulfur content.

Estimate of sulfur trioxide in combustion gas

%Sulfur in fuel oil	0.5	1.0	2.0	3.0	4.0	5.0
Excess O_2 %	Sulfur trioxide expected in gas (ppm)					
1	2	3	3	4	5	5
2	6	7	8	10	12	14
3	10	13	15	19	22	25
4	12	15	18	22	26	30

The chemical reactions forming these sulfur compounds are represented as follows:

$$S + O_2 = SO_2 \text{ (sulfur dioxide)}$$

$$2SO_2 + O_2 = 2SO_3 \text{ (sulfur trioxide)}$$

One interesting note about the second reaction in particular is that the presence of hot iron surfaces and vanadium slags are required to make the reaction go in the direction indicated. If there were no surfaces to actively catalyze this reaction, the temperatures found in furnaces would effectively limit the formation of sulfur trioxide by forcing the reaction in the reverse direction.

2.2.1 Cold end corrosion

The formation of sulfur dioxide or trioxide is not a problem within the furnace (although this can be a problem outside the furnace when in the atmosphere- acid rain). The problem with sulfur trioxide is that it condenses with water vapor (formed from the combustion of hydrocarbons in the presence of oxygen) to form sulfuric acid according to the following reaction.

$$SO_3 + H_2O = H_2SO_4 \text{ (sulfuric acid)}$$

(Sulfur dioxide has a similar reaction, but the resulting acid normally does not condense at furnace temperatures.). The formation of sulfuric acid is a problem because when temperatures are low enough, the acid can condense on metal surfaces causing a severe corrosion problem. This cold end corrosion normally occurs in air preheaters where temperatures can be low enough- after heat is removed to warm incoming air- so that the acid condenses on the metal surfaces. Sulfuric acid corrosion can also occur on stack walls and particularly on any metal tops of stacks.

The formation of acid in air preheater equipment- often black in color- can also act as a trap for fly ash. This leads to deposits that can interfere with the transfer of heat in the air preheater as well as a corrosion problem. Acid smuts that leave a stack are particles of fly ash with adsorbed acid on them. When these smuts float to the ground and deposit on automobiles, they can cause additional problems.

2.2.2 Opacity and acid plumes

Opacity has many causes from unburned fuel through dispersions of metal oxides that carry through the furnace and out the stack. The cause we are concerned with is due to the same condensation of sulfuric acid discussed above. Very fine droplets of acid form after the exhaust gases leave the stack and are cooled to the acid dew point. This leads to a visible plume often blue-white in color (can be reddish or yellow-brown depending on sun angle). The most obvious trait of this type plume is its persistence. An acid plume will carry a long way before dissipating.

3. Counter Measures to Problems

There is one nonchemical method that can solve all these problems available to the furnace operator. That is to reduce the amount of excess air going to the furnace. Excess air is defined as the amount of air above that needed to exactly support combustion. This works by limiting the formation of vanadium oxides to those of a lower "valence" state. These compounds have higher melting points. Also since the formation of sulfur trioxide is dependent upon oxygen being present, by reducing the excess air to the absolute minimum, there is no oxygen present to form sulfur trioxide.

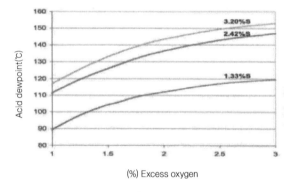

Acid dew point vs. excess oxygen (with fuel oil of different sulfur levels)

The reason this method becomes impossible to attain is long before this minimum excess air level can be reached, incomplete combustion would result in severe, unacceptable furnace smoke. However, reducing the excess air to as low a level as possible is always desirable for minimizing the formation of both hard slags and sulfur trioxide.

Experimentation on chemical means to solve these problems has been conducted since the 1950's. Early in this work magnesium was found to offer the most cost-effective method to control slag, high temperature corrosion, cold end corrosion, and some opacity problems. And just as the problems are caused by chemical reactions, the solution of the problems using magnesium can be described by chemical reactions too.

3.1 Solution to hot side problems

Magnesium combines with vanadium and sodium (chiefly vanadium) to form higher melting compounds. When magnesium is combusted in the presence of air, it forms magnesium oxide (MgO). The magnesium oxide combines with vanadium pentoxide to form among other compounds, magnesium orthovanadate.

$$3MgO + V_2O_5 = 3MgO.V_2O_5$$

Whereas vanadium pentoxide has a melting point from 675°C and lower, magnesium orthovanadate has a melting point of 1,243°C. This temperature is well above the typical operating temperature of a furnace so the ash is no longer sticky.

3.1.1 High temperature slag

When the melting points of any vanadium compounds formed are above the operating temperature of the furnace they will no longer be molten or sticky. If no longer sticky, there will be no build up on furnace surfaces and heat transfer will not be impeded.

3.1.2 High temperature corrosion

Similarly, if the vanadium compounds formed are no longer molten, they will no longer dissolve the protective oxide coatings of the tube metal. If they no longer dissolve the oxide coatings, they will no longer cause corrosion.

3.2 Solution to cold end problems

Magnesium is also used for eliminating or at least greatly reducing cold end problems. In this instance magnesium is used to coat the internal surfaces of the furnace. Recall that sulfur trioxide formation requires the catalyzation by hot iron surfaces or vanadium slags.

3.2.1 Cold end corrosion

When the catalytic surfaces are coated with the fine particles of magnesium oxide that result from the combustion of oil soluble magnesium products, they are rendered passive. Magnesium orthovanadate does not catalyze the reaction of sulfur dioxide to sulfur trioxide. Thus the formation of sulfur trioxide is greatly reduced. Also, magnesium oxide acts to neutralize any acid or sulfur trioxide that may still form by the following reactions.

$$MgO + SO_3 = MgSO_4$$

$$MgO + H_2SO_4 = MgSO_4 + H_2O$$

Magnesium sulfate is not corrosive and water vapor already exists in the stack plume. By the coating action and neutralization, formation of sulfur trioxide can be substantially reduced or eliminated. This eliminates cold end corrosion problems.

3.2.2 Opacity

Opacity that is the result of sulfuric acid formation in the plume can be controlled in the same manner with the use of magnesium additives. When sulfur trioxide is reduced or eliminated, sulfuric acid can no longer be formed so there is no acid plume.

4. Heater and Coke

Coke formation was initiated by asphaltene precipitation from unstable crude (West Texas, New Mexico, Ohio/ Pennsylvania and Alberta). The TMT (tube metal temperature) increases as coke lays down on the inside of the tube. With rising TMTs, heater firing must decrease or the TMTs will progressively escalate until their limit is reached. The heater must then be shut down to remove the coke. Rapid coke formation is caused by a combination of high oil film temperature, long oil residence time and inherent oil stability. In the most cases where atmospheric heater coking occurs, the root cause is high average heat flux, high localized heat flux or flame impingement. Oil thermal stability depends on crude type. For example, some Canadian and Venezuelan crude oils have poor thermal stability and begin to generate gas at heater outlet temperature 360℃. At outlet temperatures much above 371℃, these same crudes begin to deposit sufficient amounts of coke to reduce heater runs to two years or less.

Another form of oil instability is asphaltene precipitation. As the oil is heated, the asphaltenes become less soluble, depositing in low velocity areas, fouling crude preheat exchanger, heater tubes or atmospheric column internals. When asphaltenes separate from the crude oil, the material deposits inside the tubes. This increases heat transfer resistance, raising asphaltene temperature and TMTs.

Furthermore, when asphaltene deposits are widespread in the convection or radiant sections, heater firing must increase to meet the targeted heater outlet temperature. This leads to a higher localized heat flux, further raising the temperature of the asphatenes deposited on the inside of the tubes. The temperature of these asphatene eventually exceeds their thermal stability, resulting in coke formation and even higher TMTs, because the coke layer has lower thermal conductivity than asphaltenes. Heater TMTs eventually exceed metallurgical limits, requiring a heater shut down to remove the coke. In this example, heater run lengths were as low as 90 days between piggings.

The oil velocity inside the heater tubes was only 5.5-6.0 ft/sec prior to the oil vaporizing, which corresponds to a 250 lb/ft^2 sec oil mass flux. Crude exchanger and heater tubes should be designed for oil velocity of 8-10 ft/sec or higher. Since many designers set the maximum allowable pressure drop through exchangers and the heater as design criteria, low velocities are often the result of meeting pressure drop.

A refinery revamped its existing atmospheric heater and installed a new parallel "helper" heater. Only reducing heater firing by the helper would not have improved the heater run length. That is because the oil velocity of existing heater will be reduced to less than 4 ft/sec after installing the helper. Total fired heater "absorbed duty" is the sum of the convection and radiation duties. Maximizing the convection duty minimizes the radiant duty, which lowers the oil film temperature, reducing the rate of coking.

So, the convection section was replaced with a similar design, except some of the tube fins were upgraded from carbon steel to 11-13 Cr to avoid damage from high temperature and to maintain performance through the run. And the radiant section was completely retubed with smaller diameter. Bulk oil velocities were increased from, 5.6-6.0 ft/sec to almost 10 ft/sec, which resulted in an oil mass velocity of 460 lb/ft^2 sec (twice before revamp).

The smaller tube diameter dramatically increased the pressure drop, requiring a larger pump impeller and motor in the flashed crude pump. Crude charge has been increased by 14% and the heater outlet temperature has risen from 335-354℃. The heaters has not been pigged and the TMTs have shown very little rise since startup.

5. Heater Problems & Maintenance

5.1 Burner

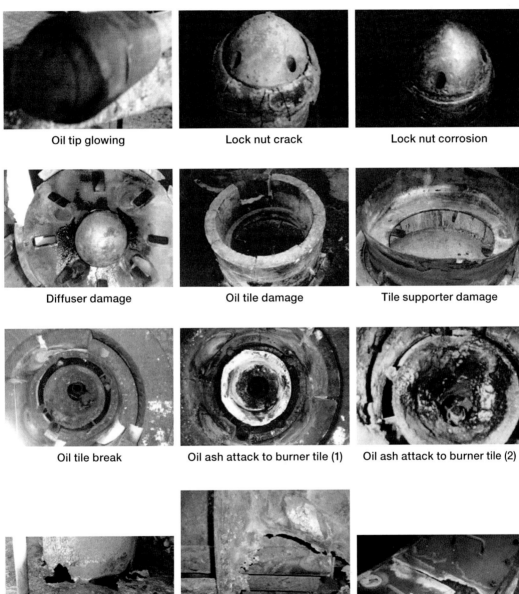

Oil tip glowing	Lock nut crack	Lock nut corrosion
Diffuser damage	Oil tile damage	Tile supporter damage
Oil tile break	Oil ash attack to burner tile (1)	Oil ash attack to burner tile (2)
Burner casing damaged (1)	Burner casing damaged (2)	Heater casing damage

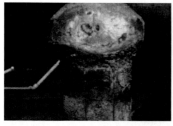  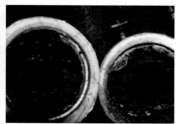

Stage gas tip thermal damage   Gas tip riser corrosion

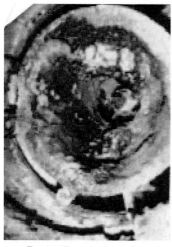  

Regentile ash corrosion   Burner tile damage (1)   Burner tile damage (2)

Refractory brick collapsed   Burner tile cracked & broken   Burner tile damage 3

5.2 Efficiency improvements

Efficiency improvement 1

Efficiency improvement 2

Efficiency improvement 3

Efficiency improvement 4

Efficiency improvement 5

Efficiency improvement 6

Efficiency improvement 7

Good damper control 8

Fire box sealing 9

Tube added 10

Before & after burner cleaning 11

5.3 Heater & burner accidents

Heater explosion

Heater explosion   Heater collapse   Furnace explosion

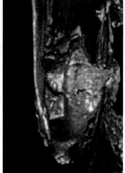

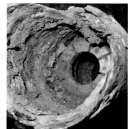

Pilot & main burner burnt

Tube rupture

5.4 Monitoring & troubleshooting

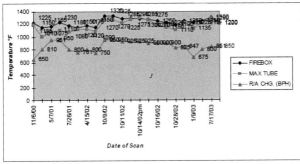

Infrared scan

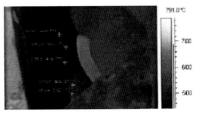

Infrared scan

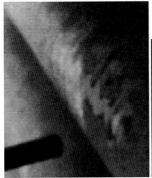

Fuel oil ash removal

Fuel gas coalescer

Fuel oil strainer

Pig cleaning

CDU heater coke

CDU heater coke

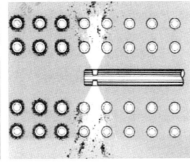

Retractable soot blower

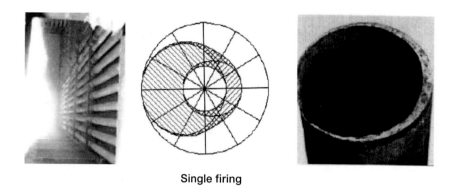

Single firing

5.5 Tube rupture step

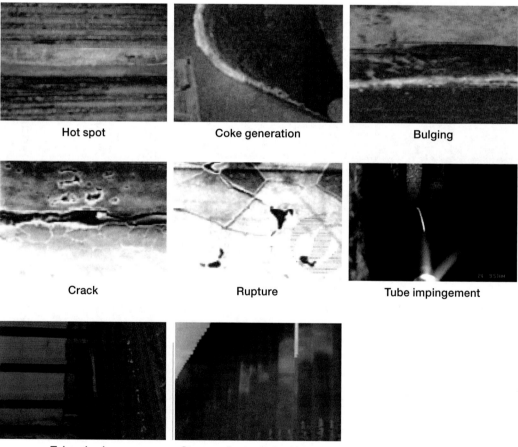

Hot spot	Coke generation	Bulging
Crack	Rupture	Tube impingement
Tube glowing	Steam reformer tube hot spot	

5.6 Radiant tube leak accidents

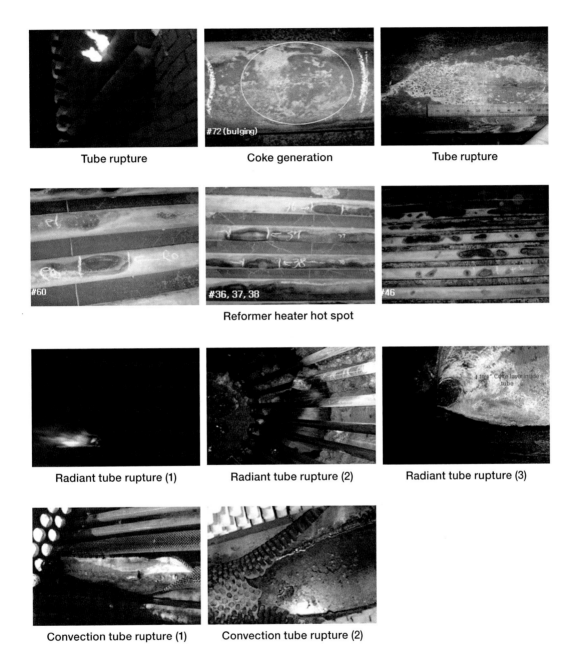

Tube rupture   Coke generation   Tube rupture

Reformer heater hot spot

Radiant tube rupture (1)   Radiant tube rupture (2)   Radiant tube rupture (3)

Convection tube rupture (1)   Convection tube rupture (2)

5.7 Burner & other accidents

Burner oil leak

Fuel gas manifold steam purging

Oil gun steam flushing

Burner oil leak

Burner oil dripping & burned

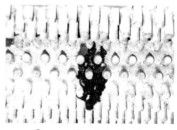
Convection tube leak

Convection inspection door

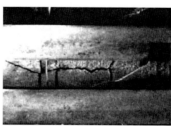
Wall refractory crack (1)

Castable refractory crack (2)

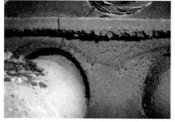
Tube sheet crack (1)

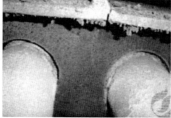
Tube sheet crack (2)

Refractory detached

5.8 Pass flow coking

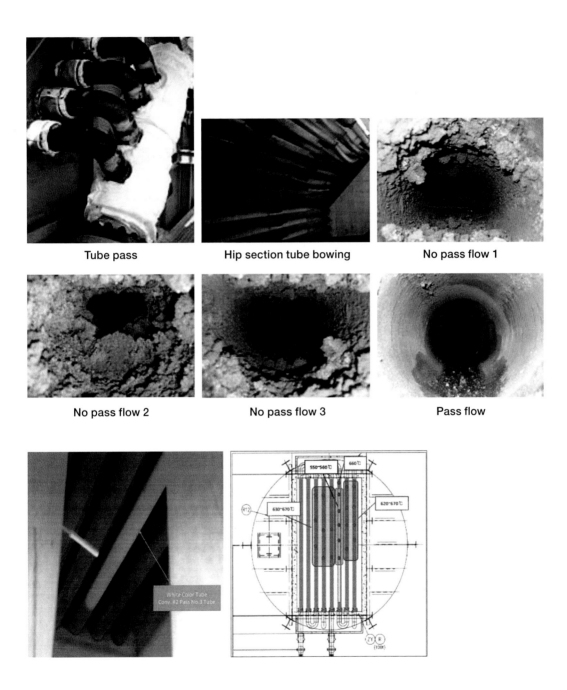

Tube pass   Hip section tube bowing   No pass flow 1

No pass flow 2   No pass flow 3   Pass flow

Circle is flue gas temperature thermocouple to check bridge wall temperature (BWT). Rectangular box indicates the convection shield tube with white color that has lower temperature by 100℃ in infrared scan than adjacent tubes.

Tube bowing

5.9 Burner plugging & coking

Flame out

Over duty

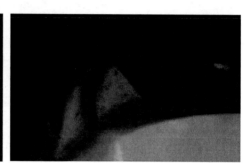

Tile coke generation

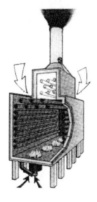

Afterburning

Flashback

Lift-off

Gas tip coking (1)

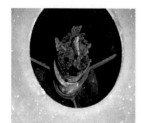

Gas tlp coking (2)

Gas tip plugging

Gas tip corrosion

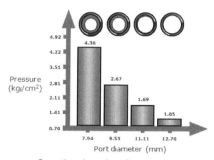
Gas tip plugging & pressure

5.10 Safety & check equipments

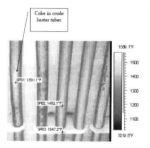
Infrared survey

Burner problem

Arch heat flux probe

Pilot burner flame detector

FLIR camera

Flue gas analyzers

Cyclops Pyro-gun

6. Corrosion in Flexible Burner Hoses

6.1 Flexible hoses in burner systems

Flexible hoses have several specific advantages when used in the design of pipe work. Among the advantages are the hoses' ability to absorb vibration and operate effectively under high pressure. However, the most important advantage is the ability of flexible hoses to be adjusted easily. When employed in burner piping for fuel oil, atomizing steam or fuel gas, flexible hoses are generally used for the purpose of burner-gun positional adjustment. Flexible hoses permit a more economical installation compared to rigid piping in difficult locations- when connected to flexible hose, it is relatively easy to adjust the elevation or orientation of a burner gun without any mechanical modifications in burner piping.

6.2 Piping leaks cause a fire

The process details were as follows: operating fuel gas pressure of 1.2-1.5 kg_f/cm^2: operating fuel gas temperature of between 30-60℃; fuel gas composition of 69% H_2, 10% ethane, 8% C_3H_8, and 13% other components, but no critical toxic components. The system was designed to operate with either oil or gas, but could not use both fuels simultaneously. In general, because burner tips are custom-designed in number, size and in the angle of the tip holes for specific applications, the damaged burner tip will result in undesired flame characteristic, including length and size, as well as low performance in operation, such as higher NO_x emissions.

Flexible hose

Burner tip

Damaged burner tip

6.3 Failure analysis

Attention then turned to corrosion as ultimate cause of the accident. Because the corroded pits were found on the corrugate tube, it was necessary to carry out EDX (Energy-Dispersive X-ray spectroscopy) in an effort to identify the component of corrosion from the deposit scale on the inside of the tube. EDX is an analytical technique used for the elemental analysis or chemical characterization of a sample. The EDX studies were carried out to determine the elemental compositions of the matrix and the deposits and scales on the failed tubes. The below EDX spectrum of the failed tube shows iron (Fe), sulfur (S) and chromium (Cr) in very high concentration.

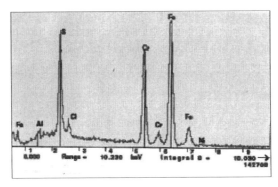

EDX spectrum

The results of EDX studies indicate that there is substantial incorporation of sulfur compounds in the corrugated tube inside during operation. Sulfur and chloride result in corrosion, while Cr, Fe, Ni resulted from the corrosion. The pitting corrosion is caused by the effects of sulfur and chloride, especially when they are present in hydrous solutions. Attack on the material is affected by chemical concentration, temperature and the type of material which the corrugated tube is manufactured.

Stagnation of fuel gas condensate during heater operation may increase the corrosivity of the environment, and reduce stability of the protective surface films and increase susceptibility to metal loss. Most stainless steels form a protective film of stable oxides on the surface when exposed to oxygen gas. The rate of oxidation is dependent on temperature. At ambient temperatures, a thin film of oxide is formed on the stainless steel surface. In accordance with the corrosion resistance charts published by NACE (National Association of Corrosion Engineers), it is not recommended for Type 304 to be used with sulfuric acid and sulfurous acid. The sulfur content in fuel gas is only 10 ppm, so the fuel gas condensate is not likely to have caused the corrosion. Tube failure due to fuel gas was ruled out.

6.4 Dew point corrosion

Given the evidence of sulfur from the EDX, what is the source of the sulfur? In approaching the corrosion issue, it is necessary to look into the flue gas side, as well as fuel gas itself, in order to find out the source of the sulfur. Also it is crucial to understand the mechanism of flue gas acid dew point corrosion. It is very important not to cool the flue gas below its acid dew point because the resulting liquid acid condensed from the flue gas can cause serious corrosion problems for the equipment. During oil firing, the gas burner is not in operation, however, the gas guns are placed in the burner and the gas tips are exposed to the hot flue gas in the radiant box. One of the most striking features of this combustion process is that the flue gas penetrates through the idling fuel gas tip holes, and collects inside of the corrugated tube.

To explain the flue gas flow mechanism, and why flue gas enters the burner gun, it is helpful to use Charles' Law of gas volume- at constant pressure, the volume of a given mass of an ideal gas increases or decreases by the same factor as its temperature on the absolute temperature scale. The hot flue gas continuously flows into the burner gun and into the corrugated tube due to the gas volume difference between the hot burner tip area and cold flexible hose area. Once the flue gas stays inside the corrugated tube, then the gaseous flue gas becomes condensate when the temperature drops below the dew point. The fuel oil contains sulfur at a concentration of 0.3 wt% and the flue gases of combustion may also contain small amount of sulfur oxides in the form of gaseous sulfur dioxide (SO_2) and gaseous sulfur trioxide (SO_3). The gas phase SO_3 then combines the vapor phase H_2O to form gas phase sulfuric acid (H_2SO_4), and some of the SO_2 in the flue gases will also combine with water vapor in the flue gases and form gas phase sulfurous acid (H_2SO_3):

$$H_2O + SO_3 -> H_2SO_4 \text{ (sulfuric acid)}$$

$$H_2O + SO_2 -> H_2SO_3 \text{ (sulfurous acid)}$$

The collected flue gas (gaseous acid) in the flexible hose between the gas tip and the isolation valve will continuously condense into liquid acid, because the burner piping located outside the furnace cools down to atmospheric temperature, which is far below the sulfuric acid dew point of flue gas (about 120℃ at 0.3 wt% fuel oil, depending upon the concentration of sulfur trioxide and sulfur dioxide). Eventually, the liquid phase sulfurous and sulfuric acids lead to severe corrosion. The explanation may be somewhat confusing, because it is generally thought that the amount of flue gas flowing through the small gas tip holes is negligible. However, in actual field operation, especially during cold weather, it is observed more than 50mL of condensate inside the 1in. flexible hose when the flexible hose is dismantled after one week of operation with fuel oil firing only (no fuel gas firing). Therefore, there is no doubt that the failure was the result of corrosion by flue gas condensation. See below figure detail 'A' for illustration.

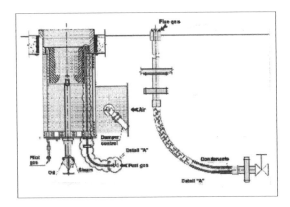

Flue gas condensate

6.5 Burner tip plugging

Liquids, particulate matter, unsaturated hydrocarbons and H_2S in fuel gas can cause most plugging problems. In order to identify the material causing the tip plugging, the fuel gas analysis and the design review of the knockout drum that removes liquids from the fuel were carried out. However, there were no out-of-specification instances in the above listed items. Nonetheless, the focus needed to be on the condensate from flue gas. It is important to recognize that the collected condensate will be carried over to the gas tip as soon as fuel gas is pressurized and serviced. Under continuous fuel gas firing operating conditions, this may not be a problem because the tips are cooled enough by the high velocity fuel gases flowing through gas tips. Upon switching from oil firing to gas firing, the condensate, which stays inside the flexible hose, will automatically be delivered to the hot gas tip. This will lead to abrupt evaporation of liquid inside of the hot gas tip, and then result in plugging due to hydrocarbon coke build up, and finally to melting of gas tip. Overheating the burner tips can cause the carbon in the fuel to thermally crack, giving rise to severe coking inside the tips, which leads to plugging of the holes.

6.6 Recommendations

Considering the above, it is highly recommended that the fuel gas piping for combination type burners that could possibly have flue gas condensation be designed with rigid piping (size 1 in. Sch. 40, 3.4mm thickness) instead of flexible hose. The rigid piping is about 13 times thicker than flexible bellows tube's thickness, as depicted in Fig. below.

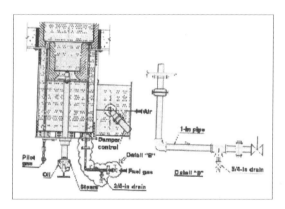

Burner rigid piping

In real world industrial practice, little is known about corrosion failure of rigid burner piping that may experience dew point corrosion from flue gas condensation. It is possible that the thicker walled piping could prolong pipe lifetime. For gas firing burners, the use of rigid piping is also recommended in the case of intermittent gas firing burners that use high-sulfur fuel gas. If the flexible hose is not avoidable, then the material of the bellows tube should be Inconel 625, which is properly resistant to sulfur corrosion, discolored from chemical attack and begin to fracture. In order to prevent fuel gas tip damage due to liquid carryover, a drain system at the nearest point from the burner gun should be provided at the lowest point of fuel gas piping between the first block valve and burner tip (see above Fig.).

Also, it is necessary that the activity of the liquid drain before gas firing should be strictly specified on the burner operation manual. In case of the API RP535 2nd ed. (Burners for Fired Heater in General Refinery Services), it is highly recommended that the detail requirement for preventing "flue gas acid dew point corrosion" should be clearly specified, in addition to the current mechanical requirement for flexible hoses (Flexible hoses require special attention to avoid failure due to kinking) or stainless steel lined with a PTFE (polytetrafluoroethylene) liner and flared end fittings.

Periodic soap bubble tests on the surface of the flexible hoses can eliminate the potential of fire accident. Also, close visual monitoring can allow earlier identification of possible failures. During inspection, corrosion of a flexible corrugated metal hose can be spotted by looking for signs of chemical residue on the exterior of the assembly, or by pitting of the metal hose wall.

7. Fired Heater Safety Practice

7.1 Feed control valve (field)

Provide steady and controllable feed rates. Provide feed control valve required for each separate feed to furnace.

Exceptions include: (1) Hydrotreaters and Rheniformers where the recycle H_2 is controlled by the recycle compressor operation (2) Flashing liquids from high pressure separators (3) Manual valves acceptable for reboiler loops and BFW circulation

(Example) Feed control valve (FC) installed

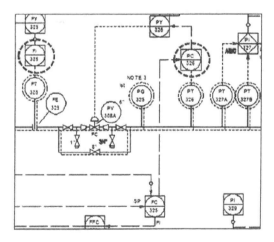

7.2 Fuel control valve (field, panel)

Control fuel to prevent overfiring. Fuel indicator in the control room helps operators troubleshoot furnace operating problems such as bogged firebox. Provide fuel control valve for each furnace and provide fuel flow indication in control room.

(Example) Fuel gas flow control valve (PC)/fuel gas flow indication (FI) provided

7.3a Total feed low flow alarm (panel)

Total feed FI/FAL prevents a decrease or loss of feed from causing coking, overheating of the tubes, or tube rupture. Need independent instrumentation to prevent an instrument malfunction from causing both the upset condition and loss of alarm. Provide total feed flow indication and provide total feed low flow alarm. The flow FI/FAL can share a flow element (orifice) with the FRC, but they must have separate orifice taps and transmitter. Separate orifice taps and transmitter will be necessary for safety shut down system.

Exceptions include: (1) Furnace with multiple feeds can have separate FAL's on each feed stream in place of total feed (2) Separate FI/FAL on recycle hydrogen feed to furnace is required for hydrotreater, hydrocracking, and Rheniformer heaters and optional for the hydrocarbon feed to these furnaces (3) Separate FI/FAL on the steam feed to H_2 reformer furnace and optional for the hydrocarbon feeds to these furnaces (4) Manually controlled reboiler loops can have a software FI/FAL (5) Multi-pass furnaces with pass balancing and redundant FI/FAL on each pass

(Note) Panel alarms brought into the DCS or Triconex SSD system and displayed as a common trouble panel alarm for each furnace may be used in place of a separated panel alarm, i.e., Low feed FAL, High Tube Skin TAH, etc., provided the operator can call up the alarm on the DCS or Triconex alarm log, identify it, and the panel alarm will not clear until the alarm is cleared in the field.

7.3b Individual pass low flow alarm (panel)

Prevent loss of flow in an individual pass from causing coking, tube overheating or tube rupture. Need independent instrumentation to prevent an instrument malfunction from causing both the upset condition and loss of the alarm. Provide pass flow indication on multi-pass furnaces with single phase flow at the inlet and provide pass low flow alarm on multi-pass furnaces with single phase flow at the inlet.

Exceptions are Hydrogen & Catalytic Reformers with many passes. Furnaces with 2 phase flow at the inlet will be instrumented with skin TI's, since flow measurement will not be accurate. These furnaces should also have symmetrical inlet and outlet piping and pass configuration.

(Note) An alternative to provide redundancy for alerting to low pass flow is to provide alarms on control valve positioners instead of separate FAL transmitter. However, the separate FAL transmitter is preferred. FAL's can be combined into a common alarm.

(Example) Individual pass flow indication (FI), pass flow low alarm (FAL), and pass flow low low alarm (FALL) installed

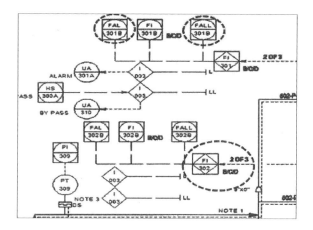

7.4 Furnace outlet TI/TAH (panel)

Pass outlet TI/TAH give indication of a decrease in pass flow or a heat input imbalance between passes. TI/TAH on the combined outlet indicates decrease of total flow to the furnace or overfiring. Also needed to prevent exceeding temperature limits of downstream facilities.

Provide temperature indication on combined process outlet. Combined outlet TI needs to be separate from the TRC control signal. Provide high temperature alarm on combined process outlet. Intent can be met by TAH on individual pass outlets. Provide temperature indication on individual pass outlets. This includes convection banks in hydrocarbon service not in series with the radiant section.

Exceptions include Rheniformers and H_2 Reformers which have many multiple passes. Provide high temperature alarm on individual pass outlets. This includes convection banks in hydrocarbon service not in series with the radiant section. Pass outlet TAH's can be combined into a common trouble alarm.

(Example) TI & TC installed separately, TI installed in each pass outlet of radiant and convection section, and HI alarm (TAH) & HI HI alarm (TAHH) applied to DCS.

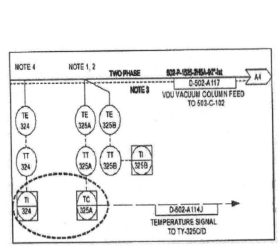

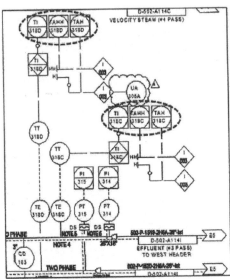

7.5a Tube skin TI/TAH (panel, field)

Ensure tubes operate within temperature limits. Give indication of internal coking or decrease in flow through the tube. Provide temperature history for tube life assessment.

Provide a minimum of 3 tube skin TI's per radiant pass. Single pass heaters require more than 3 for adequate coverage. Provide convection section/shock tube TI's: (1) one TI per pass if in series with radiant pass in hydrocarbon service (Min. of 2 for the bank) (2) two TI's per pass on convection section tubes if not in series with radiant pass in service (3) two TI's per convection section shock bank in BFW/STM service, if there is no other indicator of loss of flow.

(Exceptions) In Rheniformer heaters, radiant TI location and number should be determined with thermo graphic survey data. H_2 Reformer tube temperatures are periodically monitored with hand held pyrometers and IR scans. All tube skin TI's are to be Gayesco Refractopads or Extractopads. Set appropriate alarms on all tube skin TI's (consistent with SIS). Number and placement of TI's should be determined by technical review considering inspection history of tubes, infrared scans, and modeling results if available. TAH's can be combined into a common trouble alarm. Increase IR monitoring in the event of loss of all skin TI's on a pass.

(Example) Total 40 TI's installed- 5 TI's per each radiant pass, 1 TI per each convection pass (shock tube)

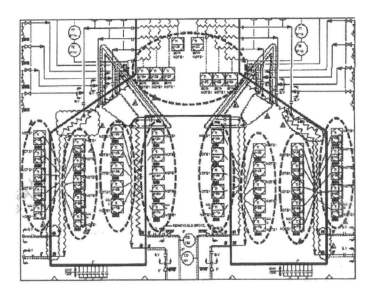

7.5b Tube temperature monitoring by IR scan (documentation)

Give indication of internal coking or decrease in flow through the tube. Perform routine infrared surveys on all furnaces with the frequency set by severity of operation and criticality of the furnace. Typical frequency is quarterly for all furnaces and more often for coking services.

Cyclops Pyro-gun

(Example) Priority places on tube color by visual (naked eye) and then it is checked using a laser thermometer (pyrometer) or a thermograph by infrared camera.

7.6 Stack and arch inlet temperature alarms (panel)

Give indication of tube leak or afterburning in convection section. Keep flue gas temperature below limits of convection section tube supports, refractory, stack damper, etc. Provide indication for stack flue gas temperature. Set stack TI alarm based on limits of stack damper or stack refractory, which ever governs (consistent with SIS). Provide indication for arch (bridge wall) flue gas temperature. Set arch TI alarm based on limits of convection section tube supports or refractory, which ever governs (consistent with SIS). Consider multiple arch TI's for large or multi-cell furnaces.

7.7 Minimum fire bypasses (field)

Prevent a firebox explosion caused by the fuel control valve closing and re-opening unexpectedly. Provide Minimum Fire Bypass (MFBP) or Pilots with independent fuel source (fuel takeoff upstream of TRC or separate fuel source). Minimum fire bypass is preferred over pilots. Acceptable MFBP's are the following:

(1) Manual globe valve car sealed in position. (2) Valve setting must be reset periodically or when number of burners changes to ensure it is set above the minimum pressure of the burners (3) Minimum pressure by-pass valves (i.e. pressure regulators) are acceptable if the following concerns are properly addressed: (1) Testing procedures to ensure speed of opening on demand (2) Self-draining inlet piping (3) Inlet filters with adequate changeout procedures.

(Example) Pilot burners applied- fuel gas & combustion air operated independently, minimum fire bypass is to mean the by-pass of regulator on pilot line, and regulator valve setting periodically checked (local PI) and reset if abnormal. However, a new heater is control with PCV and monitoring P & PCV opening possible (setting DCS alarm). In case of filter change, By-pass open and PCV isolated.

7.8 Firebox purge system and snuffing steam (panel, field)

The purge system is used to ensure a combustible free atmosphere in the furnace before start-up and after shutdown. The snuffing steam is used to control a tube rupture fire and prevent reignition of flammable vapors after the fire has burned out.

Provide means to adequately purge firebox before light-off (purge steam, draft fan, steam injection at bottom of stack, etc.). Provide snuffing steam system with control at least 15m from furnace. (The purge steam and snuffing steam systems are typically the same system.) Ensure purge/snuffing steam system traps are operational (hot). Provide permanent labels for purge/snuffing valves.

(Note) Either purge steam or draft fans must be provided for use in purging firebox. Snuffing steam should not be used in high pressure furnaces during a furnace tube rupture incident, due to release of H_2 (which can easily reignite) and H_2S (which is a personnel hazard). Steam traps and self draining piping/weep holes will be provided to ensure free water does not accumulate in the purge piping when it is not in use.

(Example) Snuffing LP steam, 2"x 8 points, are installed on the bottom floor of furnace. The manipulating steam valve is prepared at least 15m away from furnace and marked outstandingly. And prior to the valve, condensate trap exists at its low point in order to use always if needed.

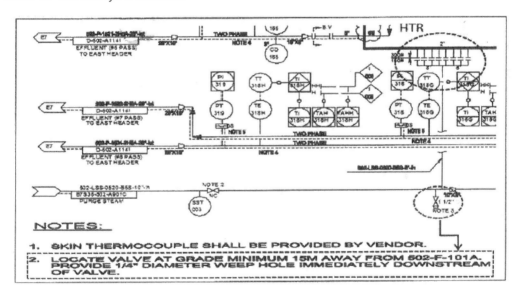

7.9 Firebox combustible gas test points (field)

These points are used to test for combustible gas prior to start-up. Locate and label combustible gas sample points at burners, inspection doors, and ports so all stagnant areas can be tested. Ensure gas sample points are consistent with description in startup procedure. Gas sample ports will be provided at burners, peepholes, and identified stagnant areas. They will be clearly labeled on the furnace.

(Example) Prior to fire off of pilot burner, it is to see whether combustible gas exists or not at any stagnant area of furnace. In general, the gas test is mainly focused on the peep doors both near burner and near hip section.

7.10 Burner isolation (field)

Positive isolation is required at each burner to prevent fuel from leaking into the firebox when the burner is out of service. Provide isolation at each burner. Acceptable systems are: (1) Double block & bleed (2) Block valve with operator blind (3) Single block with dropout spool and cap

Provide the same isolation at each pilot.

(Exception) Hydrogen furnaces with many small burners may have double block as an acceptable option.

(Example) Double block valve (gate valve + globe valve) is used for burner isolation and bleeder valve exists at low point near burner. Pilot burner is the same system as main burner also.

7.11 Protection system (minimum Class C)

Provide a minimum level of protection in the interim period before a Class A or B system is installed. Provide operator/computer controlled TCV from board/console.

Computer closes TCV valve automatically to minimum fires when conditions dictate (e.g. loss of feed, Hi skin TI, etc.). One of the minimum requirements for a furnace used on startup only (Class C) and for all other furnaces until a Class A or B safety shutdown system is installed.

(Example) Fuel increase and decrease are controlled by TCV, when tube outlet temperature rises promptly, Low & High temperature alarm happens and it is informed to operator. Feed loss is adjusted to shift of manual mode and High Tube Skin temperature follows the flame adjustment after checking field.

7.12 Safety shutdown system Class A or B (field)

Safeguard the furnace by shutting down fuel when a condition is identified which could lead to a serious incident such as a tube rupture or a firebox explosion. Safety Shutdown Systems are required for all continuous operating furnaces: (1) Class A [triconex-triple redundant] for Critical furnaces (e.g. Coker, Crude unit, Rheniformers, etc.) (2) Class B [PLC based] for smaller, less critical furnaces (e.g. distillation, little downstream impact, etc.) where false trips can be tolerated. Grouping less critical furnaces into one Triconex may be economic. Guidelines for shutdown trips (e.g. low process flow, low FG pressure, etc.) shall be determined by the associated team. Special shutdown considerations are required for hydrogen reforming furnaces, as rapid chopping could quench the tubes and damage them. Consider ramping these types of furnaces down in a controlled manner prior to chopping feed and fuel completely. Provide means to test safety shutdown system on-line.

(Example) ESD interlock logic is made up. In case of tube rupture or fire box explosion, fuel is cut due to arch draft HI HI. But pilot burner runs. All logic is managed from PLC.

7.13 Burner header fuel gas PI & PAL (panel)

Prevent low fuel gas pressure at the burner from causing a flame-out. This is normally an SSD trip as well. Provide panel indication of fuel pressure downstream of TRC. Provide low fuel pressure alarm above trip point. Provide fuel chop just above burner's low pressure limit. Fuel gas PI/PAL will be provided, with control room readout. PI tap will be located downstream of all TRC/PRCs on fuel gas to burners.

(Example) On all individual burner line behind TRC signal, PG is installed. On the pressure above the set point of ESD interlock logic, double alarms (LO & LO LO) were consisted of.

7.14 Burner header fuel gas PAH (panel)

Prevent high fuel gas pressure at the burner from blowing the flame off the burner and causing a flame-out. This is normally an SSD trip as well. Provide high fuel pressure alarm below trip point. Provide fuel chop just below burner's high pressure limit. Fuel gas PAH will be installed, with control room readout. PAH tap will be located downstream of all TRC/PRCs on fuel gas to burners. PRD or high burner pressure shutdown will be installed if burners can be over pressured at F/G supply pressure.

(Example) PAH alarm installed. In case of premix nozzle type in pilot burner, flame out due to flame blowing doesn't happen.

7.15 Fuel gas pressure control (field)

Provide a stable fuel gas supply pressure to the furnace temperature control valve resulting in stable furnace operation. Provide stable fuel gas pressure to each furnace. Usually a fuel gas pressure control valve is required. It may serve more than one furnace within a plant. Provide fuel pressure control valve for pilots, if used.

(Example) Fuel gas pressure valve installed. Fuel pressure valve for pilot burner installed as well.

7.16 Draft indication (panel, field)

Prevent running any part of the furnace under a positive pressure which can cause structural damage and a personnel safety hazard. Provide panel indication of draft at arch (bridge wall). Provide field indication of draft at arch (bridge wall).

Locate indication at grade or lower deck of furnace. Set draft alarm. Alarm is typically set at zero inches of water at arch (bridge wall). Field draft indication will be provided. Locate in arch for natural draft furnaces. Locate at minimum draft point for forced/induced/balanced draft furnaces.

(Example) At arch (bridge wall), draft DCS indicator is installed. Source points are arch, fire box bottom floor, and stack.

7.17 Alarm for loss of FD or ID Fan (panel)

Give warning so that operators can compensate for loss of combustion air preventing a potential flame out. Provide alarm for loss of FD or ID fan. Procedures for loss of fans vary. Some furnaces chop fuel and some transition to natural draft operation at reduced rates. Board indication/alarm of fan speed and driver steam rate (turbine) or power draw (motor) will be provided in plants having ID/FD Fans.

(Example) Alarm applied

7.18 Valve closure prevention for pass balanced furnaces (field)

Prevent pass balancing valves from closing unexpectedly causing loss of pass flow and potential tube rupture. Pass control valves must have mechanical stop/hand jacks to prevent operating below minimum flow rates.

(Alternative) The intent may be met by a combination of fail-open pass balancing valves and armed safety shutdown system, i.e., fuel trip based on redundant flow indication.

(Example) Not installed

7.19 Emergency isolation valves (field)

Allow access to feed, fuel and snuffing steam valves in the event of a furnace emergency. Emergency isolation valves for process feed and fuel should be located a minimum of 15m from the furnace for safe operator access during a furnace emergency. Isolation valves may be at plant waterfall. Provide clear labels for emergency isolation valves. Emergency isolation valves for feed, fuel gas, and purge/snuffing steam (i.e., 15m blocks) will be provided. Locations of these valves will be clearly marked in the field.

(Example) Hand switch and fuel total valve situated at 15m away from furnace

7.20 Fuel gas KO pot (panel, field)

Prevent liquid hydrocarbon from being carried into burners causing burner plugging, rapid variations in furnace firing, and, in severe cases, spillage and burning of fuel outside of the firebox. Provide fuel gas knock out pot for each plant. Provide high level alarm on KO pot. Provide sight glass for level verification in the field. One common KO pot may serve multiple plants that are adjacent to each other. Filter with liquid traps are not adequate as a KO pot to prevent carryover of liquid that may be trapped in lines upstream of the filters. System should be reviewed for cold weather operation as appropriate. Fuel gas knock-out pots, with local LI/LG and control room LAH, will be provided.

(Example) Fuel gas Knock Out drum, Knock Out drum high level alarm, and field sight glass for level check were all installed.

7.21 Sight ports for tube inspection (field)

Peep holes allow visual and thermographic inspection of the tubes and of the general firebox/burner condition. Provide sight ports that allow viewing a high percentage of the radiant section tube surface area. Add sight ports as needed to optimize on-line tube inspection considering furnace history and criticality. Provide safe access to sight ports.

(Radiant sections) (1) Provide peep holes for visual and infrared inspection of at least 90% of the radiant tubes to detect hot spots in the firebox (2) Locate peep holes and skin TI's so TI can be monitored by thermography to confirm skin TI readings (3) Supplement existing peep holes as needed to provide for optimum tube inspection considering furnace design obstructions.

(Convection section) (1) Provide peep holes to see as many shock tubes as practical (2) Locate peep holes so shock tube TI's can be monitored by a pyrometer to confirm skin TI reading (3) Provide safe access to the peep holes (4) Infrared survey frequency will be determined separately for each furnace, based on local operating history and in consultation with furnace expert and hydroprocessing advisor.

(Example) Observation door installed

7.22 Waste gas system (field)

Waste gas is another source of fuel to a furnace. These provisions require the same level of protection for the waste gas system as for the other fuel system. Waste gas can also contain air. The flame arrestor is required to prevent flashback in the waste gas system. Provide waste gas isolation. Provide chopper valves for waste gas stream. Provide min 15m emergency isolation valve. Provide flame arrestor in waste gas stream. For furnaces with existing waste gas system, waste gas system will have proper burner isolation, SSD, emergency isolation valve, and flame arrestor.

(Example) To use waste gas (VDU off gas) was already reflected to burner design.

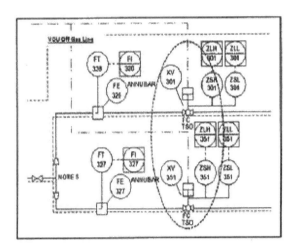

7.23 Placards (field)

Reminds the operator to use the furnace light-off procedure when lighting a burner. Placards should document light-off sequence consistent with s/u procedure. Less preferable alternative is placard which directs the operator to the correct furnace light-off procedure.

(Example) No placard but procedure prepared

7.24 Start-up checklist (documentation)

A high percentage of furnace incidents occur on startup. This provision provides a detailed step-by-step procedure for starting up the furnace to avoid conditions which could lead to accumulating unburned fuel in the fire box before light-off, or a tube rupture from low flow or overfiring on startup. Provide a startup checklist with sign off for each step. Startup procedure to include: (1) Testing critical instrumentation (e.g., skin TI's, SSD, alarms) prior to S/U (2) Fuel system preparation including draining low spots (3) Establishing and ensuring stable process flows (4) Drawing or detailed description of gas sample points (5) Firebox purging and gas testing (6) Burner lighting sequence (7) Minimize time between purge/sniff and light-off (8) Heatup steps/ milestones based on furnace limits and downstream process equipment (9) Monitoring of critical process variables during heatup (10) Relatching chopper system on hot restart after trip

(Example) All relevant procedures are documented and used practically during heater startup.

7.25 Start-up monitoring (documentation)

Recognizing that a high percentage of furnace incidents occur during start-up, this provision requires additional monitoring of the furnace until it is at normal operating conditions. Provide extra monitoring of critical furnace/plant parameters (e.g. flow rates, temperatures) during srartups.

This can be in the form of: (1) Startup run sheets, or (2) DCS printouts or (3) Recording of critical variables in a check list as part of the startup procedures

Critical variables need to be reviewed periodically by the Plant Supervisor or Head Operator during srartup to assure that conditions are within limits.

(Example) All things are specified on startup procedure.

7.26 Emergency procedures (documentation)

Provide written emergency procedures directed at shutting down and isolating the furnace safely to (1) protect the furnace and personnel, and (2) mitigate the size of an incident. Develop emergency shutdown procedures for power outage, loss of feed, etc. Develop emergency tube rupture procedures which address the following: (1) Specify tripping the fuel (2) Specify blocking in the process stream and deinventorying hydrocarbon from downstream equipment, if appropriate, to minimize the tube rupture incident (3) Specify when to use snuffing steam during a tube rupture incident (Many plants have decided to allow a tube rupture fire to burn itself out before introducing snuffing steam.)

(Example) Emergency procedure prepared and training it. Tube pin hole and tube rupture is separate in procedure. Snuffing steam uses when pilot burner flame is off.

7.27 Safety instruction sheet (documentation)

Document operating limits which cannot be changed without MOC. Each furnace shall have an updated Safety Instruction Sheet (SIS) showing the following limits: (1) Maximum tube skin temperature and basis (2) Maximum stack flue gas temperature and basis (damper metallurgical limits or stack refractory limits) (3) Maximum arch (bridge wall) flue gas temperature and basis (convection section tube support limits or convection section refractory limits)

(Example) Specified on monitoring program sheet and applicable alarm on DCS on the basis of design value

7.28 Flue gas analyzers (O_2, CO) (panel, field, documentation)

Provide necessary flue gas analyzers to (1) monitor the combustion process for potential bogging and (2) optimize the furnace efficiency. Each furnace shall be equipped with a reliable O_2 analyzer.

The number and location of sample points shall be specified to cover separate cells or each end of large cells. The system shall be designed to minimize sample and response time. When O_2 analyzers are out of service, O_2 shall be monitored each shift with a portable analyzer. Provide a low O_2 alarm above the point of CO breakthrough. Provide O_2 indication in field. Ensure O_2 analyzer is certified as ignition-safe (internal heating element will not ignite a combustible atmosphere in the furnace).

A list of ignition-safe analyzer models is available. Each continuously operating furnace shall be equipped with a reliable CO or combustibles analyzer. When CO/combustibles analyzers are out of service, CO/combustibles shall be monitored quarterly with a portable analyzer.

Provide high CO or combustibles alarm. Alarm can be based on loss of combustion air (500 or 1000 ppm) or local CO permit level or set to indicate loss of combustion air (500 or 1000 ppm). Provide CO/combustibles indication in field. Ensure analyzer is calibrated routinely and that calibration frequency is adequate to provide operators with reliable analyzer readings.

Reliable O_2 and CO analyzers will be provided on each furnace. Combustion analyzers will be installed to ensure rapid sampling and response time. The number and location of O_2 analyzers will be determined locally, based on a detailed O_2 survey of each furnace box. Portable O_2/CO analyzers will be available for use in the event the primary analyzers and out of service (or their functioning questioned), Automatic Combustion Control systems, when installed, will be configured to prevent them from bogging the furnace.

(Example) CO, O_2, and combustible analyzers installed. CO analyzer is laser type and situated on stack. With portable analyzer, operating guide of excess O_2 provided

7.29 Combustion control dampers (panel)

Provide good control of combustion air flow and draft in furnace. Combustion control dampers shall be fully operable without sticking. Remotely operated combustion control dampers shall have position indication on panel. Combustion control dampers shall have hardware minimum stop at approximately 5 to 10% open to prevent sticking on full closure. Exception is dampers that require tight shutoff during operation. Min stop requirement is met by designs with annular gap between damper and stack or duct of 5 to 10% of damper area. All furnaces will have dampers operable over full range with accurate position indications. Dampers will have a minimum stop to prevent full close.

(Example) Damper position monitored on DCS

7.30 Inspection plans/records (documentation)

Provide necessary information to predict furnace condition in order to avoid unexpected failures. Provides information to properly set and maintain the burners. Each furnace shall have an inspection file (on-line or hard copy). Inspection files shall contain history of shutdown inspections and repairs. Inspection files shall contain tube thickness history adequate to predict end of tube life.

(Example) Having inspection file

7.31 Monitoring and tracking (documentation)

Monitor key process variables, efficiency and any compliance. Ensure key process variables are tracked in order to detect any variation in operation or mechanical condition of the furnace and burners. Implement an automated efficiency tracking system. Audit furnace periodically for any compliance.

(Example) Monitoring key process variables currently

7.32 Safe & efficient furnace training (documentation)

Provide operator training ongoing basis including trouble shooting and emergency procedures. All furnace operators are to be trained in the elements of safe and efficient furnace operation. Ensure all furnace operators are given periodic refresher training on procedures, including emergency procedures. Ensure all furnace operators are given periodic refresher training on furnace tuning for efficiency.

(Example) Periodical heater training is accomplished.

7.33 Furnace surveillance (documentation)

Insure furnace operation is diligently monitored day to day. Provide outside operators with at least one of the following guidance resources to ensure furnace surveillance is adequately conducted in the field: (1) Training, Job Aid, Routine Duties, or Outside Reading Sheet (2) Outside furnace surveillance is to be performed at least twice per shift (3) Outside surveillance items include: (1) No high level in fuel KO pot (2) No positive draft (sufficient draft) (3) Sufficient combustion air (4) No abnormal flame patterns (plugged burner tips) (5) No flame impingement on tubes (6) No debris blocking burners (7) No pinched burners (8) No tube leaks (9) No hot spots on tubes (10) No excessive tube vibration (11) No sagging tubes (12) All tube supports in place

(Example) Outside routine check carried out 2 times per shift having routine check sheet.

Note 1. Check of fuel gas line up before startup (preparation)

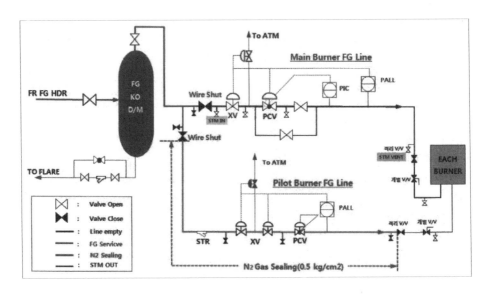

Note 2. Check of fuel oil line up before startup (preparation)

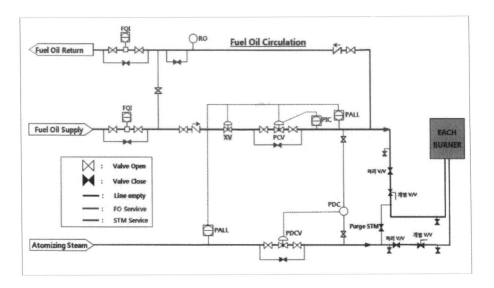

Note 3. Pilot burner firing at cold startup

$0.5 kg_f/cm^2$ N_2 sealing after steam out and under condensate removal

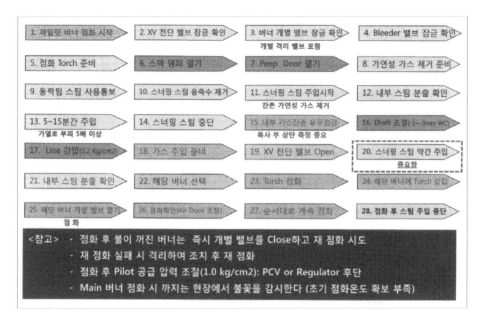

Note 4. Main gas burner firing at cold startup

Gas burner firing procedure for initial firing of heater under firing only pilot burner

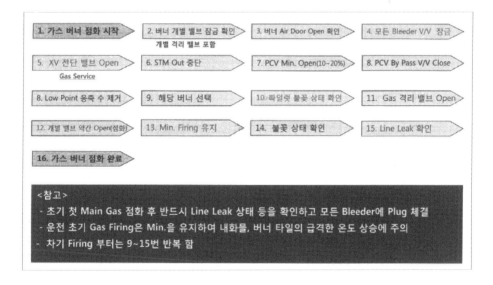

Note 5. Main oil burner firing at cold startup

Oil burner firing procedure of oil/gas burner under gas firing after firing pilot burner

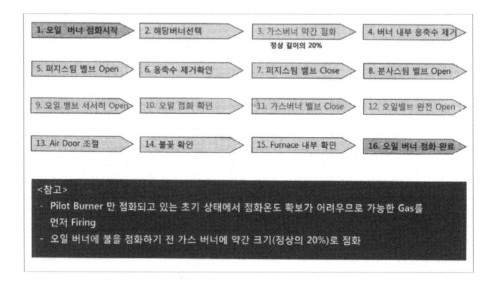

Note 6. Pilot burner firing at hot startup

In case of trouble occurring by inter lock logic work, firing after confirming fire out of all burners

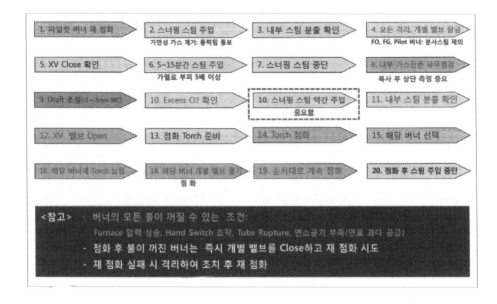

Note 7. Change in firing from oil burner to gas burner (routine job)

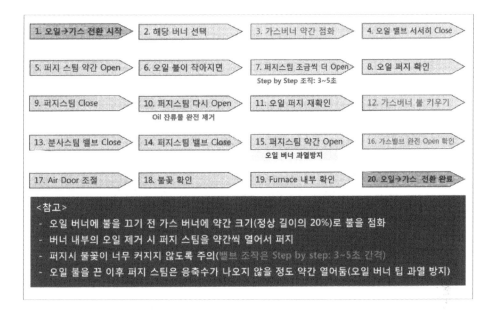

Note 8. Change in firing from gas burner to oil burner (routine job)

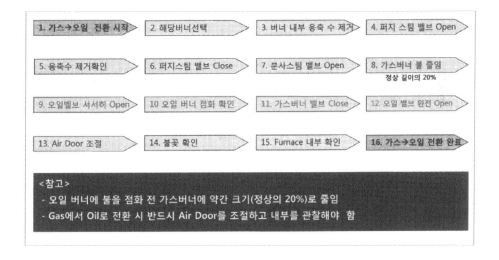

Note 9. Logic diagram for tuning furnace flame of fired heater (routine check)

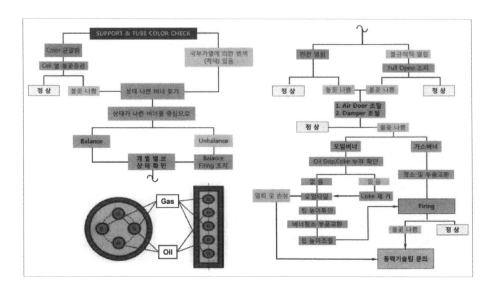

Note 10. Logic diagram for tuning balanced draft heater (routine check)

Excess O$_2$ and draft control

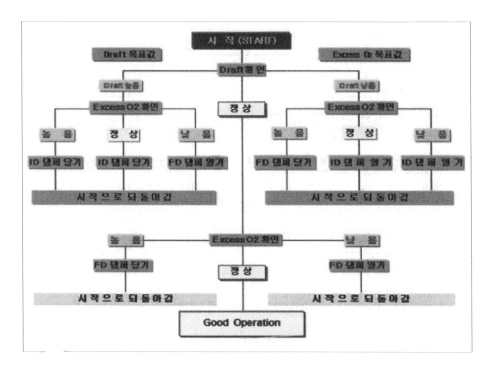

8. Heater Scale and Fouling Substances

8.1 Alkylation (SAR) furnace floor, hard coke (2015, Korea)

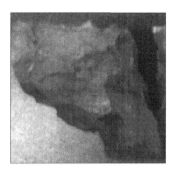 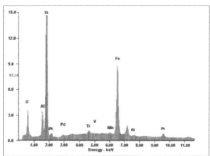

Element	O	Al	Si	Ti	V	Cr	Mn	Fe	Ni
Wt%	20.9	15.9	37.3	1.9	0.3	0.2	0.7	22.3	0.5

8.2 Steam boiler furnace bottom, hard coke (2005, Korea)

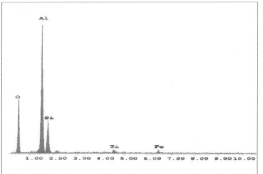 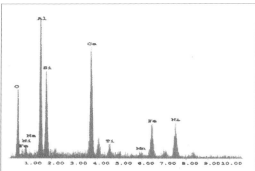

Refractory   Coke

Element(wt%)	Al	O	Si	Ti	Fe	Ni	Ca	Mg	Na	V
Refractory(oct.)	37.0	46.8	13.4	1.2	1.6					
Coke (Oct.)	22.0	34.0	13.9	1.5	6.2	8.7	11.8	0.8	1.3	
Coke (May)	28.1	37.3	11.9		4.2	4.3	10.8		2.2	1.3

Note: Firing 100% oil with 0.3% sulfur (Bunker-C treated acid crude)

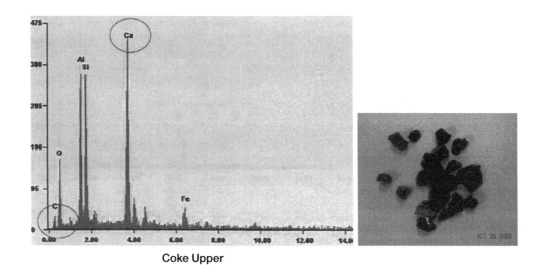

Coke Upper

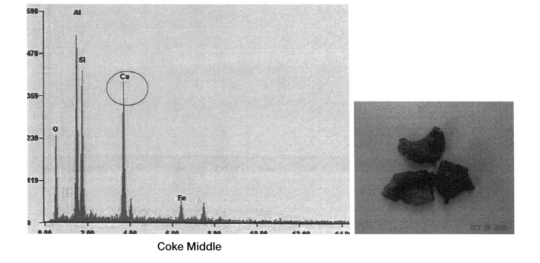

Coke Middle

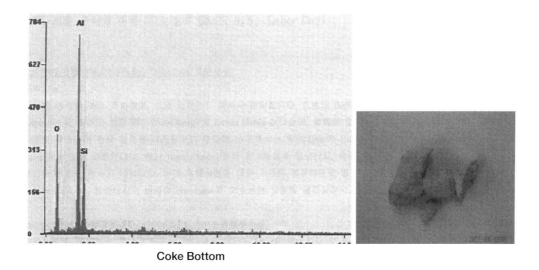

Coke Bottom

Element (wt%)	C	O	Al	Si	Ca	Ti	Fe
Coke Upper	21.1	32.3	13.5	13.1	14.9	1.9	3.3
Coke Middle		42.6	22.1	19.1	12.9		3.4
Coke Bottom		48.9	32.4	18.7			

8.3 Steam boiler, EP ash (2005, Korea)

Element	O	Na	Mg	Al	Si	S	Cu	Ln	Ca	Co	Ni	Mn
Wt%	45.7	6.4	1.8	0.7	1.0	19.2	2.5	1.4	15.4	2.2	3.4	0.8

8.4 APH ID Fan suction, thin yellow scale (2007, India)

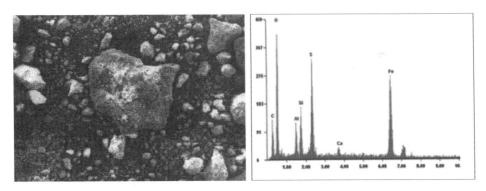

Element	C	O	Al	Si	S	Ca	Fe
Wt%	30.4	43.2	3.1	3.3	6.3	0.7	13.0

8.5 Xylene column reboiler, APH bottom, hard coke (2010, Korea)

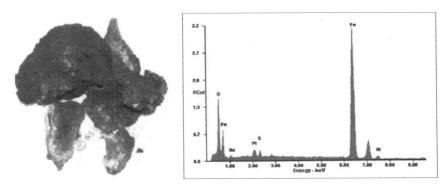

Element	C*	S*	Na	S	Fe	Ni
Wt%	12.5	16.8	4.1	2.4	91.1	2.3

Note: * measured by CS analyzer; firing 30% oil

8.6 CCR/Platformer heater, gas burner tip, coke (2015, Korea)

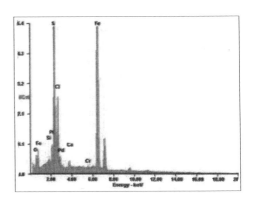

Element	*LOI	**C	**H	**N	**S	O	Si	S	Cl	Ca	Cr	Fe
Wt%	48.6	11.9	2.6	3.6	16.1	2.6	0.3	22.3	10.0	0.3	0.4	64.2

Note: * Loss On Ignition @810℃ ** measured by CHNS (organic) analyzer

8.7 CCR/Platformer heater, fuel gas line, green oil (2015, Korea)

Element	C	H	N	S	Cl
Wt%	16.0	8.4	6.6	0.1	134 wtppm

Note: Green oil is a liquid having viscosity to form at fuel gas line under chloride and olefin i.e., organic chloride and its polymerization with aromatics.

Distillation analysis (SIMDIST D6352)

Yield (wt%)	IBP	5	10	20	30	40	50	60	70	80	90	95	FBP
Temperature(℃)	217.6	227.0	233.4	242.4	250.4	257.6	264.2	271.0	278.0	286.6	413.8	441.2	478.8

Note: D90% 413.8℃ and having a range of HSR plus a little K/D product

8.8 RFCC CO boiler, scale/fouling substances (2005, Korea)

Element(wt%)	C	O	Cr	Na	Mg	Al	Si	S	Mo	Ti	Ca	Ba	V	Fe	Ni
Oil gun	61.9	18.7			1.1	6.6	7.0		2.5		0.2	0.7	0.3	0.8	0.5
Furnace		38.7			6.8	21.9	26.0		2.8	1.1			0.8	1.8	
Gas tip		35.6	3.3	14.8	3.9	13.7	14.3	11.7		0.5				1.7	0.4
Flue gas duct	100														

8.9 WCN charge heater, center fuel gas gun, coke (2005, Korea)

(Analysis) 100% Carbon

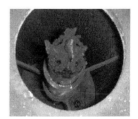

8.10 VDU heater, external tube surface, slag coke (2015, Korea)

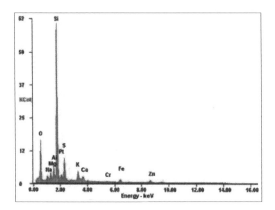

Element	LOI	O	Na	Mg	Al	Si	S	K	Ti	Ca	Cr	Fe	Ni	Zn
Wt%	12.4	33.5	2.0	2.2	2.7	42.8	5.6	2.7	0.1	0.6	0.4	2.9	0.5	4.0

Note: only gas firing, including rough off-gas

8.11 CDU heater, external tube surface, scale (2005, Korea)

Element	O	Na	Si	S	Ca	V	Fe	Co	Ni
Wt%	40.3	12.3	1.8	19.0	2.4	1.7	3.2	0.9	18.4

8.12 CDU heater, fuel gas line, sludge (2015, Korea)

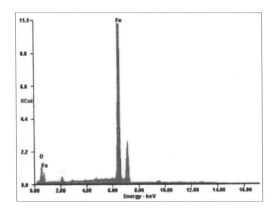

Element	LOI	C	H	N	S	O	Fe
Wt%	10.5	3.1	1.2	0.9	7.6	7.5	92.5

8.13 CDU heater, gas tip riser, coke (2015, Korea)

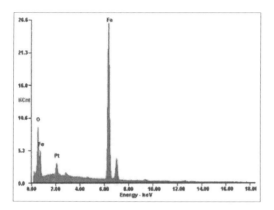

Element	**LOI**	C	H	N	S	O	Fe
Wt%	**10.5**	2.1	0.4	0.7	15.7	11.3	88.7

8.14 CDU R/C preheat exchanger, tube/shell, tank sludge (2011, Korea)

Element(wt%)	LOI	C	H	N	S	O	Na	Mg	Al	Si	P	S	Ca	Ba	V	Fe	Ni
#2 CDU *	81.7	57.3	3.4	0.9	12.1	10.8	2.3	0.9	1.7	2.8	3.1	1.8	1.3	1.8	0.8	75.2	0.8
#3 CDU**	86.0	52.1	7.3	2.2	6.3	22.1	2.2		0.8	1.2		0.8	0.3			72.6	

Note: * Service chemical: H company ** Service chemical: HP company

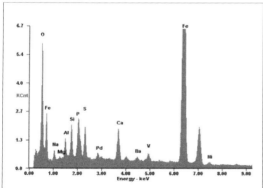

CDU R/C preheat exchanger, tube/shell, tank sludge

8.15 Chemical Company, steam boiler, coke (2006, Korea)

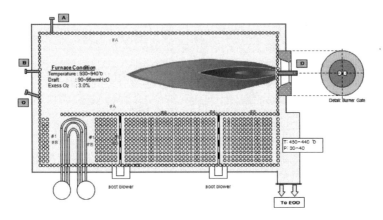

Element (wt%)	C	O	Na	S	Ca	V	Fe	Ni	Si
#A, Side wall		40.4	14.5	21.2	7.7	1.3	3.1	10.6	1.2
#B, Super heater, up		37.9		23.8	28.1		5.3	4.8	
#C, Super heater, down		35.5	8.7	24.4	12.9	1.7	4.6	10.7	1.4
#D, Further down	84	9.1	1.2	2.5	1.6		0.7		

Note: Using AR that treated a high acid crude, Doba (Ca 112, V 10, Na 6, Fe 10, Ni 12, K 1, Al 4, Zn 0.1 wt%)

9. Inspection of TA Shutdown

9.1 Convection section of CDU heater

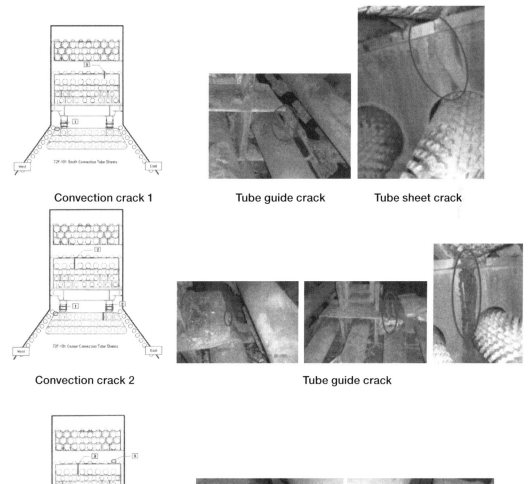

Convection crack 1   Tube guide crack   Tube sheet crack

Convection crack 2   Tube guide crack

Convection crack 3   Tube guide crack

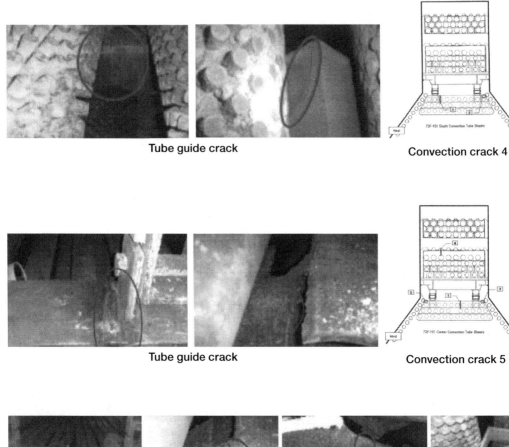

Tube guide crack

Convection crack 4

Tube guide crack

Convection crack 5

Tube support crack

Tube guide crack

Tube guide crack

Convection crack 6

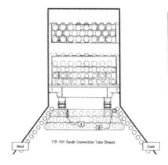

Tube guide crack

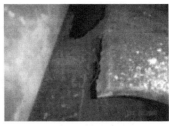

Tube guide crack

9.2 Vacuum residue hydrocracking reactor feed heater 1

Tube status (good)

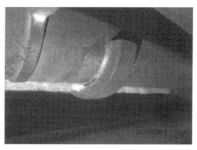

TST pad PT (good)

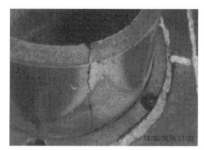

Burner tile crack

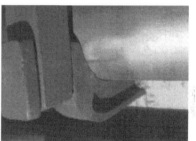

Tube scratch (0.3mm)

Bare tube status (good)

Fin tube status (good)

Castable peeled (1mm)

Tube support castable (damaged)

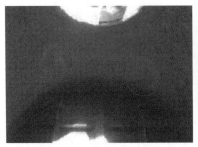

Damper upper (erosion)

Air preheater tube fouling

Air preheater tube sheet scale

9.3 Vacuum residue hydrocracking reactor feed heater 2

Tube status (good)

TST pad PT (good)

Bare tube status (good)

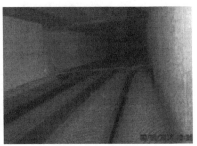

Serrated fin tube (good)

Castable peeled (1mm)

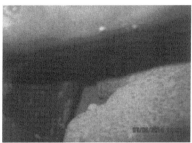

Tube support castable (damaged)

Damper upper (erosion)

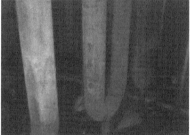

Tube scratch (0.3mm) & tube surface scale

Pyro-block damaged

TST pad PT (good)

Tube support PT (good)

Air preheater tube fouling

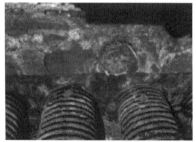

Air preheater tube sheet scale

9.4 Vacuum residue hydrocracking A tower heater

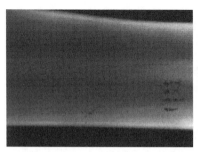

RT film measurement (coke 10mm)

Tube inside coke

Stud pig used

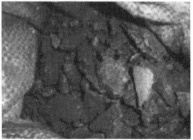

Coke removed

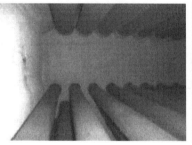

Convection inside refractory (good)

Convection inside fin tube (good)

9.5 Vacuum residue hydrocracking V tower heater

Tube lifted (1)

Tube lifted (2)

Tube lifted 185mm (3)

Tube abrasion

Tube sleeve installation

Hip section casting support

Casting support & sleeve installation

Support hanger status

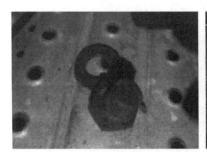

Broken bolt

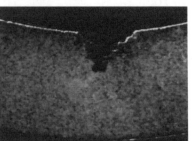

Reboiler tube pitting

Reboiler tube length direction pitting

10. Accidents & Problem Troubleshooting

10.1 CDU APH water washing

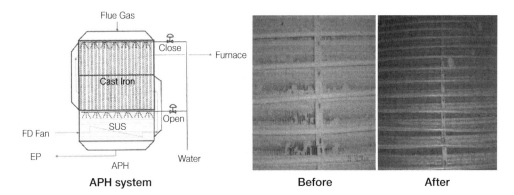

APH system   Before   After

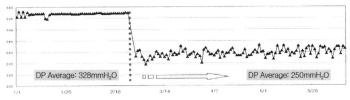

ID Fan differential pressure (DP)

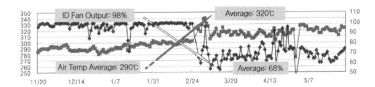

Combustion air temperature (℃) & ID Fan output (%)

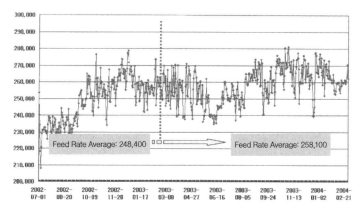

Feed flow rate (Bbl/day)

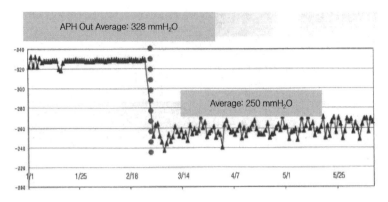

Pressure drop

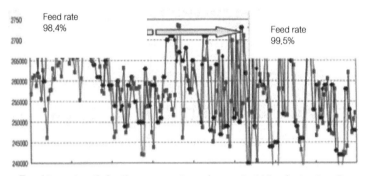

Feed target satisfaction percentage (expected blue/actual red)

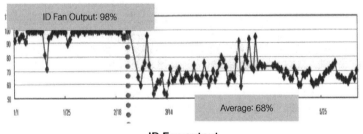

ID Fan output

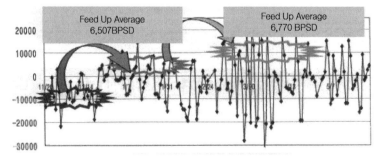

Feed throughput increase (actual/expected gap)

10.2 Xylene column reboiler APH soda ash cleaning

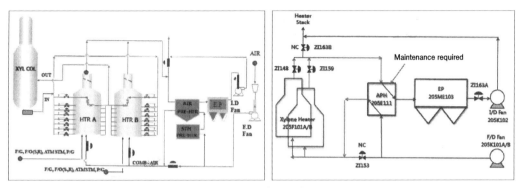

Process schematics

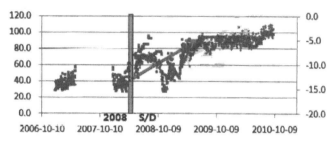

Red-EP damper opening (%) & Blue-draft (mmH$_2$O)

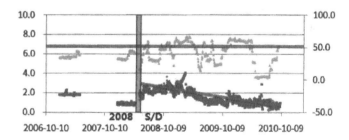

Red/blue-heater A/B excess O$_2$ (%) & Green-heater fuel oil ratio (%)

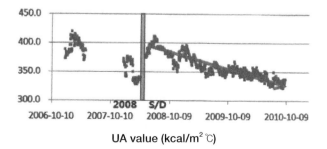

UA value (kcal/m^2 ℃)

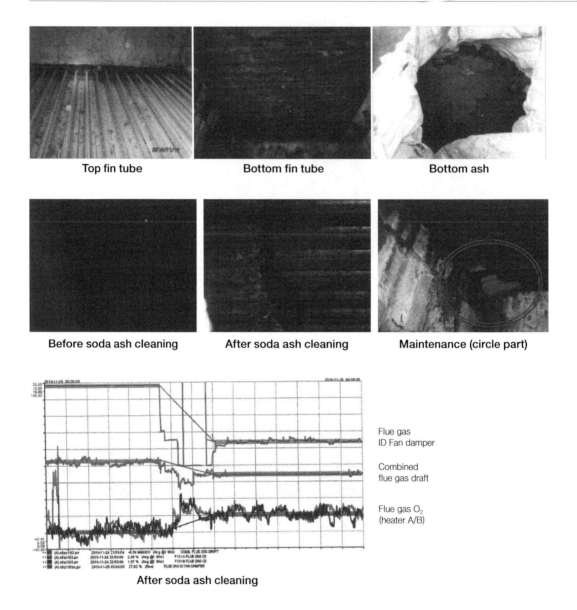

Top fin tube   Bottom fin tube   Bottom ash

Before soda ash cleaning   After soda ash cleaning   Maintenance (circle part)

Flue gas
ID Fan damper

Combined
flue gas draft

Flue gas O_2
(heater A/B)

After soda ash cleaning

Note 1. Before and after soda ash cleaning of top side in APH

Before   After

Note 2. APH shutdown for soda ash cleaning, fuel firing & draft mode

(1) During APH shutdown: Natural draft (-13mmH$_2$O) under running forced draft fan, normal feed, only gas firing, consequently using 10% fuel more

(2) Normal operation: Balanced draft, combined oil/gas firing in a burner

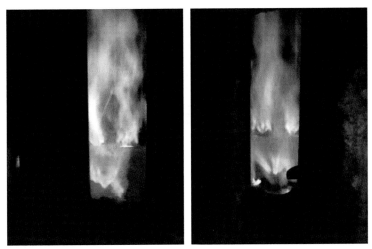

Only gas firing (@APH shut down)   Oil/gas 50%/50% firing (@operation)

10.3 Xylene column reboiler APH cleaning

(1) Before cleaning

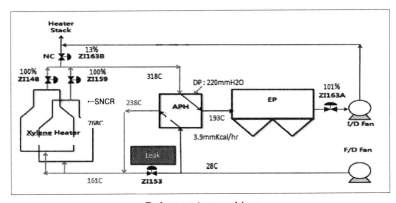

Before water washing

Top fin tube | Bottom glass coated tube | Bottom ash

(2) After cleaning

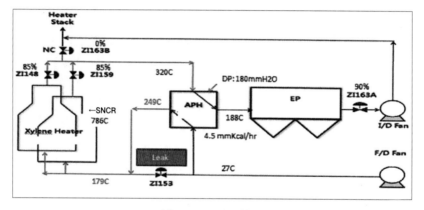

After water washing

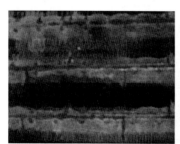

Top fin tube | Bottom glass coated tube | Washing water distributor

Item	Unit	Design	Revamp	Before cleaning	After cleaning	Difference
APH DP	mmH$_2$O	155	155	220	177	-43
Damper open	%	-	-	100	85	-15
Flue gas temp to APH	℃	336	360	318	320	2
Flue gas temp from APH	℃	170	186	193	188	-5
Combustion air temp from APH	℃	213	213	161	179	18
APH heat exchange's amount	MMkcal/hr	6.5	7.2	3.9	4.5	0.6

Results before/after cleaning

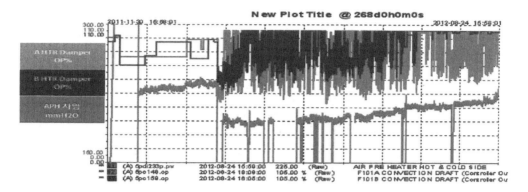

Trends before and after cleaning

Item	Operation			Performance		
	Heater efficiency (%)	Fired duty (MMkcal/hr)	Combustion air quantity (kg/hr)	Hot air temp (℃)	UA value (kcal/hr ℃)	APH duty (MMkcal/hr)
Initial design	90.40	69	133,966	123.0	47,090	6.5
Revamp calculation	88.91	83	148,771	147.0	45,621	7.2
Apr 2008 (30 days)	89.31	64	96,388	150.1	21,832	3.7
Apr 2009 (30 days)	89.96	64	107,059	160.4	24,200	4.1
Apr 2010 (30 days)	90.55	64	110,362	170.0	25,238	4.3
April 2011 (30 days)	90.04	63	96,749	162.8	21,321	3.7
Jan 2012 (30 days)	90.96	73	112,813	170.8	21,829	3.9
'12.2 before 1st cleaning	90.90	71	115,337	176.5	22,183	4.0
'12.2 after 1st cleaning	91.01	72	118,024	150.7	27,817	4.6
'12.8 before 2nd cleaning	90.77	75	119,571	157.4	24,246	3.9
'12.8 after 2nd cleaning	91.00	68	119,969	141.0	29,745	4.5

APH performance summary

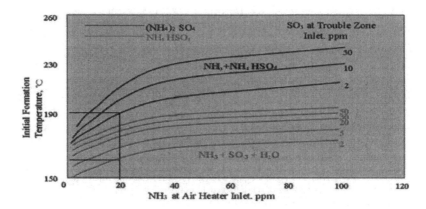

Fouling possibility by SNCR, Ammonium BiSulfate [(NH$_4$)HSO$_4$]/Ammonium Sulfate [(NH4)$_2$SO$_4$)]

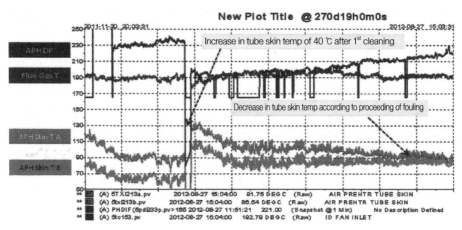

APH tube skin temperature

APH operation's details are as follows.

	Item	Unit	Initial design		Revamp calculation	Before revamp	Before 1st cleaning	After 1st cleaning	Before 2nd cleaning	After 2nd cleaning
			Operating	Maximum		Jan 2011	Feb 1-8	Feb 19-26	Aug 13-20	Sep 1-3
Process	Fuel type	-	100% FO	100% FO	100% FO	61% FO	27% FO	30%FO	47% FO	38% FO
	Excess O_2	%	4.8	4.8	4.8	2.0	1.2	1.5	0.9	1.0
	Excess air	%	25	25	25	9.3	5.4	6.9	4.2	48
	Heat absorption	MMkcal/hr	62.6		74.1	56.6	64.0	64.6	67.2	60.8
	Heat release	MMkcal/hr	69.3	95.3	83.3	62.6	71.2	71.8	74.7	67.6
	Heater efficiency	%	90.4		88.9	90.4	90.9	91.0	90.8	91.0
	Feed flow rate	kg/hr	1,147,176		1,224,700	1,391,082	1,381,002	1,381,920	1,356,183	1,348,866
APH	Combustion air quantity	kg/hr	133,966	154,958	148,771	96,374	109,374	111,922	113,389	113,671
	Flue gas quantity	kg/hr	133,314	166,643	158,754	103,133	110,431	110,431	110,431	106,223
	Radiant out temperature	℃	799		863	744	797	762	768	786
	Radiant tube skin temperature	℃		400	400	312	316	318	312	315
	Convection out temp (hot in)	℃	336	343	360	339	321	314	318	320
	Hot gas APH out (hot out)	℃	170	170	186	193	189	185	193	188
	Air APH in (cold in)	℃	16	16	16	1	2	7	28	27
	Air APH out (cold out)	℃	213	213	213	160	144	164	161	179
Draft	Draft below convection	mmH$_2$O		-2.5	-2.5	-4.8	-1.6	-2.5	-2.0	-2.5
	Heater Individual damper Op	%				76	94	75	98	85
	Draft above convection	mmH$_2$O			-15.0	-25.2	-24.8	-27.3	-21.7	-23.7
	Common damper Op	%				89	98	93	101	85
	APH pressure drop (Dp)	mmH$_2$O	138	150			236	194	217	177
Performance	UA	kcal/hr ℃	47,090		45,621	21,321	22,183	27,817	24,246	29,745
	APH duty	MMkcal/hr	6.5		7.2	3.7	4.0	4.6	3.9	4.5
	Air absorbed duty *	kcal/kg air	48		49	38	35	39	33	37

Note: * Due to APH duty depending on air flow rate, to get APH performance, APH duty is divided by air flow rate

10.4 The corrosion problem of APH of plate type

(1) Toluene reboiler heater: Only gas firing (2-3ppm H_2S in fuel gas)

(2) EDAX analysis of scale: 16.6 O, 0.3 Al, 0.7 Si, 0.1 S, 14.4 Cl, 0.6 Ca, 67.2wt% Fe

   Note: The Cl source is off gas from Debutanizer column of CCR/Platformer process.

(3) Estimated corrosion period: hot side outlet temperature to have 100-110℃ (220-230℃ inlet)

(4) Action: At present to keep the outlet temperature at more than 125℃ with air bypass

10.5 Depentanizer reboiler heater (natural draft) explosion

(1) Process

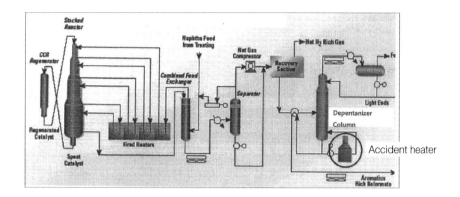

(2) Fired reboiler heater (fuel control system)

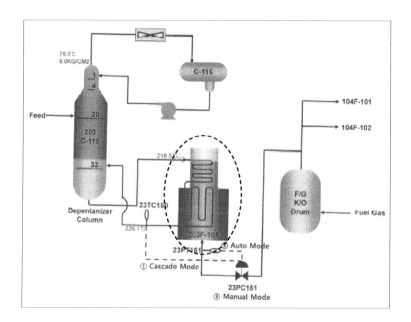

1) Cascade mode: By TC150 (auto) setting value, PC151 (cascade) adjusted

2) Auto mode: According to PT151 setting value, PC151 output adjusted

3) Manual mode: Operator adjusting PC151 output value intentionally

(3) DCS schematic

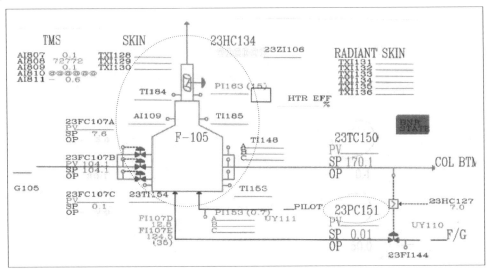

(4) Operation variable and cause of accident

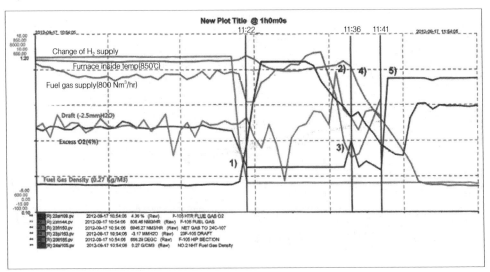

H_2 rich fuel gas supplies just before accident. 2) Closing H_2 rich gas source valve and opening fuel gas source with high heating value. 3) Inside furnace, excess O_2 reaches zero(0) and incomplete combustion occurs. 4) Burner flame is unstable, furnace draft is positive, and then air supply cut/fire off happen. 5) Excess O_2 increases, furnace (fire box) temperature decreases, and an operator increases fuel gas again not recognizing fire off. Consequently, after fire off, the built-up combustible gas around hip section/ remaining heat inside furnace cause the explosion under the bad weather of typhoon.

(5) Fuel gas fluctuation

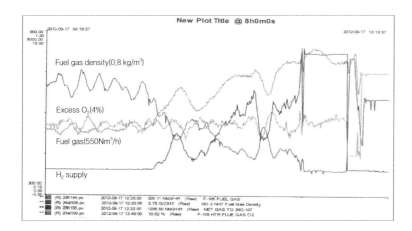

(6) Operation log sheet

Date Time	Tag Name	Parameter	Old Value	New Value	Point Description	Eng Units	Unit Name	Port
오전 11:33:03	23PC151	MODE	CAS	MAN	F-105 REB HTR F/G HDR	Kg/Cm2	23	2CCR
오전 11:33:05	23PC151	OP	81.3092	76.3092	F-105 REB HTR F/G HDR	Kg/Cm2	23	2CCR
오전 11:33:43	23PC151	MODE	MAN	AUTO	F-105 REB HTR F/G HDR	Kg/Cm2	23	2CCR
오전 11:34:15	23PC151	SP	2.2143	2.1543	F-105 REB HTR F/G HDR	Kg/Cm2	23	2CCR
오전 11:34:41	23PC151	SP	2.1343	2.1043	F-105 REB HTR F/G HDR	Kg/Cm2	23	2CCR
오전 11:34:42	23PC151	SP	2.1543	2.0943	F-105 REB HTR F/G HDR	Kg/Cm2	23	2CCR
오전 11:34:51	23PC151	MODE	AUTO	MAN	F-105 REB HTR F/G HDR	Kg/Cm2	23	2CCR
오전 11:34:55	23PC151	OP	74.4840	72.0000	F-105 REB HTR F/G HDR	Kg/Cm2	23	2CCR
오전 11:35:07	23PC151	OP	72.0000	70.0000	F-105 REB HTR F/G HDR	Kg/Cm2	23	2CCR
오전 11:35:19	23PC151	OP	70.0000	68.0000	F-105 REB HTR F/G HDR	Kg/Cm2	23	2CCR
오전 11:35:51	23PC151	OP	68.0000	66.0000	F-105 REB HTR F/G HDR	Kg/Cm2	23	2CCR
오전 11:36:20	23PC151	OP	66.0000	64.0000	F-105 REB HTR F/G HDR	Kg/Cm2	23	2CCR
오전 11:38:17	23PC151	OP	64.0000	65.0000	F-105 REB HTR F/G HDR	Kg/Cm2	23	2CCR
오전 11:38:49	23PC151	OP	65.0000	66.0000	F-105 REB HTR F/G HDR	Kg/Cm2	23	2CCR
오전 11:39:21	23PC151	OP	66.0000	68.0000	F-105 REB HTR F/G HDR	Kg/Cm2	23	2CCR
오전 11:39:59	23PC151	OP	68.0000	70.0000	F-105 REB HTR F/G HDR	Kg/Cm2	23	2CCR
오전 11:40:24	23PC151	OP	70.0000	71.0000	F-105 REB HTR F/G HDR	Kg/Cm2	23	2CCR
오전 11:41:01	23PC151	OP	71.0000	73.0000	F-105 REB HTR F/G HDR	Kg/Cm2	23	2CCR

(7) Fuel gas composition

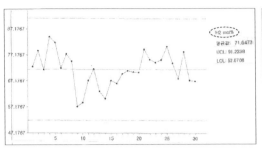
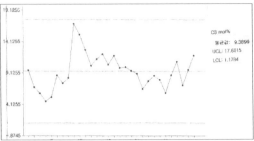
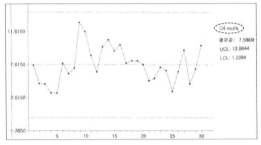

(8) Flame observation

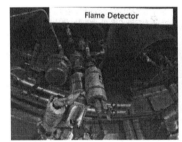

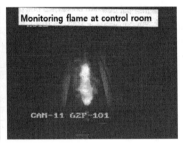

(9) Fuel gas density control constraint

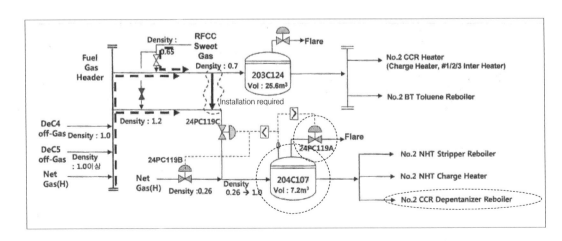

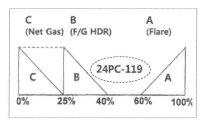

C (Net Gas)	B (F/G HDR)	A (Flare)

204C-107 pressure control scheme

(10) Fuel gas heating value

SP.GR.	0.4779		①												
EFO Factor	0.5018	L/NM3	②												
Components	MOL %	M.W	Liq. Den.	Mol	Mass	Mass	Liq. Vol.	Liq.Vol	Gas	Relative	HHV/LHV	HHV	LHV	LHV성분별	EFO KL
	(VOL %)		(kg/m3)	Fraction	(kg/kgmol)	Fraction	(m3/kgmol	Fraction	Gravity	Gravity	Ratio	Kcal/Kmol	Kcal/Kmol	Kcal/Kmol	/ Kmol
H2S	0.00	34.08	786.6	0.00	0.00	0.00	0.00	0.0	1.18	0.0	1.086	134.412	123,800	0.0	0.000
H2	71.66	2.02	69.9	0.72	1.44	0.12	0.02	0.5	0.07	0.0	1.187	68,632	57,630	41435.2	0.004
C1	1.96	16.04	299.4	0.02	0.31	0.02	0.00	0.0	0.55	0.0	1.112	213,364	191,906	3761.2	0.000
C2=	0.00	28.05	383.2	0.00	0.00	0.00	0.00	0.0	0.97	0.0	1.067	337,349	316,300	0.0	0.000
C2	5.87	30.07	359.5	0.06	1.77	0.13	0.00	0.1	1.04	0.1	1.093	372,983	341,400	20040.2	0.002
C3=	0.00	42.08	523.3	0.00	0.00	0.00	0.00	0.0	1.45	0.0	1.069	492,598	460,600	0.0	0.000
C3	9.39	44.10	507.7	0.09	4.14	0.30	0.01	0.2	1.52	0.1	1.086	530,710	488,800	45898.3	0.005
I-C4=	0.00	56.11	598.2	0.00	0.00	0.00	0.00	0.0	1.94	0.0	1.069	650,105	608,000	0.0	0.000
I-C4	0.00	58.12	562.8	0.00	0.00	0.00	0.00	0.0	2.01	0.0	1.083	686,930	634,000	0.0	0.000
N-C4	7.59	58.12	584.5	0.08	4.41	0.32	0.01	0.2	2.01	0.2	1.082	688,045	635,700	48249.6	0.005
I-C5	0.00	72.15	626.7	0.00	0.00	0.00	0.00	0.0	2.49	0.0	1.081	843,784	780,500	0.0	0.000
N-C5	1.12	72.15	630.6	0.01	0.81	0.06	0.00	0.0	2.49	0.0	1.081	846,142	732,400	8782.9	0.001
C6+	0.47	86.18	666.3	0.00	0.41	0.03	0.00	0.0	2.98	0.0	1.080	1,003,709	929,400	4368.2	0.000
O2	0.00	32.00	1135.0	0.00	0.00	0.00	0.00	0.0	1.11	0.0				0.0	0.000
CO	0.00	28.01	799.4	0.00	0.00	0.00	0.00	0.0	0.97	0.0	1.000	67,640	67,640	0.0	0.000
CO2	0.00	44.01	826.8	0.00	0.00	0.00	0.00	0.0	1.52	0.0		-		0.0	0.000
N2	1.95	28.01	806.4	0.02	0.55	0.04	0.00	0.0	0.97	0.0		-		0.0	0.000
Results	100.00			1.0	13.836	1.0	0.04	1.0		0.478		191502.3	172515.6	172515.6	0.0180

	0.6175537				8549.2	7701.6
	③ Density (kg/m3)				③ HHV (Kcal/Nm3)	④ LHV (Kcal/Nm3)

(11) Density and EFO factor (confirmation)

Date	① Sp.Gr 24al105.psi	② EFO Factor	③ Excess O2 %	④ FG XO Press. kg/cm2	⑤ FG Nm3/hr	⑥ H3/scmp Used/Reqd	⑦ * LHV Kcal/Nm3	⑧	⑨ FG	⑩ H.R. mmkcal/hr	⑪	⑫ COT ℃	⑬ COT ℃	⑭ ex ROT ℃	⑮ ex COT ℃	⑯ Stack Damper %	⑰ Draft mmH2O	⑱ FG EFO B/D
2012-39-17 10:55	0.2709	0.5378	4.3	3.6	807	0.357	5,348	1.371	140	8.85	70	216	229	668	288	61	-2.9	140
2012-39-17 10:56	0.2712	0.5383	4.4	3.6	804	0.357	5,349	1.303	140	8.83	70	216	229	667	288	61	-3.6	140
2012-39-17 10:57	0.2700	0.5385	4.6	3.6	801	0.554	5,333	1.669	140	8.79	71	216	229	666	288	61	-3.0	145
2012-39-17 10:58	0.2697	0.5382	4.4	3.6	798	0.355	5,330	1.664	140	8.76	70	216	229	665	288	61	-2.3	139
2012-39-17 10:59	0.2695	0.5384	4.3	3.6	795	0.356	5,332	1.665	139	8.75	70	216	229	664	287	61	-2.8	139
2012-39-17 11:00	0.2697	0.5385	4.5	3.6	791	0.355	5,329	1.664	139	8.71	71	216	229	663	287	61	-2.5	139
2012-39-17 11:01	0.2688	0.5349	4.5	3.6	790	0.554	5,319	1.049	139	8.69	71	216	229	661	287	61	-3.2	139
2012-39-17 11:02	0.2678	0.5336	4.4	3.6	790	0.555	5,306	1.027	139	8.69	73	216	229	660	287	61	-3.6	138
2012-39-17 11:03	0.2680	0.5338	4.5	3.6	789	0.554	5,306	1.039	139	8.68	79	216	229	659	287	61	-2.6	138
2012-39-17 11:04	0.2677	0.5334	4.5	3.7	780	0.551	5,304	1.051	139	8.70	73	216	229	659	286	61	-2.4	138
2012-39-17 11:05	0.2659	0.5309	4.4	3.7	788	0.550	5,281	1.057	138	8.68	73	216	229	659	286	61	-3.1	138
2012-39-17 11:06	0.2661	0.5312	4.8	3.7	787	0.547	5,284	1.061	138	8.67	73	216	229	659	286	61	-3.1	138
2012-39-17 11:07	0.2671	0.5325	5.1	3.7	790	0.552	5,296	1.094	139	8.71	73	216	229	660	287	61	-2.9	139
2012-39-17 11:08	0.2664	0.5330	4.7	3.7	792	0.554	5,288	1.096	138	8.72	73	216	229	661	287	61	-3.1	139
2012-39-17 11:09	0.2672	0.5329	4.4	3.6	791	0.558	5,298	1.061	138	8.70	73	216	229	662	287	61	-2.7	138
2012-39-17 11:10	0.2685	0.5349	4.4	3.6	790	0.554	5,315	1.047	139	8.67	71	216	229	662	287	61	-2.9	138
2012-39-17 11:11	0.2693	0.5395	4.4	3.6	790	0.555	5,324	1.247	139	8.62	71	216	229	662	287	61	-3.3	138
2012-39-17 11:12	0.2679	0.5340	4.6	3.6	790	0.553	5,307	1.054	139	8.66	71	215	229	661	287	61	-2.0	138
2012-39-17 11:13	0.2670	0.5340	4.6	3.6	794	0.556	5,295	1.068	139	8.73	73	215	229	661	287	61	-4.2	139
2012-39-17 11:14	0.2678	0.5336	4.6	3.6	793	0.551	5,306	1.059	139	8.73	73	215	229	660	286	61	-5.3	139
2012-39-17 11:15	0.2655	0.5304	4.5	3.6	790	0.551	5,277	1.063	139	8.69	73	215	229	660	286	61	-3.0	139
2012-39-17 11:16	0.2654	0.5302	4.5	3.7	789	0.550	5,275	1.064	139	8.69	73	215	229	660	286	61	-3.5	139
2012-39-17 11:17	0.2661	0.5310	4.6	3.7	794	0.551	5,310	1.073	139	8.79	73	215	229	662	286	61	-3.4	140
2012-39-17 11:18	0.2697	0.5355	4.6	3.7	792	0.559	5,329	1.064	140	8.80	73	215	229	665	287	61	-2.5	140
2012-39-17 11:19	0.2705	0.5373	4.4	3.7	793	0.557	5,340	1.070	141	8.84	73	215	229	664	287	61	-2.8	140
2012-39-17 11:20	0.2697	0.5362	4.3	3.8	797	0.205	5,330	1.072	141	8.80	72	215	229	667	287	61	-2.5	141
2012-39-17 11:21	0.2698	0.5362	4.1	3.8	795	0.356	5,331	1.084	141	8.90	73	215	229	665	288	61	-2.5	141
2012-39-17 11:22	0.2798	0.5494	3.1	3.8	796	0.366	5,452	1.344	142	8.92	72	215	228	658	288	61	-2.7	140
2012-39-17 11:23	0.4072	0.7085	3.0	3.7	774	0.741	7,107	1.551	147	9.27	72	215	229	680	287	61	-2.2	133
2012-39-17 11:24	0.6651	1.0935	8.0	3.6	754	1.108	10,559	1.204	168	10.46	72	215	230	682	288	61	-3.2	129
2012-39-17 11:25	0.9159	1.4279	8.0	3.6	768	1.465	14,023	1.530	182	12.06	73	216	229	600	266	61	-5.2	137
2012-39-17 11:26	1.0277	1.5915	8.0	3.6	785	1.624	15,548	939	203	12.90	79	216	228	637	262	61	-4.4	142
2012-39-17 11:27	1.0276	1.5918	8.0	3.6	795	1.625	11,580	929	203	13.06	72	216	228	623	278	61	-4.2	144
2012-39-17 11:28	1.0278	1.5915	8.0	3.6	806	1.625	15,589	919	203	13.12	72	216	228	615	275	61	-8.2	145
2012-39-17 11:29	1.0278	1.5919	8.0	3.6	804	1.725	15,589	902	203	13.22	72	215	228	613	274	61	-7.6	145
2012-39-17 11:30	1.0278	1.5915	8.0	3.6	807	1.625	15,590	869	203	13.27	76	214	228	613	273	61	-3.9	146
2012-39-17 11:31	1.0278	1.5918	8.0	3.6	811	1.625	15,590	869	203	15.35	76	213	227	617	273	61	-2.9	147
2012-39-17 11:32	1.0561	1.6415	8.0	3.6	821	1.606	15,426	870	205	15.42	71	213	227	624	273	61	-1.8	148
2012-39-17 11:33	0.9568	1.4816	8.0	3.6	824	1.325	14,594	835	191	15.11	76	213	227	629	274	61	-1.0	147
2012-39-17 11:34	0.7799	1.3711	8.0	3.6	808	1.406	13,498	928	182	12.41	97	213	227	634	274	61	-1.9	142
2012-39-17 11:35	0.8176	1.2212	8.0	3.6	777	1.319	12,658	852	181	11.56	94	213	228	640	275	61	0.9	135
2012-39-17 11:36	0.7674	1.2206	8.0	3.6	730	1.207	11,961	860	169	10.62	95	213	228	652	276	61	2.0	126
2012-39-17 11:37	0.7146	1.1499	2.5	3.6	667	1.171	11,296	814	154	9.73	67	213	227	650	276	61	-1.2	117
2012-39-17 11:38	0.6862	1.1227	2.5	3.6	679	1.140	11,032	870	152	9.54	64	213	222	590	271	61	-2.6	116

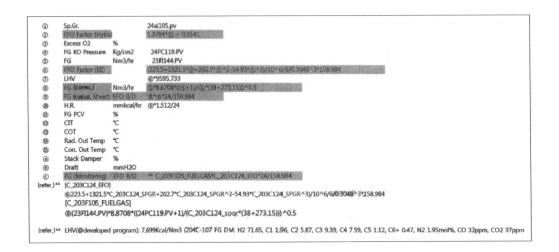

①	Sp.Gr.		24ai105.pv
②	EFO Factor (Hysis)		1.3794*① + 0.1641
③	Excess O2	%	
④	FG KO Pressure	Kg/cm2	24PC119.PV
⑤	FG	Nm3/hr	23FI144.PV
⑥	EFO Factor (EII)		(223.5+1321.5*②+202.7*①^2-54.93*①^3)/10^6/6/0.3048^3*158.984
⑦	LHV		⑥*9595.733
⑧	FG (correc.)	Nm3/hr	⑤*8.8708*((⑥+1)/(⑥*(38+273.15)))^0.5
⑨	FG (calcul. Sheet)	EFO B/D	④*⑥*24/158.984
⑩	H.R.	mmkcal/hr	⑨*1.512/24
⑪	FG PCV	%	
⑫	CIT	℃	
⑬	COT	℃	
⑭	Rad. Out Temp	℃	
⑮	Con. Out Temp	℃	
ⓐ	Stack Damper	%	
ⓑ	Draft	mmH2O	
ⓒ	FG (Monitoring)	EFO B/D	** C_203F105_FUELGAS*C_203C124_EFO*24/158.984

(refer.)** [C_203C124_EFO]
⑥223.5+1321.5*C_203C124_SPGR+202.7*C_203C124_SPGR^2-54.93*C_203C124_SPGR^3)/10^6/6/0.3048^3*158.984
[C_203F105_FUELGAS]
⑧(23FI144.PV)*8.8708*((24PC119.PV+1)/(C_203C124_spgr*(38+273.15)))^0.5

(refer.)** LHV(@developed program): 7,699Kcal/Nm3 (204C-107 FG DM: H2 71.65, C1 1.96, C2 5.87, C3 9.39, C4 7.59, C5 1.12, C6+ 0.47, N2 1.95mol%, CO 32ppm, CO2 37ppm

10.6 BT reboiler heater draft high high alarm & trip

(1) Air preheater system

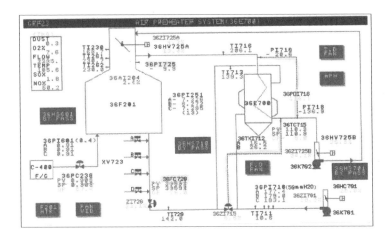

(2) BT column reboiler heater tube skin/pass flow

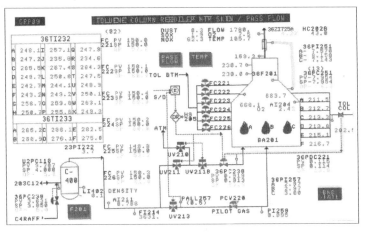

(3) Process alarm journal

```
'AGE 002    S-02/20/14 20:00  E-02/21/14 04:00    PROCESS ALARM JOURNAL
---------------------------------------------------------------------------
tCK 21:46:35   36TI116     PVLO              LOW   306C-101 TRAY #26      36
iLM 22:08:00   36XL123     CHNGOFST          LOW   306C110 DRUM MIXER     36
tCK 22:08:05   36XL123     CHNGOFST          LOW   306C110 DRUM MIXER     36    RUN
tTN 22:09:09   36TI116     PVLO    105.000 LOW   306C-101 TRAY #26      36    108.776
iLM 22:14:08   36TI116     PVLO    108.000 LOW   306C-101 TRAY #26      36    107.984
tCK 22:14:16   36TI116     PVLO              LOW   306C-101 TRAY #26      36
tTN 22:27:59   36TI116     PVLO    108.000 LOW   306C-101 TRAY #26      36    109.002
iLM 23:18:24   36PI251B    PVHI     10.000 LOW   F201 TOL REB DRF PRESS B 36    12.607
iLM 23:18:24   36PI251C    PVHI     10.000 LOW   F201 TOL REB DRF PRESS C 36    14.499
iLM 23:18:24   36PX251C    OFFNORM           LOW   F201 TOL  REB DF PRESS C 36    HIGHHIGH
iLM 23:18:25   36PI251B    PVHI     10.000 LOW   F201 TOL REB DRF PRESS A HH 36   17.247
iLM 23:18:25   36PX251A    OFFNORM           LOW   F201 TOL REB DF PRE A HH 36    NIGHHIGH
iLM 23:18:25   36PX251B    OFFNORM           LOW   F201 TOL  REB DF PRESS B 36    HIGHHIGH
iLM 23:18:25   36PHH251    OFFNORM           LOW   F201TOL COL REB DF PREHH 36    HIGHHIGH
iLM 23:18:25   36PC251     PVHI     3.000 LOW   306F-201 DRAFT         36    3.487
iLM 23:18:25   36PC251     BADPV             LOW   306F-201 DRAFT         18
iLM 23:18:26   36PI251A    BADPV             LOW   F201 TOL REB DRF PRESS A 36
tTN 23:18:26   36PI251A    PVHI     10.000 LOW   F201 TOL REB DRF PRESS A 36
iLM 23:18:26   36PI251B    BADPV             LOW   F201 TOL REB DRF PRESS B 36
tTN 23:18:26   36PI251B    PVHI     10.000 LOW   F201 TOL REB DRF PRESS B 36
```

(4) Operator process change journal

```
PAGE 001    S-02/20/14 20:00  E-02/21/14 04:00    OPERATOR PROCESS CHANGE JOUF
---------------------------------------------------------------------------
    02/20/14
    20:38:43  36FC108    306E-104B  OUTLET       SP       23.016%
    21:11:50  36FC108    306E-104B  OUTLET       SP       22.916%
    21:12:34  36FC109    306E-104A  OUTLET       SP       24.100%
    21:47:49  36FC109    306E-104A  OUTLET       SP       24.200%
    22:09:09  36TI116    306C-101 TRAY #26       PVLOTP   108.000%
    22:09:15  36TI116    306C-101 TRAY #26       PVLOTP   105.000%
    22:09:25  36FC108    306E-104B  OUTLET       SP       23.016%
    22:10:04  36HC103B   36E-403C/D LOUVER CONTRL OP      30.000%
    22:10:08  36HC103D   36E-403G/H LOUVER CONTRL OP      30.000%
    22:14:22  36HC103B   36E-403C/D LOUVER CONTRL OP      25.000%
    22:14:25  36HC103D   36E-403G/H LOUVER CONTRL OP      25.000%
    23:17:27  36HC105A   36E-207 LOUVER CTRL      OP       20.000%
    23:19:00  36PC238    FUEL GAS PRE CONTROL     MODE     CAS
    23:19:12  36PC238    FUEL GAS PRE CONTROL     OP       75.650%
    23:20:45  36PC238    FUEL GAS PRE CONTROL     MODE     MAN
    23:23:26  36FC204    BZ. COL. FEED            MODE     AUTO
    23:24:04  36FC204    BZ. COL. FEED            OP       75.825%
    23:24:37  36FC206B   306G-204A/B TOTAL FLOW   MODE     CAS
    23:24:40  36FC206B   306G-204A/B TOTAL FLOW   OP       55.891%
```

(5) BT fired heater trip history

Cause \ Effect	Time Delay Second	Pilot Gas Shut Down V/V UV-213	Fuel Gas Shut Down V/V UV-211, UV-211B & PV-238	Fuel Gas Vent V/V UV-210	FD Fan 306K-701	Emergeny Air Door XV-723	ID Fan 306K-702	Stack Damper PV-725A	APH Bypass V/V TV-715
Emergency Shut Down SW HS-205, 306F-201		Close	Close	Open	Stop	Open	Stop	Open	
Pilot Gas PSLL-257, 306F-201(2 Out Of 3)		Close							
306F-201 Draft, PSHH-251(2003)	5		Close	Open	Stop	Open	Stop	Open	
F-201 Heater Feed Flow LL, FSLL-221~226(2003)	10		Close	Open	Stop	Open	Stop	Open	
306C-204 O/H PSHH-241(2003)			Close	Open	Stop	Open	Stop	Open	
306G-301A/B Discharge FSLL-303(2003)	30		Close	Open	Stop	Open	Stop	Open	
306F-201 Fuel Gas PALL-601			Close	Open	Stop	Open	Stop	Open	
Combustion Air Pressure LL, PSLL-710(2003)	2				Stop	Open	Stop	Open	
FD Fan Stop, XL-701	2				Stop	Open	Stop	Open	
ID Fan Stop, XL-702	2				Stop	Open	Stop	Open	

(6) The trend of operation variables before & after flame out trip

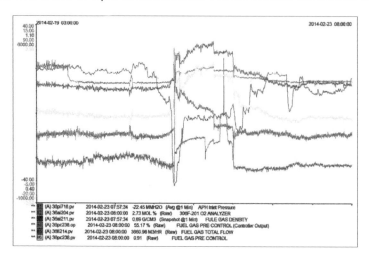

(7) Individual operation variables trend

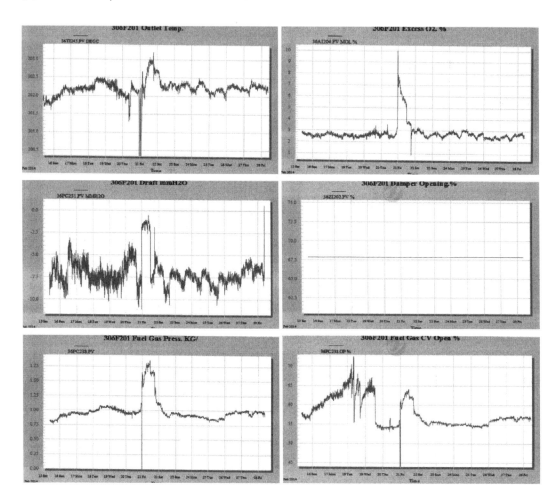

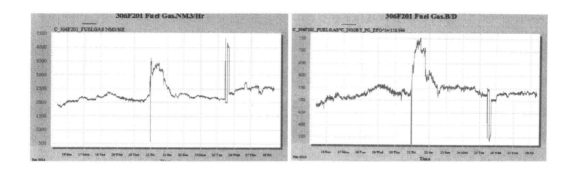

(8) Abrupt increase in the density of fuel gas

BT fired heater low density

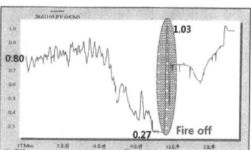

NHT fired heater low density

(9) Balanced draft heater

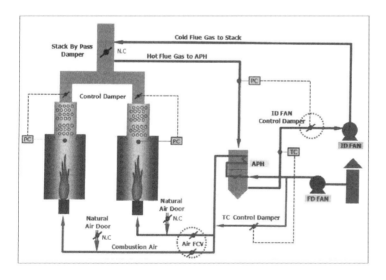

(10) Draft interlock logic status

No.	HTR	Set Point	Time Delay	No.	HTR	Set Point	Time Delay	No.	HTR	Set Point	Time Delay
1	02F-101	-10	10 Sec.	28	98F-101	+20	10 Sec.	55	107F-101	+13	X
2	02F-102	-10	10 Sec.	29	98F-103	+20	10 Sec.	56	107F-102	+13	X
3	03F-101	-10	10 Sec.	30	87F-201	+25	10 Sec.	57	306F-201	+13	5 Sec.
4	04F-101	-10	10 Sec.	31	103F-101	+13	10 Sec.	58	309F-101	+13	5 Sec.
5	04F-102	-10	10 Sec.	32	103F-102	+13	10 Sec.	59	309F-103	+13	5 Sec.
6	07F-301	-10	10 Sec.	33	103F-103	+13	10 Sec.	60	305F-101	+25	5 Sec.
7	52F-101	-5	10 Sec.	34	103F-104	+13	10 Sec.	61	305F-151	+25	5 Sec.
8	52F-201	-5	10 Sec.	35	103F-205	+13	10 Sec.	62	512-F101	+10	10 Sec.
9	54F-101	+10	X	36	103F-206	+13	X	63	512F-102	+10	10 Sec.
10	54F-201	+10	X	37	104F-201	+13	X	64	512F-103	+10	10 Sec.
11	62F-101	-8	10 Sec.	38	104F-102	None	None	65	512F-104	+10	10 Sec.
12	62F-151	-8	10 Sec.	39	104F-202	None	None	66	512F-105	+10	10 Sec.
13	62F-201	+15	X	40	105F-101A	+13	X	67	522F-101	+10	10 Sec.
14	63F-101	-8	10 Sec.	41	105F-101B	+13	X	68	522F-102	+10	10 Sec.
15	63F-102	-8	10 Sec.	42	109F-101	+13	X	69	522F-102A	+10	10 Sec.
16	63F-103	-8	10 Sec.	43	109F-102	+13	X	70	502F-101A	+10	10 Sec.
17	63F-104	-8	10 Sec.	44	203F-101	+13	X	71	502F-101B	+10	10 Sec.
18	64F-101	-8	10 Sec.	45	203F-102	+13	X	72	532F-101	+3	3 Sec.
19	64F-102	-8	10 Sec.	46	203F-103	+13	X	73	612F-101	+10	10 Sec.
20	64F-103	-8	10 Sec.	47	203F-104	+13	X	74	612F-102	+10	10 Sec.
21	67F-101	+15	10 Sec.	48	203F-105	+15	X	75	612F-201	+10	10 Sec.
22	73F-103	+25	20 Sec.	49	204F-101	+13	X	76	612F-202	+10	10 Sec.
23	72F-101	+25	10 Sec.	50	204F-102	+15	X	77	612F-301	+10	10 Sec.
24	72F-151	+25	10 Sec.	51	205F-101A	+30	5 Sec.	78	612F-401	+10	10 Sec.
25	77F-101A	+25	8 Sec.	52	205F-101B	+30	5 Sec.	79	632F-101	+3	3 Sec.
26	77F-101B	+25	8 Sec.	53	209F-101	+30	5 Sec.	80	654F-101	+15	3 Sec.
27	97F-201	+20	10 Sec.	54	209F-102	+30	5 Sec.				

10.7 C10 column reboiler heater problem- high excess O$_2$

This heater needs more tight operation in the view of excess O$_2$ control. Due to the low throughput of upstream, the heater is also operated below minimum load (high excess O$_2$ of 6-8%) as follows.

	Design	Normal	Minimum	Actual
Duty (MMkcal/hr)	16.83	15.3	7.65	Less than 7.65

Heater itself can be operated below minimum duty but needs higher safety guards such as higher excess air control or wider damper open (higher draft) at the load near turndown. So, it is not recommended to operate the tight excess O$_2$ control at low duty case and the higher excess O$_2$ may be a proper operation of heater for safety. And also the main obstacle to get tight O$_2$ control is due to ID fan which is based on 125% flow margin and 156% static pressure margin.

ID Fan can't control smooth flow rate below 15' because of damper actuator mechanism. Minor change of ID Fan will cause huge draft turbulence as well as it makes O_2 control difficult. Heater maker recommends to hold higher excess air control at low duty case again. If user wants to get tight O_2, the best way is to control the damper by hand. The purpose of ID Fan is to control heater draft and O_2 is controlled by FD Fan at normal duty, but by burner air register (damper) at turndown duty. It is not clear how much burner damper is open in such case but it shall be tight close position to get tight O_2 control. However, since this control is sensitive to the change of heat duty, whenever any circumstance changes, field operator should adjust it at all times.

10.8 CDU fired heater's coke generation on inside of burner regentile

(1) Coke generation

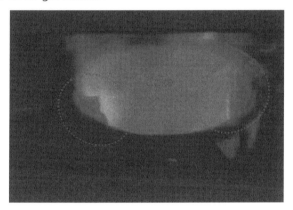

(2) Original gas tip

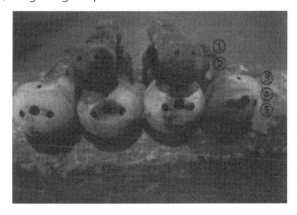

Upper: 2 primary gas tips (total 3 ports/tip)
① Crossover Ports (2)
② Firing Port (1)
Lower: 4 secondary gas tips (total 4 ports/tip)
③ Ignition Port (1)
④ Crossover Port (1)
⑤ Firing Ports (2)

(3) Original gas tip orientation and ports number

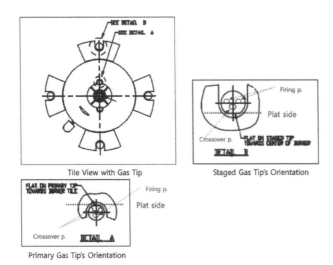

Tile View with Gas Tip

Staged Gas Tip's Orientation

Primary Gas Tip's Orientation

(4) Test gas tip

1) 1st/2nd gas tip revision for test

The firing ratio of primary gas tip and staged gas tip per burner depends on burner capacity, but it will be 10-20% for primary and 88-90% for staged. In case of short flame design, it will be 18-20% for primary. However, technical review of total balance of burner capacity is required. And normally single port flame length is longer than multi ports if total firing capacity is same. 1st trial gas tip flame has improved to prevent high skin temperature of heater tube. NO_x will be increased due to lower capacity of staged flame. A burner maker has applied same port angle as 1st test for 2nd test gas tip. But, they applied two firing ports on primary tip and this port angle is not same as 1st test. But, angle against axis of the burner is same. The burner maker is sure that two firing ports made better mixing with combustion air.

	Primary Gas Tip		Staged Gas Tip		
	Crossover Port, EA/mm	Firing Port, EA/mm	Ignition Port, EA/mm	Crossover Port, EA/mm	Firing Port, EA/mm
Original	2-Ø2.6	1-Ø2.6	1-Ø2.6	2-Ø3.2	1-Ø5.6
1st Trial Test (Feb, 2013)	2-Ø3.2	1-Ø3.2	1-Ø3.0	2-Ø3.2	2-Ø4.4
2nd Trial Test (Apr, 2013)	2-Ø2.6	1-Ø4.1	1-Ø3.0	2-Ø3.2	2-Ø4.4

2) 3rd gas tip revision for test

To avoid generating coke inside the upper regentile, a burner maker has changed gas tip design to get more short flame due to thermo couple damage on the heater tube first. Then, they increased primary firing ratio. However, air/gas mixing inside of regentile became worse, and generated coke inside of regentile. So, the burner maker changed primary gas tip design again to get more good mixing condition of gas and air. The modification of gas tip includes 2 primary ports added and port size revised to existing tip.

	Primary Gas Tip			Staged Gas Tip	
	Crossover Port, EA/mm	Firing Port, EA/mm	Ignition Port, EA/mm	Crossover Port, EA/mm	Firing Port, EA/mm
3rd Trial Test (Jul, 2013)	2-Ø2.6	2-Ø2.9	1-Ø3.0	2-Ø3.2	2-Ø4.4

Upper: 1 primary gas tip (total 4 ports/tip)
① Crossover Ports (2)
② Firing Ports (2)
Lower: 4 secondary gas tips (total 5 ports/tip)
③ Ignition Port (1)
④ Crossover Ports (2)
⑤ Firing Ports (2)

(5) Test gas tip orientation and ports number

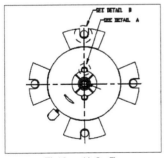

Tile View with Gas Tip

Staged Gas Tip's Orientation

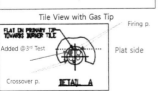

Primary Gas Tip's Orientation

10.9 Hydrotreating charge heater accident (case 1)

A refinery in Korea has experienced the tube crack and leakage in old Hydrotreating charge heater in December 2007. The cause of accident was flame impingement. After shutdown the tube replacement and the tube decoking were carried out as first action. Furthermore the refinery has modified coil inlet manifold (to "Y" type) to get even pass flow as well as the replacement of old burner.

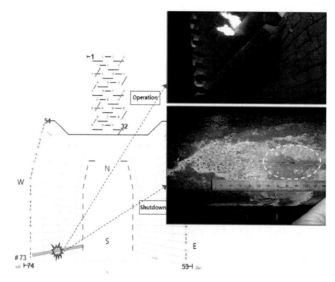

Crack and leakage

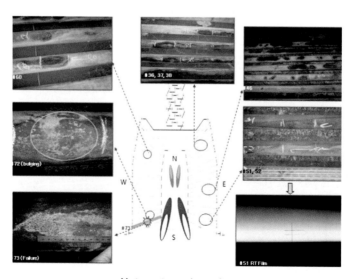

Hot spots and cracks

10.10 Platforming charge heater accident (case 2)

A refinery in Korea had tube sagging problem just after the revamp rearranging feed pass in Platforming charge heater in July 2008. The incident took place due to the pass flow depletion (3 of 4 passes) with unbalanced feed distribution at coil inlet manifold. The feed maldistribution has come from the incorrect design in manifold by heater maker. As symptom of the issue, one week after running, the color of tubes changed into white gray.

4 passes

Tube sagging

Pass 2 Coke blocked

Pass 1 No coke blocked

Pass 3 Coke blocked

Pass 4 Coke blocked

10.11 Oil tip lock nut seizing problem

The seizing problem of lock nut of oil gun in advanced burner has been solved by changing its metallurgy into stainless steel from carbon steel although the oil gun is set in the burner in no service.

Parts	Metallurgy	Grade		Cr contents	Scaling temp, deg C
Oil Tip	A276	440C(Martenstic S.)	17Cr-Mo		815 deg C/intermittent heating 760 deg C/continuous heating
Lock Nut(Existing)	SA193	B7(Carbon S.)		0.75-1.20 Cr, 0.15 - 0.25 Mo	No data was found
Lock Nut(Test)	JIS G4303	310SS(Austenitic S.)	25Cr-20Ni		1,035 deg C/intermittent heating 1,095 deg C/continuous heating
Sleeve	SA582	416SS(Martenstic S.)	13Cr-S		760 deg C/intermittent heating 675 deg C/continuous heating

Note: Thermal expansion coefficient (500°C): carbon steel 0.007mm/m vs. 310SS 0.009mm/m

PART 4

Revamp

Sub Contents
for Revamp

1. Reference Revamp

Below example has been used in many field reports to show the effects of fuel composition. The refinery was burning the high hydrogen fuel in the catalytic reforming heaters. New hydrotreating units were added and the hydrogen was removed from the fuel gas and sent to the hydrotreaters. Pipeline natural gas replaced the high hydrogen fuel. The operation group could no longer maintain the current feed rate. At the same feed rate, the heat release increased 5% and the flue gas rate increased 21%. The heater was stack limited and could not handle the large 21% increase in flue gas flow rate. The heater was revamped and an induced draft fan was added to overcome the stack limitation. Below is the result of revamp.

Fuel	High H_2 (@15% excess air)	Natural gas (@15% excess air)	Fuel oil (@25% excess air)
Flue gas rate difference	Base	121%	133%
Flame temperature, ℃	1,950	1,817	1,749
Heat release	Base	105.2%	108.4%

2. Conditions Influencing Coke Formation

Hot spots in fired heaters can lead to accelerated rates of coking and tube failure. Two examples show how integrating fireside and oil side calculations can help identify and correct conditions leading to high localized peak flux rates.

High tube metal temperatures (TMTs) and coking in fired heaters continue to reduce run lengths and in some instances cause tube failure. High TMTs are the result of non-uniform heat flux distribution in the firebox and tube side conditions that lead to accelerated rates of coke formation. Several factors including the number of burners, burner layout, burner flame length, burner-to-tube spacing, firebox height to width (L/D) and non-ideal flue gas flow patterns influence heat flux distribution. Moreover, tube side conditions that affect the rate of coke formation such as tube size, oil mass velocity and oil residence time need to be considered. When local conditions generate high TMT, determining the root cause is essential. Otherwise, the problem cannot be fixed. In some instances, it is more cost effective to improve fireside performance. In others, tube side changes are needed. Two cases will illustrate how local conditions can lead to high TMT, high rates of coke formation and short run lengths. In each example, revamp modifications increased run length by reducing heat flux variability or improving tube side operation.

2.1 Hot spots-local conditions

Hot spots exist in all fired heaters' radiant sections; the magnitude depends on the specific heater design. Since many heaters have been purchased solely on lowest initial installed cost, the low bidder is forced to supply an "inferior" design. Lowering the cost may include reducing the number of burners, decreasing the distance between the burners, lowering the distance between the burner center-line to tube face, stacking the tube passes along the wall, increasing the L/D or several other cost saving considerations. However, many of the techniques that lower cost also cause large heat flux variability throughout the firebox or degrade tube side operation. As with most process equipment, there is no free lunch. Fundamental principles cannot be violated without paying the consequences. In fired heaters, rapid coke formation and high TMTs are often the outcome.

Specific heater design will determine the severity of non-uniform heat flux distribution. Hence, typical evaluation criteria such as average radiant section heat flux or maximum duty can be misleading. For instance, design guidelines for vacuum heaters may specify that the average radiant heat flux should not exceed 10,000 Btu/ft^2-hr-°F. In one instance, a vacuum heater operating at only 9,000 Btu/ft^2-hr-°F was decoked every two years. Whereas another operates reliably for four years at an average heat flux of approximately 12,000 Btu/ft^2-hr-°F. Fireside design problems that cause very high, localized heat flux will produce elevated TMTs even when there is no coke in the tubes. This often raises the film temperature above the oil's thermal stability and causes coke to form. At very high, localized heat flux, even a small amount of coke results in very high TMTs. In these instances, tube metallurgy limits can be exceeded. Even when localized heat flux is not extremely high, coke can form because the tube side conditions result in oil film temperature and residence time above the oil's thermal stability limits. Tube size (oil mass velocity) and the location of the outlet tubes are important considerations affecting the rate of coke formation. It is not unusual to have the pass outlet tubes routed through the highest heat flux zone. Furthermore, outlet tubes for heaters such as vacuum heaters are often larger diameter. The larger diameter results in lower mass flux, higher oil film temperature and increased oil residence time, and can lead to high rates of coke formation if the tubes are located in the highest heat flux zone. In summary, localized hot spots can be caused by fire or tube side design shortcomings, or both.

2.2 Rate of coke formation

Coke forms because the oil film temperature and residence time exceed the oil's thermal stability. The oil film temperature depends on the heat flux and oil mass velocity. Raising the heat flux increased the oil film temperature at a fixed mass flux, while a higher mass flux lowers the film temperature at a given heat flux. The oil residence time depends on the oil's mass velocity and the amount of coil steam or condensate added to each heater pass (vacuum and coker heater). A lower residence time reduces the rate of coke formation.

Radiant section heat flux varies from the combustion zone where the fuel/air mixture is burned to the flue gas outlet from the radiant section. In the combustion zone, the maximum heat flux is controlled by direct flame or hot gas radiation. Direct flame radiation depends on the number of burners, flame size, distance between the burners and burner-to-tube spacing. In some instances, direct flame radiation controls the maximum heat flux in the combustion zone. Talmor's (Talmor, E, "Combustion Hot Spot Analysis for Fired Process Heaters," Gulf Publishing, 1982) combustion zone heat flux calculation should be used to identify the specific location of the hot spots.

In many heaters, the highest heat flux is outside the combustion zone. Here, the flue gas temperature determines the maximum heat flux. These hot spots can be determined by measuring the TMTs when the heater is clean, through direct heat flux measurements with a flux probe or calculation using a proprietary program such as HTRI that performs a zone analysis and considers burner type, location, flame size and distance from burner centre-line to tube face.

Coking occurs when the oil film temperature is too high for too long. The film temperature (at a fixed heat flux) depends on the tube side heat transfer coefficient. Increasing the mass flux by decreasing the tube size raises the heat-transfer coefficient and lowers the film temperature. Conversely, when the tube size is increased (such as heater outlet tubes) the heat transfer coefficient decreases, thereby increasing the oil film temperature and residence time. Furthermore, when the pass outlet tubes are routed through the highest heat flux zone, the oil film temperature and residence time often exceed the oil's thermal stability, rapidly forming coke inside the tubes. When coke accumulates inside the tubes, it acts as an insulator, raising TMTs. In some instances, TMTs exceed the maximum temperature for the tube metallurgy for extended periods, causing the tube to rupture.

2.3 Fired heater models

Various models are used to design and predict the operation of fired heaters. These make several assumptions about the actual heater's performance. Some assume that the radiant section is a well stirred isothermal zone or that the flue gas flows upwards, with the radiant section heat flux determined by hot gas radiation alone. Some models take into account direct flame radiation in the combustion zone while others ignore the specific burner system design completely. Rarely will a fired heater's model default match the operation of a specific heater design. Thus, most models fail to predict local conditions without the user manipulating them. Since actual heater design is so variable, the model user must have significant know-how concerning fire and tube side operations. Field measurements and observations are critical. Simple techniques like throwing baking soda into the burners to observe flame behavior may be used to identify flame impingement. Heater models can be powerful tools, but they can also easily give answers that do not represent true heater performance.

Understanding fired heater operation and design requires that both the fire and process side be evaluated together. Correctly predicting the oil film temperature and residence time in each coil is essential to specify design changes during a revamp to avoid problems or to establish the root cause of rapid coking when local conditions are non ideal. In one case, the heater model assumed the radiant section was a well stirred box. Thus it predicted a peak heat flux of 25,000 Btu/ft^2-hr-°F, but the actual peak heat flux in the combustion zone was closer to 42,000 Btu/ft^2-hr-°F. Each model makes simplifying assumptions that influence the model results, so the user needs to be aware of these.

Once an individual heater's tube oil film temperature and residence time are established from the heater model, the oil's thermal stability can be assessed. A lower oil residence time permits a higher allowable temperature for a given level of thermal stability. While thermal stability depends on the oil type and upstream processing, it can be represented as an exponential function of oil film temperature and residence time. For instance, even though the coker heater's operating temperatures cause thermal cracking to occur inside the heater tubes, it is the oil residence time and film temperature that dictate the rate of coke lay-down inside the tube. The TMTs will increase as coke forms inside a tube, to the point where the heater must be decoked. Reducing the rate of coke formation to increase heater run length requires either lower oil residence time or lower oil film temperature, and sometimes both.

2.4 Poor burner layout (Case study 1)

In case study 1, a vertical cylindrical hydrotreater reactor heater that had operated well for years began to have problems when the fired duty was increased. Very high, localized heat flux raised the oil film temperature above the thermal stability limit of the oil, forming a thin coke layer inside the tubes. Since the thin layer of coke raised the resistance to heat transfer at a very high heat flux, TMTs increased above the metallurgical limits, causing a tube to fail. Although tube failure was localized, several tubes in the bottom 10-12 ft of the radiant section showed damage from high temperature. Interestingly, burner flame lengths were approximately 10-12 ft.

Combustion zone problems occur when the localized heat flux from direct flame radiation exceeds that calculated from hot gas radiation. The maximum heat flux in the combustion zone depends on the number of burners, burner layout, heat release per burner, burner volume and burner- to- tube spacing. In this case, very high, localized heat flux in three areas of the combustion zone was evident from infrared TMT measurements and tube metallurgical evaluation. Infrared scans showed TMTs highest in the area predicted by Talmor's peak heat flux method for combustion zones.

Furthermore, one of the three hot zones had even higher heat flux than the other two.

2.5 Burner layout

Burner configuration plays a significant role in combustion zone heat flux and flue gas flow patterns. Burner layout, number of burners, distance between burners and heat release per burner all affect the localized heat flux. Fig.1 shows the heater floor with three burners installed on an equilateral triangular pitch. Visual inspection of the burner locations inside the heater (and from the heater drawings) shows three areas where the distance from the burner to the tube face is small. Heat flux from direct flame radiation increases as the distance from a flame to a tube decreases. Infrared measurements of TMTs showed the three areas closest to the burners had the highest TMTs and one of the three higher than the other two.

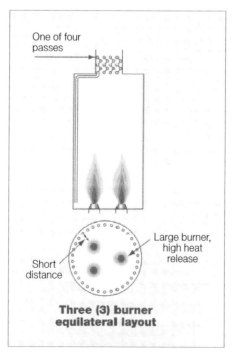

Fig.1 Burner layout- three large burners

Distributing heat in the combustion zone requires a balance between heat release per burner, distance from the burner to the tube face and distance between the burners. Ultimately, the heater designer (engineering contractor or heater vendor) determines the number of burners and the diameter of the heater floor. While more burners distribute the heat better, this requires a larger footprint. Fig.2 shows a schematic of three potential burner layouts using three, four and six burners. The hatched area represents the location of the convection section inlet relative to the different burner configurations.

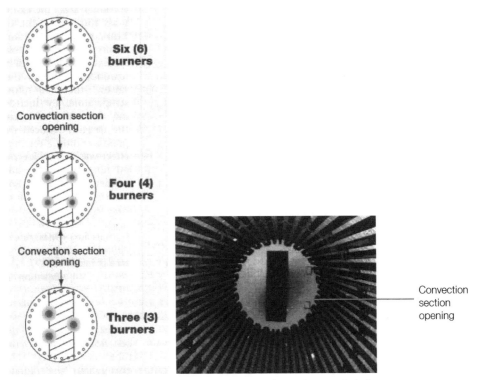

Fig.2 Alternative burner layout- three, four and six burners

In a cylindrical heater, good heat distribution is easier when six or more burners are designed in a circle, but this increases total installed cost.

Increasing the number of burners would have reduced the heat release per burner and helped distribute the heat. However, the floor diameter must be large enough to accommodate additional burners while maintaining the proper distance between them to allow for stable flame operation and the correct distance from the burner to tube face, to reduce the peak heat flux caused by direct flame radiation to an acceptable level. An acceptable level is when the heat flux from direct flame radiation is at or less than that caused by hot gas radiation alone. However, when the lowest initial installed cost is the basis for purchasing a heater, cost cutting methods such as decreasing the number of burners and reducing the floor diameter will almost always be the result. While heater OEMs are often blamed for poor heater design, it is the purchaser that is truly at fault for allowing fundamental principles to be violated. In this case, the number of burners could not be increased due to the existing floor diameter. Heat exchange from hot gas radiation can be calculated from Eq.1 (Lobo-Evans method) as a function of flue gas temperature (T_g), tube metal temperature (T_t) and the radiant section surface area (αA_{cp}).

Radiation section $Q_r = 0.173\,(\alpha A_{cp})\,(F)\,[(T_g/100)^4 - (T_t/100)^4]$ ---- (Eq.1)

Duty= Btu/hr

Although this Eq. makes several assumptions to simplify what happens in an actual heater, it highlights that radiant heat transfer is controlled by flue gas and process fluid temperature differences. Since the flue gas absolute temperature is so much higher than the process fluid, the localized flue gas temperature largely determines how much heat is transferred at any location within the firebox. Thus, large variations in flue gas temperature throughout the firebox result in large heat flux differences.

2.6 Flue gas flow patterns

Localized heat flux is also influenced by flue gas flow patterns. Flue gas does not flow uniformly upward in the firebox. Burner location relative to the convection section inlet and cold flue gas flowing downward are some of the factors influencing flue gas recirculation. In a vertical cylindrical heater, ideally hot flue gas flows upward in a circular pattern, while cold flue gas flows downward behind and around the tubes. In practice, burner layout, burner type and convection section opening play a major role in flue gas flow patterns. In the three-burner arrangement, two burners are located below one edge of the convection section, requiring flue gas from these two burners to flow toward the middle of the convection section inlet area. This pushes the flue gas from the single burner at the other edge of the convection section toward the tubes (Fig.3). Modeling this with computational fluid dynamics (CFD) shows these flow patterns with three burners. Furthermore, when baking soda is thrown into the firebox to better observe flame patterns, the two burners lined up on one side of the convection section opening were forcing the single flame closer to the tubes.

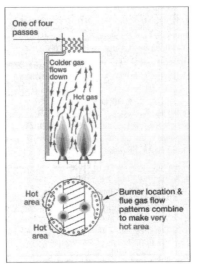

Fig.3 Flue gas flow pattern- three large burners

Tube failures occurred in this very hot area of the heater (Fig.4). Heater flue gas flow patterns have a large effect on flame interaction and burner stability.

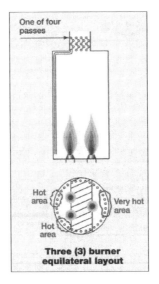

Fig.4 Hotspots- three large burners

Flame stability problems are increasing because of the use of low and ultra-low NO$_x$ burners. Thus, high heat flux variability is becoming more common. In the high flux areas, TMT and oil film temperature are high, which raises the rate of coke formation. Field measured heat flux and infrared measurements of TMTs show that flame stability and flue gas recirculation problems are increasingly common today.

2.7 Heater revamp

Ideally, six burners should have been used to better distribute heat in the firebox and to improve flue gas flow patterns. But in this instance, the burner floor was too small to maintain a proper distance between the center-line of the burner and the tube face with the six-burner layout. Increasing the distance between the burners and the tubes would have reduced the heat flux due to direct flame radiation, but the floor was not large enough in diameter. Therefore, the only practical solution was to install one large burner with multiple gas tips in the middle of the heater floor (Fig.5).

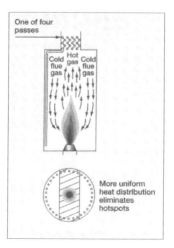

Fig.5 Reduced localized heat flux- single large burner

When this solution is implemented, it will reduce combustion zone heat flux from direct flame radiation to below that from hot gas radiation, thereby reducing peak heat flux in the bottom of the heater by 40%. This will dramatically lower the oil film temperature and rate of coke formation, but it will also raise the bridge wall temperature and the convection section heat flux. When revamping heaters, existing configurations force the revamp engineer to make compromises.

2.8 Coker heater revamp (Case study 2)

In case study 2, a coker heater had run lengths of less than two months between decokes, although the average radiant section heat flux was less than 9,500 Btu/ft^2-hr-°F. However, oil residence time was very high and the oil's mass velocity low throughout the radiant section tubes. In the larger 4" outlet tubes, mass flux was extremely low and oil residence time very high. Moreover, fireside combustion zone evaluation showed the local heat flux was highest where the radiant section tube size changed from 3.5" to 4". This was confirmed with infrared measurements of TMTs.

Fig.6 shows the cabin heater prior to making modifications. The convection section was designed with both oil and steam coils. Convection section oil coils were 4" while the steam tubes were 3.5". radiant section coils were down-flow with 4" tubes on the roof section, 3.5" tubes on the majority of the wall and the last six tubes were 4". All radiant section tubes were on two-to-one spacing except the last six 4" outlet tubes that were spaced at three-to-one. Coke was forming in the 4" tubes at the bottom of the heater, causing very short run lengths. Low oil mass flux and high oil residence time were the two main reasons for this short run length.

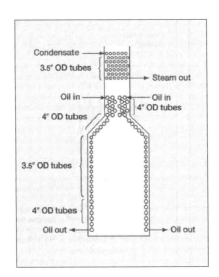

Fig.6 Coker heater- before revamp

2.9 Oil mass flux

Oil mass flux (Eq.2) is a measure of the flow of oil through the tube cross sectional area. Increasing mass flux lowers the oil film temperature and reduces the rate of coke formation. Conversely, lowering the mass flux raises the oil film temperature. By increasing the outlet tube size, the designer reduced the pressure drop. Yet the effect was to increase the oil film temperature and residence time and thereby raise the rate of coke formation.

G (mass flux)= Mass rate of oil/inside cross-sectional area of heater tube ---- (Eq.2)

Mass flux= Lb/ft^2 sec

Coke forms because the oil flowing inside the tube is cracking. The temperature throughout the radiant section of a coker heater is above the oil cracking temperature, so coke always forms. The key is to minimize the rate of formation and increase the duration between decokes. Maximizing the inside tube's heat transfer coefficient minimizes the film temperature and increases the heater's run length. Since the outlet tubes' mass velocity was well below 200 Lb/ft^2 sec, these tubes needed to be changed from 4" to 3.5". By reducing the tube size to 3.5", the process heat transfer coefficient (Eq.3) would be increased and the oil film temperature reduced.

Inside tube heat $(h_i)= (0.023) \, k/D \, (DG/\mu)^{0.8} \, (c_p\mu/k)^{0.33} \, (\mu/\mu_w)^{0.14}$ ---- (Eq.3)

Transfer coefficient = $Btu/ft^2\text{-}hr\text{-}°F$

2.10 Localized oil film temperature

Lowering film temperature reduces the rate of coke formation. Eq.4 shows how the temperature drop across the oil's film is calculated. The D_o and D_i are the outside and inside tubes' diameters, respectively. Combustion zone direct flame radiation or flue gas temperature largely determines the amount of heat transferred locally (Q_{local}).

Δt_f = Temperature drop across the oil film ----- (Eq.4)

$$= Q_{local} \, D_o/D_i \, h_i$$

$$= °F$$

For a given heat flux (Q_{local}) the temperature drop through the oil's film is set by process fluid inside the film coefficient (h_i). Reducing the tube size lowers the oil film temperature and reduces the rate of coke formation in a heater.

Fireside calculations showed that the highest heat flux (Q_{local}) occurred in the area where the tube size changed from 3.5" to 4" (Fig.7). Since the 4" tubes were spaced at three-to-one spacing, installing 3.5" tubes on the same tube hangers reduced the peak heat flux, further lowering the oil film temperature.

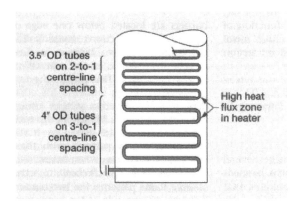

Fig.7 Location of highest heat flux

2.11 Oil residence time

Oil residence time is controlled by the radiant section's tube length, tube size, oil mass flow rate and condensate injection. Decreasing the radiant section's oil residence time lowers the rate of coke formation. In this example, the heater operated less than 60 days between decokings due to the high rate of coke formation. The radiant section's oil residence time was more than one minute and had to be reduced to less than 30 seconds to materially improve run length.

Increasing condensate injection would have been the easiest way to decrease oil residence time, but raising the rate would increase the heater's firing because condensate must be vaporized. Another alternative was to reduce all the radiant section's 4" tube sizes to 3.5". Although this reduced oil residence time, it was not enough to materially increase run length and it also raised the pressure drop so much that the charge and condensate pumps were no longer able to provide sufficient head to overcome the additional pressure drop. Therefore, other means to increase run length were needed.

2.12 Convection & radiant section duty

Fig.8 is a simplified representation of a fired heater. Total duty is supplied by the convection and radiant sections. In this instance, the majority of the convection section's surface area was used to make steam. Thus, most of the heat added to the oil was radiant section duty. It was clear that increasing the convection section's oil duty would decrease the radiant section's flux and enable some of the 4" tubes to be replaced with 3.5" tubes.

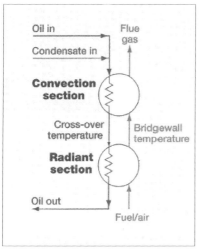

Fig.8 Simplified heater schematic

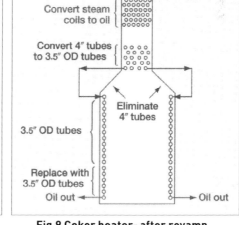

Fig.9 Coker heater- after revamp

Fig.9 shows the revamp modifications. The convection section's surface area was added for the oil, and the radiant section's surface area was reduced by removing some tubes and replacing others with smaller tubes. These changes, in addition to a higher condensate flow injection, reduced the radiant section's flux rate and reduced the radiant section's residence time to less than 30 seconds without exceeding the condensate pump's capacity. The convection section's duty was increased by converting the steam coils to oil, and the 4" convection section's oil tubes were reduced from 4" to 3.5" to reduce residence time and oil film temperature. The oil convection section's surface area was increased by more than 600%. Since the revamp needed to be completed as fast as possible, the radiant section's modifications were minimized. The 4" roof tubes were removed and the 4" outlet tubes were changed to 3.5" tubes. The heater run length has increased dramatically.

After startup, the skin temperatures were still higher in the highest heat flux section of the heater compared to the other tubes, but the lower residence time and lower film temperature have dramatically reduced the rate of coke formation. In the longer term, the convection section has unused space to add more tubes to further lower the radiant section's heat flux (Fig.10).

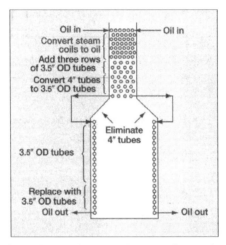

Fig.10 Coker heater- increased convection surface area

Identifying the local conditions that lead to rapid coke formation is the key to increasing heater run length. Quantifying the radiant section's heat flux variability and minimizing it is essential. Since oil film temperature and residence time have a dramatic influence on coke formation rates, properly evaluating each coil to establish the location of hot spots is essential. Fireside and oil side calculation methods must be integrated and calibrated with plant data to predict true heater performance.

3. Revamping Atmospheric Crude Heaters

The revamp of an atmospheric crude unit heater which was suffering from coking caused by asphaltene precipitation and poor burner stability, resulted in a significantly increased heater run length.

Refinery atmospheric crude heaters can experience rapid increases in tube metal temperature (TMT), requiring unplanned shutdowns for decoking. In the case of a refining company, rapid increases in the atmospheric crude heater's TMT resulted in a shutdown every three to six months. In this case, as well as others observed in the industry, coke formation was initiated by asphaltene precipitation from unstable crudes.

Industry-wide atmospheric crude heater coking is an unusual problem with some heaters operating reliably at an average radiant section heat flux of 13,000-14,000 Btu/ft^2 hr or higher. However, some crude oils including those produced from North American fields in West Texas, New Mexico, Ohio/Pennsylvania and Alberta are known to be unstable when there is asphaltene precipitation in certain areas of the crude unit. The equipment in these areas includes preheat exchangers, fired heaters and atmospheric column flash zone and stripping section internals.

3.1 Atmospheric crude heater coking

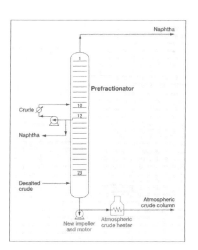

The refinery was experiencing chronic coking in its atmospheric column heater, with periodic shutdowns to remove coke at intervals as short as 90 days. Fig.1 shows the process flow scheme with the heater located downstream from the prefractionator column. Flashed crude is charged to the atmospheric crude heater, which feeds the atmospheric crude column.

Fig.1 Process flow diagram

The heater was a vertical tube hexagonal shaped four pass design with 12 burners. Prior to the revamp, the radiant section's average heat flux rate was only 9,400 Btu/ ft^2 hr with an oil outlet temperature of just 335℃. Furthermore the heater had oil mass flux rates of 250 Lb/ft^2 sec and relatively poor flame stability. This, in combination with poor asphaltene stability, caused a very short heater run length even though the heater was operating at relatively mild conditions. Some atmospheric crude heaters operate at an average radiant section heat flux of 13,000-14,000 Btu/ft^2 hr and oil outlet temperatures of 388℃ or higher while meeting four to five year run lengths.

3.2 Heater coking

Coke forms because conditions in the shock or radiant tubes cause the oil to thermally decompose to coke and gas. The TMT increases as coke lays down on the inside of the tube. With rising TMTs, heater firing must decrease or the TMTs will progressively escalate until their limit is reached. The heater must then be shut down to remove the coke. Rapid coke formation is caused by a combination of high oil film temperature, long oil residence time and inherent oil stability. In the majority of case where atmospheric heater coking occurs, the root cause is high average heat flux, high localized heat flux or flame impingement.

3.3 Oil stability

Oil thermal stability depends on crude type. For example, some Canadian and Venezuelan crude oils have poor thermal stability and begin to generate gas at heater outlet temperatures as low as 360℃. At outlet temperatures much above 371℃, these same crudes begin to deposit sufficient amounts of coke to reduce heater runs to two years or less. Another form of oil instability is asphaltene precipitation. As the oil is heated, the asphaltenes become less soluble, depositing in low velocity areas, fouling crude preheat exchangers, heater tubes or atmospheric column internals. In some cases, the asphaltenes do not drop out until they reach the bottom of the atmospheric column or inside the vacuum heater. With some Canadian crude oils, especially the bitumen based oil sands crudes, asphaltene precipitation occurs inside the vacuum heater tubes rather than in the atmospheric heater. The same heater design parameters that improve atmospheric heater performance also increase vacuum heater run length. When asphaltenes separate from the crude oil, the material deposits inside the tubes. This increases heat transfer resistance, raising asphaltene temperature and TMTs. Furthermore, when asphaltene deposits are wide spread in the convection or radiant sections, heater firing must increase to meet the targeted heater outlet temperature. This leads to a higher localized heat flux, further raising the temperature of the asphaltenes deposited on the inside of the tubes. The temperature of these asphaltenes eventually exceeds their thermal stability, resulting in coke formation and even higher TMTs, because the coke layer has lower thermal conductivity than asplaltenes. Heater TMTs eventually exceed metallurgical limits, requiring a heater shutdown to remove the coke. In this example, heater run lengths were as low as 90 days between piggings.

3.4 Asphaltene precipitation

Crude stability is a function of its source and highly variable. However, the designer can influence the process and equipment design to minimize the effect of poor asphaltene stability. In some cases, the material deposits inside the exchangers, piping heater tubes or fractionation column. The lower the velocity, the more likely it is that asphaltenes will precipitate. In this example, the oil velocity inside the heater tubes was only 5.5-6.0 ft/sec prior to the oil vaporizing, which corresponds to a 250 Lb/ft^2 sec oil mass flux. At these velocities, whether in an exchanger or heater tube, asphaltenes will likely drop out. Heat exchanger data gathered from hundreds of operating exchangers shows the rate of fouling and the ultimate fouling factor are to a large extent determined by the velocity of the crude flowing through the tubes (assuming no shell side design problems).

Asphaltene precipitation in crude preheat exchanger tubes is common. Fig.2 shows asphaltene precipitation inside the channel head and tubes in a unit processing West Texas crudes.

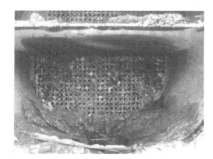

Fig.2 Fouled exchanger tubes- asphaltene precipitation

In this case, the oil velocity inside the tubes was less than 5 ft/sec and severe fouling occurred. Exchangers operating at higher velocities in the same unit had less fouling. Moreover, the atmospheric heater downstream of the fouled exchanger had short heater runs, with TMTs increasing at 0.6°C/day. This rate of TMT rise is similar to a delayed coker heater. Crudes with poor asphaltene stability are especially difficult to process and the equipment must be carefully designed.

Maintaining a high velocity in the equipment minimizes asphaltene precipitation. Crude preheat exchangers and heater tubes should be designed for oil velocities of 8-10 ft/sec or higher. Experience shows significant improvements in crude preheat and heater reliability when velocities are high. Since many designers set the maximum allowable pressure drop through exchangers and the heater as design criteria, low velocities are often the result of meeting pressure drop specifications. Crude preheat exchangers and fired heater designs should be based on a higher velocity, with the pressure drop simply a result of the design.

3.5 Heater revamp

In late 2005, a refinery revamped its existing atmospheric heater and installed a new parallel "helper" heater. The helper heater was needed because the existing heater burner spacing was increased and the number of burners decreased for improved flame stability with Ultra-Low NO$_x$ Burners (ULNB), resulting in lower design heater firing. Yet, installing the parallel heater without revamping the existing heater would have decreased oil velocity to less than 4 ft/sec in the existing heater. Moreover, the refinery wanted to increase the atmospheric heater outlet temperature from 335-354°C to unload the downstream columns because they limited the crude charge rate. Since the root cause of heater coking was asphaltene precipitation inside the heater, reducing heater firing alone would not have improved the heater run length. Furthermore, designing the new parallel heater for similar tube velocities as the existing heater would have created a second reliability problem. The revamped heater included a new convection section, completely retubed radiant section, new ULNB burners and floor. Prior to the revamp, heater performance showed the convection section was not performing well, radiant section tubes were fouled and the burners had poor flame stability. Poor convection performance was caused primarily by fin damage. Radiant section fouling was caused by asphaltene precipitation and poor burner stability resulting from the burners being too close together, causing adverse flame interaction.

Total fired heater "absorbed duty" is the sum of the convection and radiant section duties. Maximizing the convection section duty minimizes the radiant duty, which lowers the oil film temperature, reducing the rate of coking. In this example, the convection section was replaced with a similar design, except some of the tube fins were upgraded from carbon steel to 11-13 Cr to avoid damage from high temperature and to maintain performance throughout the run. The radiant section was completely retubed with smaller diameter tubes. Bulk oil velocities were increased from 5.6-6.0 ft/sec to almost 10 ft/sec which resulted in an oil mass velocity of 460 Lb/ft^2 sec. The revamped heater mass velocity is over twice the rule-of-thumb (ROT) values of 150-200 Lb/ft^2 sec that have been used to design atmospheric crude heaters. The smaller tube diameter dramatically increased the pressure drop, requiring a larger pump impeller and motor in the flashed crude pumps (Fig.3).

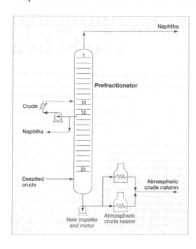

Fig.3 Increased system pressure drop

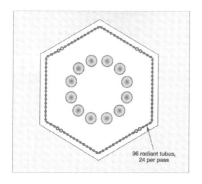

Fig.4 Burner floor before revamp

It is not uncommon for process design engineers to specify the pump design before the heater is designed because they assign a maximum allowable pressure drop to the heater. This approach expedites design but causes heater reliability problems once it is built. Prior to the revamp, the burners had poor flame stability. The heater had 12 burners that were spaced too close together (Fig.4).

A completely new floor was installed with six larger burners, which eliminated flame interaction producing a stable flame (Fig.5).

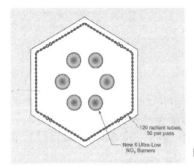

Fig.5 Burner floor after revamp

These burners were latest generation ULNB burners.

3.6 Revamp results

The revamped heater and new helper heater have been operating for 18 months without a shutdown. Crude charge has been increased by 14% and the heater outlet temperature has risen from 335-354℃, with the revamped heater operating at approximately 103% of the pre-revamp firing rate or 9,700 Btu/ft^2 hr average radiant section heat flux. The heaters have not been pigged and the TMTs have shown very little rise since startup.

[Case Study 1] Troubleshooting of Kero Splitter reboiler

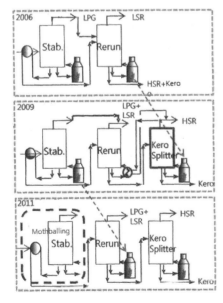

Kero Splitter Reboiler

Naphtha Splitting Unit above is process to treat imported naphtha and in 2009, the unit did a revamping to install new kero splitter to improve the quality of product HSR, accompanying to rearrange the heat exchange network and heater. Just after the revamp, the kero splitter's reboiler occurred a serious problem which due to continuously rising tube skin temperature, unscheduled shutdown for decoking is required once every 10 months (shutdown 3 times). So burner replacement and retrofit of convection breeching space were implemented but they did not help the problem solve basically. And during shut down in 2011, the unit experienced 2nd revamp that, after mothballing the stabilizer, the stabilizer reboiler was transferred to the new reboiler of rerun column, and rerun column/kero splitter operate independently. As the result of 2nd revamp, the rising tube skin temperature at last stopped and didn't show an increase any more for 16 months. After all, the serious problem became to be solved completely and got the benefit of fuel saving 40% from 350 to 200 EFO Bbl/day.

[Case Study 2] Troubleshooting of CDU heater

In 2011, those heaters did revamp to replace existing bare tubes to stud tubes in convection. With the revamp, the rate of increase in flue gas temperature was faster than before the revamp. So to mitigate the fouling rate, in 2012, 8 numbers of soot blower were installed to the provisioned convection zone of each heater under operation. Based on the installation of new soot blowers, total saving fuel by increase in efficiency (decrease in stack flue gas temperature) was about 12 EFO Bbl/day.

	#3 CDU Heater 1	#3 CDU Heater 2
Stack flue gas temp (℃)	24 down (298 to 274)	15 down (294 to 279)
Efficiency (%)	1.2 up (84.2 to 85.4)	0.5 up (84.5 to 85.0)
Fuel saved (EFO B/D)	8.3	3.9

The trend of flue gas temperature is as follows.

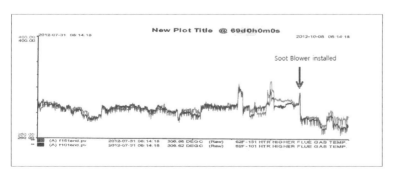

4. Examples of Revamping- Case Studies

There are situations where a refinery requires a different type of solution on the basis of process unit need. The different requirements which, in many cases, may be in combination, can be: Capacity increase, utility consumption reduction, operation reliability increase, increase or reduce steam production in the steam generation recovery unit, emission reduction, and process unit life extension. Each requires different approach to the problem and must be carefully evaluated before finding the optimum solution. A guideline to approach the various cases is given in the following points.

4.1 Capacity increase

The upgrading of the capacity of a fired heater without increasing the heat flux can be obtained by increasing the heat transfer surfaces. Often, a fired heater designed with low efficiency can be revamped by adding surfaces in the convection section and in some cases also in the radiant section. In case of process side pressure drop limitations, the replacement of a portion of the coil with tubes having bigger inside diameter can be considered.

4.2 Utility consumption reduction

The improvement of the utility consumption is normally achievable by increasing the fired heater efficiency and reducing the heat fired accordingly. The efficiency increase can be achieved by modifying the heat transfer surfaces or by installing an air preheating system to recover the heat discharged to atmosphere by preheating the combustion air to the burners. In other cases, the recovery of additional heat from the flue gas downstream of the process coil can be obtained generating steam or heating other colder fluids such as hot oil.

4.3 Operating reliability increase

The efficiency of the fired heater operating in natural draft is heavily affected by the excess O_2. If this parameter is not continuously and carefully controlled, the efficiency is low with the consequent waste of fuel fired. Since the trimming of combustion air must be performed frequently at site, even if the fired heater is in good condition, a continuous operation with the correct excess O_2 is almost impossible.

The installation of an air preheat system, with consequent increase of efficiency, has the additional advantages of providing effective ways in which to optimize the operation from the control room. Operators are easily able to act on the forced fan to adjust the combustion and air excess, also on the induced fan to provide the required draft at arch level.

4.4 Increase operational reliability

The installation of a waste heat boiler downstream of the process coil to provide steam and improve the overall unit efficiency was very common in the past. This arrangement is an acceptable solution to increase efficiency. The strong disadvantage, however, is to have steam production depending on the process unit load. In addition, steam is produced with low efficiency if compared with steam produced from a fired boiler. Replacement of the waste heat boiler with air preheat boiler with air preheat system and switching the steam production to the boiler house is very often an attractive solution to reduce the utility cost and improve the overall fired heater operability and efficiency.

4.5 Increase or reduce steam production in the steam generation recovery

Recently the flue gas emissions from a refinery have been of primary importance. The furnaces designed with old criteria (with natural draft burners and low efficiency) have NO_x, SO_2, CO and particulate emissions which often are not complying with the local regulations. In other cases, the increase in capacity or the installation of a new unit is not possible since the allowable overall amount of pollutant from the plant has already reached the maximum allowable level. The increase of the fired heater efficiency and the replacement of the burners with new low NO_x, reduce the emissions dramatically. Alternatively, it allows the plant capacity increase with very similar overall fired heater emissions as indicated previously.

4.6 Process coil life

The life of a fired heater is normally limited by the tube life, designed for 100,000 hours operation at design capacity in accordance with the API RP 530 code. If the coil is working in the elastic stress range, the rupture stress does not control the coil life. For operation in creep range the temperature impact is dramatically important, as a 25-30°C tube metal temperature increase may result in a reduction of tube life down to 20% of the original value.

Visbreaking, thermal crackers, reformers, crude and vacuum are units where creep conditions govern. In this case, a reduction of the average and/or peak heat flux by adding surface or obtaining an even distribution of the heat firing with no flame impingement on heat transfer surfaces should result in a decrease in tube metal temperature with a substantial increase of pressure parts life. An example of what can be achieved in the increasing of tube life is given in Table below.

Item	Unit	Visbreaking	
Average heat flux	kcal/ m² hr	25,000	25,000
Average peak heat flux	kcal/ m² hr	47,500	47,500
Maldistribution	%	20	0
Maximum heat flux	kcal/ m² hr	57,000	47,500
ΔT_{film}	°C	50	42
ΔT_{metal}	°C	18	15
ΔT_{coke} (3mm)	°C	42	35
Metal temperature	°C	560	542
Tube life	hours	40,000	100,000

The same concept is applicable for a fired heater with no heat flux maldistribution where a reduction of 20% of heat flux has the same effect shown above. Additional limitation of fired heater life is generally given by the tube supports design. This can be improved by reducing firing (by increasing efficiency) or by decreasing the flue gas temperature in critical areas, by providing an even heat distribution. In addition to the above, proper maintenance of the internal lining and of the external steel casing eliminates air infiltration and substantially increases the life of the fired heater.

Examples of fired heaters, subject to adequate maintenance and analysis, continue to work properly after more than half century of operation in many refineries worldwide. Further to the guideline provided above, some specific cases where revamps and upgrading have been carried out recently, are given as highlight. Modifications have been carried out in order to overcome and solve the clients' most stringent issues. In some cases the analysis and the activities made to solve such critical issues resulted in a general optimization of operating conditions of the fired heater.

Additionally, a great number of smaller studies and revamps on specific parts and components of the fired heaters have been also carried out such as: (a) Process conditions (b) Rerating based on new feeds (c) Mechanical design (d) Burners operation (e) Emission reductions (f) Pressure parts, tube supports and tube sheets (g) Heat recovery section (h) Cleaning devices (i) Pressure profiles (j) Forced and induced draft fans, dampers, and stack which have optimized various aspects of operation

5. Case Studies

5.1 Case1 European refinery

In the late 1980s a twenty-five year old topping heater was upgraded with the installation of an air preheating system; the main objective being to reduce its fuel consumption. The following is an illustration of the fired heater performances before and after modifications: Taking into account that overall project investment was in the region of 2,000,000 Euro equivalent, payback was less than two years. Due to the current market conditions and the current fuel cost which is clearly above the former level, a similar investment today is even more attractive.

Item	Unit	Before modifications	After modifications
Fired heater duty	MMkcal/hr	52	52
Fired heater efficiency	%	72	90
Fuel fired	MMkcal/hr	72.2	57.8
Fuel saving	MMkcal/hr	14.4	
Fuel saving	kg/hr	1,520	
Annual fuel oil saving	Tons/yr (8,000hrs)	12,160	
Fuel cost	Euro (equivalent)	0.1	
Annual saving	Euros (equivalent)	1,216,000	

5.2 Case2 European refinery

In the mid-nineties an Italian refinery carried out a program of firing optimization and emission reduction in order to operate according to new EU codes. Under the terms of the project, the complete refinery flue gas emissions discharged to the atmosphere through two independent common stacks, was measured and analyzed. While considering all fuel balances, the objective of the study was to identify the most effective fuel oil versus fuel gas ratio to minimize stack emissions. On this basis taking into account the original heater capacity and the target of the refinery, the refinery improved the emission level proposing low NO$_x$ burners properly designed to be installed on existing heaters with marginal modifications to the furnace floor only. After the replacement the new concentrations of pollutant were analyzed, confirming the full compliance with EU environmental requirements.

5.3 Case3 European refinery

An Italian refinery required to revamp their vacuum heater, the main objectives being to improve the operating conditions and optimize performances affected by problems with the original design. The refinery also required to increase the feed throughput to the maximum possible level. Taking into account the above and considering the crucial importance of the vacuum unit, this work was successfully carried out during a scheduled unit turnaround. The activities performed include burner replacement, convection section replacement, stack replacement without any modification to the steel structure and foundations.

5.4 Case4 South American refinery

In the middle nineties an old box type crude heater was upgraded to increase the radiant surface and modify the convection section. A summary of the fired heater performances before and after modification is as follows.

Item	Unit	Before modifications	After modifications
Unit load	BPSD	30,000	42,000
Fired heater duty	MMkcal/hr	25	35
Efficiency	%	65	82
Heat fired	MMkcal/hr	38.5	42.7
Heat flux	kcal/ m^2 hr	27,000	28,500
Radiant surface	m^2	780	930
ΔT Film peak	℃	42	34
ΔT Metal	℃	13	14

The increase of unit capacity was achieved by adding extra surface in the radiant section. The convection section was also modified by replacing the bare tubes with extended surface tubes. The mechanical design of the heater was modified and reinforcements were studied and added as needed. It has to be noted that the 40% increase in throughput in the fired heater was obtained with only a 19% increase in radiant surface and 11% increase in heat firing. The process unit was upgraded with modifications to the column and heat exchanger train to achieve the process unit capacity increase.

5.5 Case5 Middle East refinery

A hydrogen plant having 60,000 Nm^3/hr nominal capacity was built in the mid-eighties and had its fired heater revamped to increase hydrogen production and at the same time, to replace the catalyst tubes close to the end of their life. The upgrading project was limited to the replacement of catalyst tubes and improvement to the convection section heating surfaces. The original catalyst tubes in 25/35 Cr/Ni material were replaced with new ones in 25/35 Cr/Ni micro alloy having the same outside diameter but with lower thickness. The upgrading was based on a hydrogen production increase of 20%. The summary of the unit performances at two considered capacities are as follows.

Item	Unit	Before project	After project
Hydrogen capacity	Nm^3/hr	60,000	72,000
Fuel + Feed consumption	MMkcal/hr	237	298
Overall steam production	Ton/hr	120	158
Export steam	Ton/hr	63	93

Assuming: Hydrogen value of 0.07 US$/$Nm^3$, fuel cost of 14 US$ for MMkcal, export steam value of 17 US$/ton

The utilities consumption benefits is as follows.

Item	Unit	After project
Hydrogen capacity	Nm^3/hr	72,000
Delta hydrogen	Nm^3/hr	12,000
Delta feed + fuel	MMkcal/hr	51
Delta export steam	Ton/hr	30

The utilities hourly costs benefit is as follows.

Item	Unit	After project
Hydrogen capacity	Nm^3/hr	72,000
Delta hydrogen	$/hr	12,000 x 0.07 = -840
Delta fuel	$/hr	51 x 14 = +714
Delta steam	$/hr	30 x 17 = -510
Total	$/hr	-636
Saving	$/yr	636$/hr x 8,000hrs = 5,088,000

Since the overall project investment was in the region of US$6,000,000 the payout for 20% capacity increase was approximately one year. The advantages are much more significant considering that most of the project cost was due to the catalyst replacement. As the tubes were close to the end of their life, the cost of replacement was considered normal maintenance cost.

In summary, it is clear that the number of factors affecting fired heater operation is multiple and quite complex. It is also clear that there is not a single solution applicable to all cases on the basis of an already-established understanding of fired heater operation. In many cases there are knock-on effects which may lead to wrong assessments of current operations and therefore to a wrong solution. And different situations and sometimes opposite needs, must be properly evaluated on a case-by-case basis, in order to find out the most attractive solution from both an economical and technical point of view.

6. Decrease Tube Metal Temperature in Vacuum Heaters

In a refinery of Middle East, a vacuum furnace was found to be operating at a high tube metal temperature (TMT) and was forced to shut down before the normal run length of five years. The coke laydown was also found to be substantial. An evaluation revealed that the furnace outlet temperature and pressure were operating at much higher values, compared to the design values, to maintain the same flash zone conditions at the vacuum tower.

This scenario was the result of a change in furnace location, which increased the transfer line length by 200m. The increase in transfer line length resulted in an additional pressure drop of 1.631mH$_2$O, which increased the furnace outlet temperature requirement for the same flash zone conditions in the vacuum column.

Various options for reducing the TMT of the vacuum heater, such as changing the transfer line size and adding extra surface area in the radiation and convection sections, were examined. The best option generally depends on the limitations under which the heater operates with regard to the process requirements and the downstream equipment limitations and flexibility. Vendor-developed heater models are used to evaluate the various options.

6.1 System description

Below Fig.1 shows the vacuum transfer line arrangement. The existing heater is a horizontal tube, box type heater with a convection section. There are four process passes- two on each radiant wall. Process passes from the convection section enter at the top of the radiant section and exit close to the radiant floor. The heaters' absorbed duty for the design condition is 13.86 MMkcal/hr, and the fired duty is 17.14 MMkcal/hr (i.e., 100% fuel gas firing). From the outlet, radiant tube sizes are 10in., 8in., 6in. and 5in. NPS. In the convection section, there are three bare rows and seven studded rows (all 5in. NPS tubes). The convection section is for process service only. All radiant and convection tubes are 9Cr-1Mo steel material, schedule 40 thickness. Eight upward firing burners are arranged at the center of the radiant floor. The burners are of the dual firing type, although they no longer fire fuel oil.

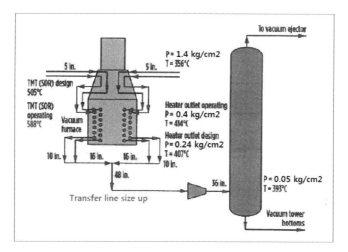

Fig.1 Schematic of a vacuum transfer line

The expected TMT under clean condition is 506℃; for a coke thickness of 3mm and a corrosion allowance of 3mm, the design TMT is 621℃. This temperature is based on a rupture pressure of 9.3 kg_f/cm^2 and a corrosion allowance of 3mm. As per the American Petroleum Institute (API) standard 530, 9Cr-1Mo steel material can withstand temperatures up to 705℃. The vacuum heater was installed in 2003 to replace the original heater. The transfer line size of the original heater was 36 in. During the replacement of the vacuum heater, the transfer line was changed from 36 in. to 48 in.; however, a nozzle size similar to the transfer line size could not be installed due to construction issues. Consequently, a 48in.-36in. reducer was installed, connecting the new and old transfer lines closer to the column.

6.2 Performance of existing operation

The feed composition of the reduced crude oil (RCO) feedstock is shown in Table1. The furnace heat release is calculated based on fuel gas pressure and burner capacity (i.e., pressure curves). This calculation has been crosschecked with the fuel gas flowrate. The fired duty was found to be 20 MMkcal/hr, while the average radiant flux, based on absorbed duty, was calculated to be closer to the design duty of 27,084 kcal/m^2 hr. Taking these calculations into consideration there is no scope for increasing the firing rate without adding more area in the radiant section.

Recovered Mass (%)	Boiling Point (℃)	Recovered Mass (%)	Boiling Point (℃)
Initial BP	303	45	557
5	379	50	579
10	413	55	602
15	438	60	628
20	459	65	656
25	478	70	693
30	497	75	731
35	516	77	750
40	536	-	-

Table1. Boiling point distribution (BP) of reduced crude oil feedstock

The heater was found to be operating at approximately 108% of the throughput, approximately 112% of the process duty and approximately 116% of the firing rate of the original heater. The higher load was presumed to be the main reason for the higher TMT and the high coking rate. The heater stack damper was found to be operating at 95% of the open position, and the arch draft was +2.5 mmH$_2$O. The excess air was back calculated from excess O$_2$ at 3.3% in the flue gas, which was found to be 18.6 vol% (10% for the design calculation). Under existing excess air condition, there is a limited margin for increasing the firing rate without needing to modify the heater stack to satisfy the draft requirement.

6.3 Improving furnace performance

Several options have been considered to raise the performance for the heater, including increasing the run length, improving the draft and increasing the firing rate without exceeding the radiant flux limit.

6.3.1 Adding radiant surface area

Adding heat transfer surface area will directly reduce the firebox temperature and the average flux density to the tubes. This, in turn, will reduce the TMT and the coking rate. The firing rate can also be reduced with higher fuel efficiency, due to the increased heating surface. The addition of eight tubes in the section between the convection and radiant sections shows that the TMT can be decreased by 8℃.

Table2 provides a comparison of other present operating case and the proposed modification mentioned above. Note: the base case has been adjusted to meet 10% excess air (design conditions) while operating with the current load. When additional draft is shown as negative, it implies that the existing stack height is sufficient, while a positive increase in draft translates to increased stack height for the same damper opening.

Property	Unit	Simulated base case (adjusted to present operation)	Proposed operating case
Flow rate to heater	kg/hr	329,000	329,000
Heater inlet/outlet temperature	℃	356/413	356/413
Heater inlet/outlet pressure	kg_f/cm^2	1.6/0.4	1.6/0.4
Total process duty	MMkcal/hr	15.6	15.6
Calculated maximum TMT	℃	592	584
Fuel efficiency	%	79.5	79.6
Extra draft	mmH_2O	-3 (less)	-3 (less)

Table2. Comparison between present and proposed operating cases

6.3.2 Adding convection surface area

Adding two rows of convection tubes has been studied as a means of shifting part of the process duty from the radiant section to the convection section. Table3 provides a comparison of this option with the base case scenario.

Property	Unit	Simulated base case (adjusted to present operation)	Addition of convection tubes option
Flow rate to heater	kg/hr	329,000	329,000
Heater inlet/outlet temperature	℃	356/413	356/413
Heater inlet/outlet pressure	kg_f/cm²	1.6/0.4	1.6/0.4
Total process duty	MMkcal/hr	15.6	15.6
Calculated maximum TMT	℃	592	589
Fuel efficiency	%	79.5	81
Extra draft	mmH₂O	-3 (less)	+1 (more)

Table3. Comparison of convection tubes option with base case scenario

Since the addition of convection surface area increases the pressure drop on the flue gas side, the required draft is higher. However, the required draft is already limited by the stack height, and the damper opening is close to 95%, so this option as not pursued further.

6.3.3 Extending convection surface tubes

The lowest three rows (shield rows) of the 12 tubes in the convection section are bare tubes. The present operating maximum stud temperature is already close to the design temperature. Since adding studs will further increase the temperature, this option was not pursued.

6.3.4 Converting natural draft furnace to forced draft

Adding a forced draft fan in the furnace will neither reduce the firing, nor provide more draft. Moreover, this option requires extra ducting and additional plot area. As a result, this option was not investigated in detail.

6.3.5 Converting natural draft furnace to induced draft

Adding an induced draft fan in the furnace will provide the necessary draft at the heater and allow for the possibility of more firing, although it will not reduce the TMT. This option requires extra ducting, arrangement of space and arrangement of the steel structure; consequently, this option is not investigated in detail.

6.3.6 Converting natural draft furnace to balanced draft

Introducing a complete air preheat system has the following consequences: For the same process duty, it will significantly increase the fuel efficiency, therefore reducing fuel gas firing. However, due to the higher combustion air temperature, it will increase the firebox temperature and raise the TMT. Since the furnace is already limited by high skin temperature, and since this option requires extra ducting, plot space and arrangement of the steel structure, this method is not recommended.

6.3.7 Increasing the transfer line size

Since the pressure and temperature in the flash zone is fixed for a given feed, the enthalpy is also fixed. There is no enthalpy change in the vacuum transfer line; therefore, the enthalpy at the furnace outlet should be the same as in the flash zone. Increasing the transfer line size reduces the pressure drop, which calls for a lower furnace outlet temperature for the same enthalpy. This decrease in bulk temperature lowers the TMT, along with the coking rate. Coking is caused by high temperature and low residence time for a given feed. By increasing the transfer line size and raising the column nozzle size to 60in., the TMT is expected to be lowered. Table4 provides an overview of the projected results.

Property	Unit	Simulated base case (adjusted to present operation)	Increased transfer line size option
Flow rate to heater	kg/hr	329,000	329,000
Heater inlet/outlet temperature	°C	356/413	356/409
Heater inlet/outlet pressure	kg_f/cm^2	1.6/0.4	1.6/0.25
Total process duty	MMkcal/hr	15.6	15.6
Calculated maximum TMT	°C	592	586
Fuel efficiency	%	79.5	79.5
Extra draft	mmH_2O	-3 (less)	-3 (less)

Table4. Comparison of increased transfer line size option with base case scenario

While the reduction in TMT is approximately 6°C, the heater outlet pressure is expected to be 0.25 kg$_f$/cm^2, or a reduction of 63mm, due to the increased transfer line size. This, in turn, reduces the coil outlet temperature by 6°C. In all previous cases, the vapor at the heater outlet was around 9%, and the increase in the transfer line size raised the amount of vapor to 17%.

6.3.8 Velocity steam and vacuum tower debottlenecking

Another way to reduce coke formation is to increase the velocity (and, therefore, the residence time) by adding steam. Fig.2 shows a graph of TMT vs. steam flow.

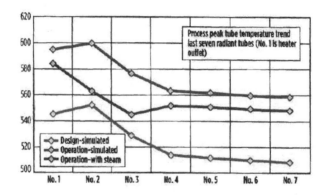

Fig.2 TMT vs. steam flow

For the purposes of this study, 800 kg/hr steam was used. This option depends largely on the availability of additional capacity on the vacuum column overhead steam jet ejector. Although this will definitely lead to a decrease in TMT, the vapor at the heater outlet will be approximately 14%. It is evident that increased transfer line size, along with extra radiation surface area and coil steam injection, provides the best solution to reduce the TMT and increase vaporization while marginally raising heater efficiency. Table5 summarizes this option as compared to the base case scenario.

Property	Unit	Simulated base case (adjusted to current operation)	Increased radiant area, plus increased transfer line, plus addition of velocity steam option
Flow rate to heater	kg/hr	329,000	329,000 + 820
Heater inlet/outlet temperature	℃	356/413	356/406
Heater inlet/outlet pressure	kg_f/cm^2	1.6/0.4	4.05/0.25
Total process duty	MMkcal/hr	15.6	15.6
Calculated maximum TMT	℃	592	574
Fuel efficiency	%	79.5	79.6
Extra draft	mmH_2O	-3 (less)	-3 (less)

Table5. Comparison of velocity steam and vacuum tower debottlenecking option with base case scenario

With this option, the TMT drops by 18°C, reducing the coke buildup within the heater tubes and extending the unit run length by about nine months. However, with the injection of steam, the pressure at the heater inlet will increase. The maximum design temperature was calculated for the maximum expected pressure of 4 kg_f/cm^2 (i.e., the pressure relief valve set pressure in the vacuum column, plus the transfer line and column pressure drop).

The calculated maximum TMT for a 3mm corrosion allowance, with a pressure of 4 kg_f/cm^2 for convection tubes of schedule 40 thickness, is 675°C. Although the TMT can be stretched from 621°C for a tube life of 100,000 hours during end-of-run conditions, caution is advised, as this circumstance depends on the actual thickness of the tubes. The run length of the heater can be further increased by one year, if allowable TMT is raised from 621°C to 650°C for the same tube life of 100,000 hours.

6.4 Summary

Reduction of TMT can be achieved with several different approaches involving modifications within the heater box, outside of the heater box, or a combination of both. In cases where the transfer length is significant, a change in the transfer line size can have a marked effect on the heater tube TMT. While the best solution in the above case involves a combination of changing the transfer line size, steam injection and the addition of tubes in the radiant zone, a detailed mechanical integrity analysis should be carried out at the vacuum column inlet nozzle area to confirm the viability of the nozzle size increase from 36 in. to 68 in.

Subsequently, as interim solution, velocity steam can be introduced to the heater, with an increase in radiant heat transfer area, thereby alleviating the TMT issue in the heater. A detailed analysis is being performed to further optimize the system and implement the most technoeconomic solution available.

7. Heater Performance Analysis Study

A refining company in Korea has #2 Crude Distillation Heaters (A/B) that were revamped by a heater maker in 2007. The heaters are two cabin with horizontal tube coil type heaters. Both the heaters are identical. Though this study refers to one heater, this assessment remains valid for both the heaters. The refinery engaged another heater maker to carry out the heater performance analysis and feasibility study for performance improvement. As outlined in client's requisition, another maker had been verified existing heater design and current operation. Based on verification results, feasibility study of revamp cases has been performed. Modification of APH has been selected among several revamp cases.

7.1 Study summary

Case1. Additional new two block installation under existing APH

Description	Unit	Before revamp		Revamp case1	
		Design data (w/o SAPH)	Operating data (w/o SAPH)	Design prediction (w/o SAPH)	Operating prediction (w/o SAPH)
Absorption duty	MMkcal/hr	58.5	69.4	58.1	69.4
Heat release	MMkcal/hr	64.7	79.7	64.1	77.7
Flue gas flow rate	kg/hr	106,848	128,916	105,840	128,268
Flue gas temp @APH inlet	℃	285	385	285	387
Flue gas temp @APH inlet	℃	161	211	144	165
Air flow rate	kg/hr	101,952	122,868	100,980	122,400
Air temp @APH inlet	℃	15.6	15.6	15.6	15.6
Air temp @APH outlet	℃	165	230	185	286
APH duty	MMkcal/hr	3.7	6.4	4.1	8.1
Payback period				SOR/EOR	SOR/EOR
Fuel saving ost	Million ₩	-	-	825/388	2,698/1,112
Equipment cost	Million ₩	-	-	1,120/1,120	1,120/1,120
Construction cost	Million ₩	-	-	1,090/1,090	1,090/1,090
ROI	yr	-	-	2.7/5.7	0.8/2.0

Note: ① Additional new APH two blocks shall be installed below of existing APH ② Glass coated block shall be installed to avoid corrosion from acid dew point ③ Pipe type APH: Less pressure drop (ID fan modification or replacement is not required)/Glass coated pipe ④ Plate type APH: Higher pressure drop (ID fan modification or replacement is required)/Glass coated plate ⑤ Due to ID fan capacity, pipe type APH is recommended ⑥ Existing three (3) block + new two (2) block ->total five (5) block APH

Case2. Bottomblock replacement of existing APH

Description	Unit	Before revamp		Revamp case2	
		Design data (w/o SAPH)	Operating data (w/o SAPH)	Design prediction (w/o SAPH)	Operating prediction (w/o SAPH)
Absorption duty	MMkcal/hr	58.5	69.4	58.4	69.4
Heat release	MMkcal/hr	64.7	79.7	64.3	78.2
Flue gas flow rate	kg/hr	106,848	128,916	106,056	129,024
Flue gas temp @APH inlet	℃	285	385	284	387
Flue gas temp @APH inlet	℃	161	211	147	175
Air flow rate	kg/hr	101,952	122,868	101,196	123,084
Air temp @APH inlet	℃	15.6	15.6	15.6	15.6
Air temp @APH outlet	℃	165	230	182	275
APH duty	MMkcal/hr	3.7	6.4	4.1	7.8
Payback period				SOR/EOR	SOR/EOR
Fuel saving ost	Million ₩	-	-	648/386	2,088/706
Equipment cost	Million ₩	-	-	1,120/1,120	1,120/1,120
Construction cost	Million ₩	-	-	1,090/1,090	1,090/1,090
ROI	yr	-	-	3.4/5.7	1.1/3.1

Note: ① Existing bottom block shall be removed and new one (1) Block shall be installed at the same place ② New block size shall be same as existing APH bottom block ③ Plate type APH block shall be replaced to increase APH performance since plate type has more heat transfer area ④ Glass coated plate type APH shall be installed to avoid corrosion from acid dew point ⑤ Existing three (3) block ->existing two (2) block + new one (1) block

7.2 Study detail

7.2.1 Scope of work

① Performance analysis: 1) Existing heater design review 2) Existing heater design analysis 3) Current operation review 4) Current operation analysis

② Feasibility study for APH modification: 1) Case study for APH modification 2)

Prediction of fired heater performance for revamp case 3) Technical review/analysis for revamp case 4) Cost estimation

7.2.2 Basis of study

① Performance analysis

Main objective of the study has been to analyze process condition for revamp study. During course of this study, some discrepancies were found between operating data and prediction based on operating case and simulation case. Details of findings have been presented in this study. However, the some differences are negligible in terms of the APH modification. These differences could be happen because different process property (past vs. current), calculation methods, fuel composition, environment, etc.

② Feasibility study for APH modification

Feasibility study of basic design package for APH modification has been carried out based on performance analysis results. This study described prediction of fired heater performance for revamp case. Also, economic has been estimated for payback period.

7.2.3 Details of software and approach to the study

The heater maker uses fired heater simulation program FRNC-5 from PFR Engineering Systems Inc. FRNC-5 is well proven to be reliable and trusted industry-wide. This program simulates performance of most types of fired heaters found in refineries and petrochemical plants.

Most part of the heater can be simulated including complex processes, multiple fire boxes, convection sections, ducts, stacks, most coil configurations, plain/extended surface tube and various extended surface types, transfer lines, manifolds and fittings. Process and other critical conditions are determined at any point entering, leaving or within the unit. This simulation tool is a combination of process simulator and comprehensive heat transfer simulator with pressure drop estimator using proven state-of-the-art methods.

The rigorous step-wise approach used by the simulator, estimates conditions for the process and flue gas streams as heat transfer takes place through each coil/furnace section and reliably allows fairly accurate estimation of furnace parameters: (1) Firebox and overall thermal efficiency (2) Bridge wall and stack entry flue gas temperature (3) Local radiant and convective average and peak heat flux (4) Peak and average tube metal film and fin tip temperatures (5) Determination of two phase flow and boiling regimes (6) Heat transfer and pressure drop in two phase flow (7) Prediction of flow regimes (8) Prediction of relative coking rates.

The heater maker used existing furnace geometry is simulated to represent current operation and compared simulation results. Usually, this is done to validate model and subsequently this model is used to predict the performance of the heater for various future operations. In this instance, significant discrepancies were found between operating data and heater capability. Details of findings have been presented in this study.

7.2.4 Heater description

#2 Crude Distillation Heaters are two cabin box with horizontal tube coil heaters originally revamped by a heater company in 2007. Revamp scope was to remove existing steam generation and BFW coils and install additional four (4) rows of crude heater, total 32 tubes (A335 P5) above MP steam. There are thirty two burners in each heater. The original burners were replaced in 2001. These burners are designed for 3.013 MMkcal/hr heat release for each gas and oil fuel. The heater is not known for any unusual heater related operational issues so far.

7.2.5 Review of existing heater (original design)

The heater has been simulated based on its existing design (2007)- per details provided by a heater maker's data sheets and general arrangement drawings. Heater outlet conditions (pressure, temperature) have been fixed while simulating the heater; process temperature has been adjusted to meet absorbed duty. Property grid is used from datasheet issued by a heater maker. Design datasheet simulation matched fairly well with data sheet. Original heater datasheet shown TBP at 760 mmHg (LV), API=29.66 as follows:

Wt%	1	5	20	30	50	70	90	96	98
Temp(℃)	45.05	72.69	114.54	220.00	326.16	453.13	665.56	753.82	889.13

Above TBP data has been used for all simulation cases.

Simulation results are shown in Table below.

Description	Unit	Design datasheet (revamp 2007)	Design datasheet (simulated)
Process data			
Process absorbed duty	MMkcal/hr	58.5	58.5
Steam absorbed duty	MMkcal/hr	0.72	0.74
Total absorbed duty	MMkcal/hr	59.22	59.24
Process			
Process feed rate	kg/hr	756,500	756,500
Inlet temperature	℃	265	269
Outlet temperature	℃	350	350
Coil pressure drop	kg/cm^2	7.5	8.3
Inlet vaporization	wt%	1.9	0.5
Outlet vaporization	wt%	42.3	43.9
Steam superheating			
Steam flow	kg/hr	13,000	13,000
Inlet temperature	℃	240	239
Outlet temperature	℃	347	350
Coil pressure drop	kg/cm^2	0.3	0.2
Combustion data (fuel oil)			
Fired duty (calculated)	MMkcal/hr	64.16	63.83
Excess air	%	20	20
Air flow rate	kg/hr	–	108,396
Flue gas flow rate	kg/hr	119,418	116,820
Bridge wall temperature	℃	870	882
APH inlet section	℃	288	287
FGT leaving APH	℃	183	184
Air temp after SAPH	℃	75	75
Air temp after APH	℃	198	198
Mechanical data			
Calculated Max TMT for process	℃	739	394
Calculated Max TMT for steam	℃	359	362

Note: One heater item is described. Two identical heaters shall be considered (fuel, air and flue gas flow rate etc.)

Major observations from analysis of above existing design data and simulation results: ① Design datasheet simulation matches fairly well with datasheet. Largest difference is in bridge wall temperature (BWT, Arch)- simulation predicts about 12 ℃ higher and could be possibly due to calculation procedure (A vs. B heater maker) ② Simulation has been carried out in absorbed duty fixed based on datasheet ③ Prediction of process condition based on existing design simulation does not have large differences. Some minor differences are found but it is usual results since thermal calculation method and consideration is not same among heater companies. ④ Refractory design has been reviewed. Existing design is as below Table.

Location	Type	Thickness (mm)
Radiant exposed wall	LHV = 1:2:4	150
Radiant shield wall	LHV = 1:2:4	125
Arch	LHV = 1:2:4	125
Floor	LHV = 1:2:4	216
Convection wall	LHV = 1:2:4	125
Hot flue gas duct	LHV = 1:2:4	75

⑤ Simulation results are used as hot face temperature to calculate cold face temperature since hot face temperature is not available from datasheet. Also, equivalent material Kaolite 2000HS is used instead of refractory type LHV 1:2:4 for calculation. The calculated cold face temperatures are indicated in the Table as below.

Location	Thickness (mm)	Existing design ($°C$)	
		Hot face	Cold face
2D shielded wall	125	578	93
Exposed wall	150	870	111
Floor	216	759	90
Roof	125	870	116
Convection wall	125	579	93
Breeching & stack	75	288	76

⑥ Cold face temperature exceeds allowable maximum temperature as per API 560 4th ed. Cold face temperature is calculated based on 27°C ambient temperature and 0 m/sec wind velocity as per API 560 4th ed.

7.2.6 Review of heater current operation

Heater operating data was reviewed. Significant differences were not observed between operating data and heater. The summary of simulation results are indicated in the Table as below.

Description	Unit	Operating data (29 Mar 2012)	Operating data (simulated)
Process data			
Process absorbed duty	MMkcal/hr	69.7	69.3
Steam absorbed duty	MMkcal/hr		1.71
Total absorbed duty	MMkcal/hr		70.98
Process			
Process feed rate	kg/hr	727,465	727,452
Inlet temperature	℃	270	257
Outlet temperature	℃		362
Coil pressure drop	kg/cm²		7.96
Inlet vaporization	wt%		0
Outlet vaporization	wt%		47
Steam superheating			
Steam flow	kg/hr	21,040	21,024
Inlet temperature	℃		280
Outlet temperature	℃	440	440
Inlet pressure	kg/cm²		9
Coil pressure drop	kg/cm²		0.56
Combustion data (fuel oil + fuel gas)			
Fired duty (calculated)	MMkcal/hr		79.0
Excess O₂ (dry)	%	2.6	2.6
Air flow rate	kg/hr		121,680
Flue gas flow rate	kg/hr	119,418	127,692
Bridge wall temperature	℃	873 (Floor, Note 1)	951
APH inlet section	℃	381	384
FGT leaving APH	℃	212	217
Air temp after SAPH	℃	31	31
Air temp after APH	℃	236	236
Mechanical data			
Calculated Max TMT for process	℃		417
Calculated Max TMT for steam	℃		468

Note: BWT is not available due to provision at arch. Floor temperature is shown instead of BWT.

Major observations from analysis of above operating data: ① Process temperature at inlet is not matched with simulation result. It looks that property grid is not same between current and design process. ② Bridge wall temperature is not available. Since provision is not existed on radiant arch. The refinery confirmed that there is a temperature measuring point at floor only. We assumed that bridge wall temperature is floor temperature +66 ℃ as per the heater maker's standard. So, estimated BWT is about 939℃. Temperature is pretty matched since it is 12℃ less than simulation. ③ Flue gas temperature at APH outlet is 212℃/217℃ at each current operation and simulation results (refer to next Table) ⓐ Fired duty is higher than design: 64.16MMkcal/hr (design) vs. 79MMkcal/hr (estimated fired duty for current operation). Fired duty is increased up to 23%. ⓑ Although air temperature is less than design, flue gas temperature at outlet is higher than design since flue gas temperature increase at inlet is higher ⓒ Flue gas temperature at APH inlet is higher than design due to more firing: 320℃(design) vs 381℃ (current operation)

Description	Unit	Design	Operation	Operation (simulated)
Air flow rate	kg/hr	242,189	-	243,360
Air inlet temperature	℃	75	31	31
Air outlet temperature	℃	236	236	236
Flue gas flow rate	kg/hr	260,663	-	255,384
Flue gas inlet temperature	℃	320	381	384
Flue gas outlet temperature	℃	183	212	217

④ Cold face temperatures are calculated based on current operation received from the refinery. Arch temperature is estimated. Cold face temperature is calculated based on 27°C ambient temperature and 0m/s wind velocity as per API 560 4[th]ed.

Location	Thickness (mm)	Existing design (℃)	
		Hot face	Cold face
2D shielded wall	125	629	98
Exposed wall	150	951	118
Floor	216	840	96
Roof	125	951	123
Convection wall	125	666	102
Breeching & stack	75	381	91

⑤ Cold face temperature exceeds allowable maximum temperature as per API 560 4[th]ed. Also, thermal images are taken and many hot spots are observed as right picture.

7.2.7 Assessment for future modifications

Even though gap between operating data and heater prediction is not close fully, impact of this difference has little bearing on this study. Main objective of this study has been to carry out the process review required for the modification of existing APH. In terms of energy saving from existing heater performance, difference between the current operation and simulation results is negligible. For APH review, APH performance review has been carried out with a APH maker. Prediction of revamp case has been analyzed based on both design case and operating case. And, glass coating has been considered where predicted minimum temperature would be less than dew point.

① **Revamp case1**

ⓐ **Concept of revamp case**

Flue gas temperature at APH outlet temperature is 212℃. New APH block is installed bottom of existing to collect waste heat from flue gas.

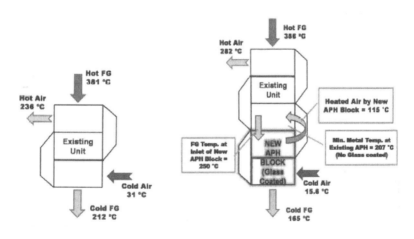

Existing APH modification

ⓑ **Prediction of minimum metal temperature**

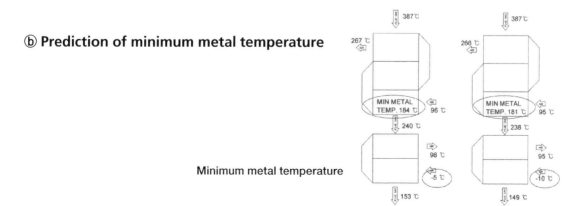

Minimum metal temperature

Estimated minimum metal temperatures are 184℃ and 181℃ for each air temperature -5℃ and -10℃. Glass coating is not required to avoid acid dew point due to enough margin.

ⓒ Prediction of revamp case (APH modification)

Refer to results for prediction of revamp case as Table below.

Description	Unit	Operating data (29 Mar 2012)	Operating data (simulated)	Prediction data (revamp case)
Process data				
Process absorbed duty	MMkcal/hr	69.7	69.3	69.4
Steam absorbed duty	MMkcal/hr		1.71	1.83
Total absorbed duty	MMkcal/hr		70.98	71.23
Process				
Process feed rate	kg/hr	727,465	727,452	727,452
Inlet temperature	℃	270	257	256
Outlet temperature	℃		362	362
Coil pressure drop	kg/cm^2		7.96	8
Inlet vaporization	wt%		0	0
Outlet vaporization	wt%		47	47
Steam superheating				
Steam flow	kg/hr	21,040	21,024	21,024
Inlet temperature	℃		280	270
Outlet temperature	℃	440	440	440
Inlet pressure	kg/cm^2		9	9
Coil pressure drop	kg/cm^2		0.56	0.54
Combustion data (fuel oil + fuel gas)				
Fired duty (calculated)	MMkcal/hr		79.0	77.5
Excess O$_2$ (dry)	%	2.6	2.6	2.7
Air flow rate	kg/hr		121,680	122,076
Flue gas flow rate	kg/hr	119,418	127,692	127,944
Bridge wall temperature	℃	873 (Floor, Note 1)	951	969
APH inlet section	℃	381	384	386
FGT leaving APH	℃	212	217	165
Air temp after SAPH	℃	31	31	15.6
Air temp after APH	℃	236	236	282
Mechanical data				
Calculated Max TMT for process	℃		417	417
Calculated Max TMT for steam	℃		468	470

The following points regarding heater performance may be noted from Table above: ① Bridge wall temperature is increased to 969°C as compared to 951°C for prediction data of current operation ② Process side pressure drop is increased slightly and under allowable value ③ Calculated fired duty is decreased to 77.5 MMkcal/hr as compared to 79 MMkcal/hr prediction of current operation. Fired duty 1.5 MMkcal/hr is decreased per one fired heater. Total 3 MMkcal/hr fired duty is decreased ④ Flue gas temperature at APH outlet is decreased to 165°C(revamp) from 212 °C(current operation)

ⓓ APH type comparison

Both plate and pipe type APH could be installed as new block. Comparison Table is as below.

Item	Plate type APH	Tube type APH
Installation area	Relatively small size and weight	Easy size matching
Maintenance	Repair	Replacement
Cost	Lower price (approximately 50%)	High price
Glass coating	Max 1,219mm x 1,524mm (one element size)	Applicable for less than 4m
Pressure loss control	Plate gap thickness	Pipe size

ⓔ Study for installation of new APH block

New APH block is installed under existing APH unit without any foundation work. Addition structure shall be installed to support new APH block weight. Below schematic drawing is installation concept to add new block.

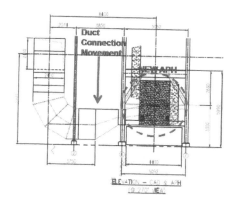

ⓕ Draft loss review

Draft loss has been estimated for both plate type and pipe type APH as below Table.

Description	Unit	Design	Revamp1 (plate type)	Revamp1 (pipe type)
Heat release	MMkcal/hr	128.3	144.7	144.7
DP @ SCR	mmH₂O	100	96.2	96.2
DP @ APH	mmH₂O	150	216.9	184
Others (duct, margin, etc)	mmH₂O	121	173.9	173.9
ID Fan suction	mmH₂O	471	487	454

Pipe type APH pressure loss is less than plate: ① For plate type APH, estimated required ID Fan suction is 487 mmH₂O and exceeds design fan capacity. ID Fan shall be replaced. ② For pipe type APH, estimated required ID Fan suction is 454 mmH₂O and less than design fan capacity. ID Fan shall cover current operation. However, fan capacity is allowable but not much.

② Revamp case2

ⓐ Concept of revamp case

Original design is to preheat combustion air before APH by steam APH. But, some of combustion air is bypassed to avoid acid dew point at cold APH block since SAPH does not preheat combustion air enough due to expensive steam cost. Bottom cold block shall be replaced with new glass coated block due to acid dew point. Then, heater efficiency would be increased since combustion air bypass is not required

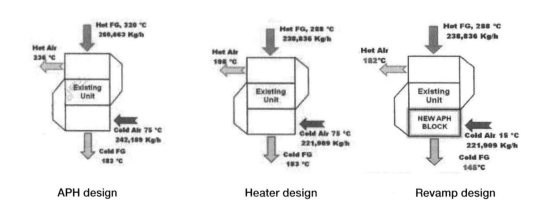

APH design   Heater design   Revamp design

Below is the summary of APH condition.

Description	Unit	APH design	Heater design	Revamp (heater design)
Flue gas flow rate	kg/hr	260,663	238,836	238,836
Flue gas inlet temperature	℃	320	288	288
Flue gas outlet temperature	℃	183	183	145
Flue gas temperature, ΔT	℃	137	105	143
Air flow rate	kg/hr	242,189	221,909	221,909
Air inlet temperature	℃	75	75	15
Air outlet temperature	℃	236	198	182
Air temperature, ΔT	℃	161	123	167

For revamp case2, air bypass is not required because of replaced glass coated new block. As the result, flue gas outlet temperature is decreased. And, overall heater performance shall be increased. However, performance review for temperature is not exact since existing and replaced new APH block types are different. Also, operation data is not provided due to site condition. The prediction of temperature is estimated based on assumption as per experience. This prediction shall be used only for information. The minimum metal temperature is predicted above 150℃ at existing APH unit with -10℃ ambient air. But, it shall be verified throughout actual operation. The minimum temperature shall be monitored for operation at existing unit area expected minimum temperature.

ⓑ Prediction of minimum metal temperature for non glass coated area

Estimated minimum metal temperatures are 204℃ and 196℃ for each air temperature 15.6℃ and -10℃ for non-glass coated area.

ⓒ Schematic drawing

New block size is matched existing size. Bottom one block is replaced with new glass coated block.

ⓓ Draft loss review

Description	Unit	Design	Revamp 2
Heat release	MMkcal/hr	128.3	126.4
DP @ SCR	mmH$_2$O	100	100
DP @ APH	mmH$_2$O	150	Existing unit 2 block + 56
Others (duct, margin, etc)	mmH$_2$O	121	121
ID Fan suction	mmH$_2$O	471	-

Existing bottom block shall be replaced with new plate type block. New block draft loss is estimated 56 mmH$_2$O. Its one block draft loss is assumed 50 mmH$_2$O for existing APH, draft loss would be increased 5 mmH$_2$O. This additional loss 5 mmH$_2$O is within Fan margin.

③ Cost estimation

For cost estimation, heater simulation has been adjusted to set same basis as below. Fuel cost saving is estimated without SAPH operation base although air is heated by SAPH at exiting operation.

Description	Unit	Before revamp			
		Design data (w/ SAPH)	Design data (w/o SAPH): cost basis	Operating case (w/ SAPH)	Operating case (w/o SAPH): cost basis
Absorption duty	MMkcal/hr	58.5	58.5	69.3	69.6
Heat release	MMkcal/hr	63.8	64.7	79.0	80.0
Flue gas flow rate	kg/hr	116,820	106,848	127,692	129,348
Flue gas temp @APH inlet	℃	287	285	384	386
Flue gas temp @APH inlet	℃	184	161	217	212
Air flow rate	kg/hr	108,396	101,952	121,680	123,300
Air temp @APH inlet	℃	75	15.6	31	15.6
Air temp @APH outlet	℃	198	165	236	231
APH duty	MMkcal/hr	3.25	3.69	6.07	6.44

This cost estimation is estimated based on prediction of heat recovery as per APH performance. Actual payback period would be varied depending on heater operation.

ⓐ Revamp case1

Fuel cost saving is calculated for both design and operating case. Operation is considered 340 days per year. For APH type, pipe type APH is considered since ID Fan shall be replaced for plate type APH. Below shows cost estimation.

Description	Unit	Before revamp		Revamp case1	
		Design data (w/o SAPH)	Operating data (w/o SAPH)	Design prediction (w/o SAPH)	Operating prediction (w/o SAPH)
Absorption duty	MMkcal/hr	58.5	69.4	58.1	69.4
Heat release	MMkcal/hr	64.7	79.7	64.1	77.7
Flue gas flow rate	kg/hr	106,848	128,916	105,840	128,268
Flue gas temp @APH inlet	℃	285	385	285	387
Flue gas temp @APH inlet	℃	161	211	144	165
Air flow rate	kg/hr	101,952	122,868	100,980	122,400
Air temp @APH inlet	℃	15.6	15.6	15.6	15.6
Air temp @APH outlet	℃	165	230	185	286
APH duty	MMkcal/hr	3.7	6.4	4.1	8.1
Payback period				SOR/EOR	SOR/EOR
Fuel saving ost	Million ₩	-	-	825/388	2,698/1,112
Equipment cost	Million ₩	-	-	1,120/1,120	1,120/1,120
Construction cost	Million ₩	-	-	1,090/1,090	1,090/1,090
ROI	yr	-	-	2.7/5.7	0.8/2.0

(Reference) Scope of work and supply for cost estimation

Considered scope of work and supply for revamp case: ① Exist ducts & steel structure demolition ② Exist steel structure modification ③ New duct & steel structures ④ Installation of new APH block ⑤ Painting ⑥ Engineering ⑦ CFD Modeling ⑧ Civil & foundation (foundation work N/A)

Out of scope of work and supply for revamp case: ① Instruments and electrical work ② Fan replacement

ⓑ Revamp case2

Below shows cost estimation.

Description	Unit	Before revamp		Revamp case2	
		Design data (w/o SAPH)	Operating data (w/o SAPH)	Design prediction (w/o SAPH)	Operating prediction (w/o SAPH)
Absorption duty	MMkcal/hr	58.5	69.4	58.4	69.4
Heat release	MMkcal/hr	64.7	79.7	64.3	78.2
Flue gas flow rate	kg/hr	106,848	128,916	106,056	129,024
Flue gas temp @APH inlet	℃	285	385	284	387
Flue gas temp @APH inlet	℃	161	211	147	175
Air flow rate	kg/hr	101,952	122,868	101,196	123,084
Air temp @APH inlet	℃	15.6	15.6	15.6	15.6
Air temp @APH outlet	℃	165	230	182	275
APH duty	MMkcal/hr	3.7	6.4	4.1	7.8
Payback period				SOR/EOR	SOR/EOR
Fuel saving ost	Million ₩	-	-	648/386	2,088/706
Equipment cost	Million ₩	-	-	1,120/1,120	1,120/1,120
Construction cost	Million ₩	-	-	1,090/1,090	1,090/1,090
ROI	yr	-	-	3.4/5.7	1.1/3.1

(Reference) Scope of work and Supply for cost estimation

Considered scope of work and supply for revamp case: (1) Exist ducts & steel structure demolition (2) Exist steel structure modification (3) Replacement of new APH block (4) Painting (5) Civil & foundation (foundation work N/A)

Out of scope of work and supply for revamp case: (1) Instruments and electrical work

7.2.8 Recommendation for operation

For current operation, air bypass has been operated to avoid acid dew point at APH. If cold block shall be replaced during TA, air bypass is not only required but also SAPH operation is not required. This operation could be begun immediately. Then, payback period will be moved up.

Reference

1. PROCESS HEAT EXCHANGE EDITED by Vincent Cavaseno and the Staff of Chemical Engineering, McGraw-Hill Publications Co., New York, N.Y.

2. 2011 ENGINEERING DESIGN SEMINAR- Fired Heaters, UOP

3. Fired Heater School, Fall Class, 26th - 28th October 2010

4. Fired Heater and Waste Heat Recovery, Chevron Corporation

5. FIRED HEATER AND WASTE HEAT RECOVERY MANUAL, CHEVRON RESEARCH AND TECHNOLOGY COMPANY, RICHMOND, CA

6. The John Zink Combustion HANDBOOK, CHARLES E. BAUKAL, JR. EDITOR, CRC PRESS

7. BP107 PROCESS BURNER FUNDAMENTALS, Tulsa, Oklahoma, John Zink, Sep 27-28, 2010

8. BP108 ADVANCED PROCESS BURNERS, Tulsa, Oklahoma, John Zink, Sep 29-30, 2010

9. http://en.wikipedia.org/wiki/Petroleum_refining_processes

10. OxyFuel Combustion for Process Heaters, Chris Leger, Praxair, Refinery Operation.com, 2011.03.16

11. Furnace Operations, third edition, Robert D. Reed, Gulf Publishing Company

12. LMGI (Liquid Minerals Group Inc.)-Magnesium Presentation

13. Market conditions encourage refiners to recover by-product gases, oil & gas journal, Oct.7.2013

14. Presentation Material for Fireside Technology, GE-Betz, 25 Jul 2003

15. ptq Refining Gas Processing Petrochemicals, Q1 2013, p 69-71

16. Revamp & Operations- ptq special report, 2004, p21-25

17. ptq Revamps, 2007, p27-29

18. CHEMICAL ENGINEERING MARCH 2013, p64

19. Hydrocarbon Processing JANUARY 2014, p73

20. AFPM "2012 Q&A and Technology Forum" The New API RP 556- Best Practices for Instrumentation, Control, and Protective Systems for Gas Fired Heater